The Conflict Over Environmental Regulation in the United States

Frank T. Manheim

The Conflict Over Environmental Regulation in the United States

Origins, Outcomes, and Comparisons
With the EU and Other Regions

 Springer

Frank T. Manheim
George Mason University
Fairfax, VA, USA
fmanhei1 @gmu.edu

ISBN 978-0-387-75876-3   e-ISBN 978-0-387-75877-0
DOI 10.1007/978-0-387-75877-0

Library of Congress Control Number: 2008933276

Printed on acid-free paper

Corrected at 2nd printing, 2009

springer.com

To Lucy

Prologue

This book describes how the current polarized political conditions in the United States evolved. Along with euphoria and positive developments, a series of problems grew and interacted to create tensions after World War II. After a decade of rising Congressional impatience with growing pollution during the 1960s, the environmental movement collided with industry at the Santa Barbara offshore oil spill of 1969. An atmosphere of crisis and loss of confidence in government triggered passage of groundbreaking new environmental laws during the 1970s. These laws created new roles for Congress and the courts. I show that whereas the new US environmental management system led to rapid improvement in environmental conditions in the United States, politicization of environmental policy created a rift in US society. It also opened the floodgates to serendipitous and intrusive lawmaking in areas where Congress has no inherent expertise or oversight capability.

The rift widened with time into the current political gridlock. I suggest that the US environmental regulatory system can be best understood by examining history and comparing United States with foreign experience – especially in the EU (European Union), where evolution of an extraordinarily flexible and cooperative political system has evolved since World War II. The EU has not only taken the lead in global climate change action. It includes nations that are world leaders in environmental and energy policy, and which are simultaneously maintaining robust industrial and economic development.

The original antagonists in the United States, the environmental movement and the energy industry, continue to be at the center of the today's conflicts. The contending groups and political supporters regularly tied the US Congress in knots during the past 30 years. Each side has been able to block the other – while not gaining its own goals. This book argues that we cannot afford this damaging internal conflict any longer. It has led to systematic national deterioration under the surface of GDP and stock market indices. It has led to erosion of infrastructure and radicalization of factions. Without change, our conflict and disorganized politics will not achieve the cooperation needed to succeed in policies of significant reduction in greenhouse gases, now committed to by presidential nominees of both political parties, but will instead produce further economic problems.

US environmental policy has been intensively studied by academic scholars, law programs, as well as think tanks, environmental and industry associations,

Congressional committees, and independent authors. A query on "environmental policy" alone brings up 30,000 books in the *Amazon.com* web site's book sales section. These key words in the *Google Scholar* search engine bring up over 2 million scholarly articles and books. And this does not take into account the enormous mass of material in the web sites, blogs, popular media, and Congressional hearings and reports. But it is almost impossible to get a solid understanding of any major policy issue in the United States in part because of the overload of superficial or partisan treatments in popular media and political campaigns. Much of the scholarly litera-ture is fragmented through attempts at theoretical analysis of "wicked problems" (multidimensional issues), while underrating nonquantifiable but important issues. Social scientists themselves have acknowledged that academic approaches have had meager influence on practical policymaking.

I suggest that a clearer picture emerges from following the story across the divid-ing lines and from many perspectives in an historical approach, moving from early US history up to the 1960s and from the present backwards to meet at the 1960s. My own professional scientific career and postretirement engagement with public policy roughly equals the span of time during which the conflict began and evolved to its present state. I had relevant industrial background as a student before the 1960s, when I worked for an international oil company before doctoral studies in Sweden. I subsequently worked for the federal government as an ocean and earth scientist, with special interest in ocean sediments, mineral resources, and coastal-estuarine environmental issues. I drew on background in both environmental science and policy activity to participate in studies of the Hurricane Katrina catastrophe.

My earlier avocational interest in science policy was fired when I noticed distor-tions in scientific and regulatory politics as well as social attitudes in the early 1970s while working in the internationally known ocean-science center at Woods Hole, MA. Federal employees like me were subject to the constraints against dabbling in policy issues. However, I participated in academic, interagency, and National Academy of Sciences panel studies and compiled shelves full of documents. During two leaves of absence, I led academic ocean policy research projects at academic institutions.

My postretirement association with public policy at George Mason brought me to intensive study of social science disciplines as a "second language." Like many persons from different perspectives, I experienced some of the widespread cynicism about the US conflict and gridlock. One recognized political scientist has stated in print that developing a rational environmental policy in the United States is impossible.

But new insights emerged after I updated background from my earlier doctoral studies in Sweden and explored comparative policies in Europe, Canada, Australia, and Japan. Research trips in Scandinavia, Germany, and other European nations provided eye openers. How did the Germans go from "red" Rudi Dutschke and violence over "green policy" earlier in Germany's post-World War II history to its present leadership in both industrial sectors and renewable energy development? Why do not nations in the EU have our gridlock? Why have they passed us in environmental performance, and achieved consensus on actions relevant to global climate change? How have they maintained traditional industries while we have lost a significant part of ours. And why are insights from abroad not better known and used in the United States? These are questions taken up in this book.

I have tried to keep up candor on subjects that are otherwise avoided. But I have also discovered that identifying problems need not degenerate into a blame game. The historical approach shows that attitudes and behaviors rarely emerge spontaneously, and that there may be complex reasons for actions we do not like. There are positive developments; even the potential forthcoming problems may create incentives for the United States to "get serious" and move away from confrontation and toward cooperation.

Part I is designed to be a readable – if sometimes fact-heavy account with tables and illustrations. Supplementary documentation, extended notes and case examples, and policy discussion are placed in Part II. Notes and references come after Part II, a style that has been adopted by other books that want to engage a broad readership while documenting points made both for interested nonspecialists and professionals.

Acknowledgments

My most important acknowledgement is to my wife, Lucy McCartan Manheim. Her combination of a Ph.D. geology background with broad peripheral knowledge, help in editing, organizing, and graphics, gimlet eye for obscure or bad text or too much academic caution, plus dedication to our research theme, made this book possible. Next, the major questions that underlie this book were launched with the help of insights and balance (indispensable for such a controversial topic) of Don E. Kash, the last Chief of the USGS Conservation Division, and John T. Hazel Sr. and Ruth D. Hazel Professor of Public Policy at the School of Public Policy. I could not have had a better academic setting for this work than the School of Public Policy at George Mason University, under the leadership of Dean Kingsley Haynes, who also provided key suggestions at critical points in my work. I must mention Greg Fuhs, an exceptionally diligent and skilled graduate student in environmental policy, who assisted in preparation of the environmental law database that provided important data for various parts of this book. I am deeply grateful to the many persons who assisted in this work through their knowledge, suggestions, discussions, and help in a variety of ways. It would be hard to know where to stop were I to try to enumerate them all, but their assistance is not forgotten. I will here mention only a few of the pioneers in the "environmental 1970s" and later, whose comments at the very end of preparation of this book led to breakthroughs in understanding of the still poorly documented origin of the major environmental laws and the Clean Air Act Amendments of 1990: Russell E. Train, former Chairman of the Council on Environmental Quality (White House) and EPA Administrator; John C. Whitaker, former White House staff member and Undersecretary of the Interior; Richard E. Ayres, cofounder of the Natural Resources Defense Council and a leading national expert on air pollution; Leon Billings, former administrative assistant to Senator Edmund Muskie on the Senate Committee on Air and Water Pollution, and an expert on both environmental law and Senator Muskie; and William K. Reilly, former EPA Administrator. William van Ness, former chief counsel of Senator Henry Jackson's committee, provided insights on the formation of NEPA. Of course, none of the named persons has responsibility for interpretations made in the book. Finally, I express my debt to Barbara Fess, Springer editor whose interest led to completion of this book, and with whom I had many useful discussions. Her place is now taken by Jon Gurstelle, aided by Gillian Greenough, whose patience I have surely tried, and whose cooperation and help I warmly acknowledge.

Contents

Part I
The Story

Chapter 1
Our Current Conflict

1.1 Politics in America

Politics in America has often been characterized by challenge and clashes. But in the past decade, political polarization has grown to levels not matched in most observers' memory. The current, deeply imbedded hostility among warring political factions in the United States is not limited to Congress and the White House. Ideological conflict extends into leadership circles of powerful sectors of society such as the environmental movement versus industry; social liberals, media, and civil libertarians against conservatives and religious groups; and groups polarized around immigration. Although the conflict intensified during the George W. Bush presidency, a list of titles and dates of books on the standoff makes it clear that the rift did not begin with the current administration (Table 1.1).

> *Polarized Politics*
>
> "In earlier days the end of policy battles meant that both sides accepted compromise because the game had been played the way the Framers had intended. [Beginning in the 1980s] politics in Congress and the White House became much more partisan. This time there were few congratulations offered to tough-but-respected opponents for a hard-fought struggle played by the rules. No! Today every new struggle seems to be viewed as another sortie in a partisan war, with each party seeking to use its control of a seat of government power to demonize the other for partisan advantage. These battles reflect a more general change: the emergence of *polarized politics* (Bond & Fleisher, 2000)."

Political ratings of US Senators and Representatives have been compiled for many years by the liberal group, Americans for Democratic Action (ADA), and by the American Conservative Union (ACU). In earlier years, these indexes, based on individual Congressional members' votes on key issues, showed that each party had a significant fraction of its members that took centrist positions or leaned toward the

F.T. Manheim, *The Conflict Over Environmental Regulation in the United States*, DOI 10.1007/978-0-387-75877-0_1, © Springer Science+Business Media, LLC 2009

Table 1.1 Book titles
evoke societal conflict

Book title	Year
Fight Club Politics	2006
Divided States of America	2006
Stalemate	2003
Polarized Politics	2000
The Dysfunctional Congress?	1999
Divided Government	1996
Out of Order	1993
Divided We Govern	1991

Source: Published books selected by
F.T. Manheim

opposite party. This kind of diversity within the parties has now almost disappeared (Fig. 1.1).

On both sides of the aisle, many members of both houses of Congress have "perfect" (100%) voting records – i.e., always vote with their party leadership – whereas formerly, 100% records were rare. Crossover voting records (i.e., more than 50% of a member's votes favoring the other party) as shown in Fig. 1.1 are now represented only by Representatives and Senators of one party who represent a state or district in which voters are primarily of the opposite party. Independents are grouped with Democrats as shown in Fig. 1.1.

When a party no longer has diversity, further polarization tends to take place, because formal resistance to the lobbying efforts of the most zealous or ideologically driven members is lessened. Outside groups can also add pressure to move the party toward ideological goals.

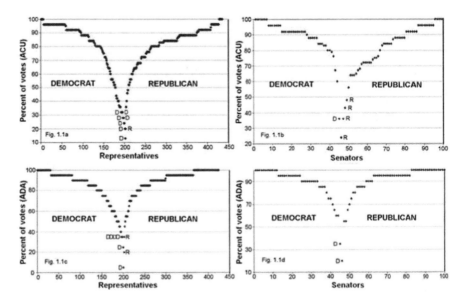

Fig. 1.1 Voting patterns for members of the U.S. House and Senate for 2006
Source: Figs. 1.1a and b, ACU; Figs. 1.1c and d, ADA. D, Democrat (including Independent); R, Republican

A notable exponent of pressure tactics to gain party conformity was former Republican House majority leader, Tom Delay. Delay was labeled "The Hammer" for his heavy hand in enforcing Republican Party discipline. This included punishing Republican Party members who opposed him (Grunwald, 2006). Delay resigned after Tony C. Rudy, his former deputy chief of staff, pleaded guilty to conspiracy and corruption charges related to enterprises run out of Delay's office (Weisman & Cillizza, 2006). However, even in his farewell speech to his colleagues in the House, Delay extolled fierce partisanship (Delay, 2006).

Tom Delay, an Apostle of Partisan Battle

"Over the course of 22 years, I've probably worked with and against almost everyone in this chamber at least once. I have scraped and clawed for every vote, every amendment, for every word of every bill.... and if given the chance to do it all again, there's only one thing I would change: I would fight even harder (Delay, 2006)."

On the Democratic side, Senator Joe Lieberman, the Democratic Party's Vice Presidential nominee in the Presidential campaign of 2000, also tasted the lash. In spite of his long and strong identification with the party, his resistance to conforming to the dominant Democratic position on the Iraq war cost him the Democratic senatorial nomination in Connecticut in 2006. Lieberman was only able to hold onto his Senate seat by running as an Independent.

Differences between the Senate and the House of Representatives have also increased. For example, from 1947 to 1970 the percent of bills that passed both houses but failed to gain agreement in conference was only 1.8%. By the late 1990s, a quarter of bills were killed at the conference stage (Binder, 2003). Other effects reported to be fostered by partisan polarization include delay in federal appointments (Mackenzie, 2001), decreased administrative efficiency, reduced minimum wages, and ability to respond to adverse economic conditions. Outside the Congress, political polarization has affected litigiousness, activism by judges of both lower and higher courts, and cynicism on the part of the general public.

Senior political figures deplore polarization:

"Partisan polarization, if allowed to continue, will destroy our economic, military, social and moral influence in the world, and it will ultimately destroy the fabric of our own country itself."[1]
"I regret that we have fostered a political culture that rewards the extremes, a culture in which dogmatic belief is deemed a virtue and openmindedness a weakness, and sarcasm and slanderous attacks frequently drown out intelligent discussion (Albright, 2006)[2]."

Concerned about polarization, the mayors of the United States' two largest cities, Republican Mayor Michael Bloomberg of New York City and Democratic

Mayor Antonio Villaraigosa of Los Angeles, co-convened a conference, *Ceasefire! Bridging the Political Divide*, in June 2007 (USC_Annenberg, 2007). Even political scientists, who love to challenge conventional wisdom, have not debunked the reality of the existing partisan stalemate.[3]

The impasse has persisted so long that many observers have come to accept it as inevitable or unavoidable. Younger people, who have only known this conflict, cannot be blamed for assuming that it has always been with us – which is not the case.

A political scientist, Morris Fiorina (Fiorina, Abrams, & Pope, 2005), cited poll statistics showing that the general US population was not polarized.[4] Many people do not understand why we need conflict. According to polls, their feelings of frustration about Congressional gridlock contribute to record low job ratings for Congress. The noted social scientist, Seymour Martin Lipset, pointed out that in 1966, 41% and 42% of citizens expressed "a great deal of confidence" in the executive branch and Congress, respectively. By 1994, the numbers had plunged to 12% and 8%, respectively, according to Louis Harris polls (Lipset, 1997). Clearly, the opinions of the general public do not have much effect on the polarization of those leadership groups whose thoughts and actions carry weight in the political affairs of the nation.

1.2 Environmentalists Versus Industry: A Collision Between Two Post-World War II Movements

Nowhere is the split more acrimonious than between environmentalists and the US petroleum industry, and especially the offshore oil industry. The standoff between the environmental movement and the offshore oil industry started when an offshore oil spill triggered events that became a major source of future conflict (see Chapter 2). Don E. Kash[5] has referred to the Santa Barbara offshore oil spill in 1969 as a "collision between two major post-World War II movements."[6] In short, battles over offshore oil go back 35 years and have involved six presidential administrations. The subtitle of a book summarizing the United States' offshore oil industry, *Offshore Development, Conflict, Gridlock* (Gramling, 1996), remains an apt description of an unremitting antagonism that is more intense today than in the 1970s (Bamberger, 2003).

The energy positions of Democratic and Republican Party leaders during the administration of President George W. Bush have modeled the opposite poles of environmentalists and industry. Both President Bush and Vice President Richard Cheney[7] came from Texas oil industry backgrounds; distrust of potential Democratic and environmentalist participation in the development of energy policy was displayed early and prominently soon after President Bush's inauguration in January 2001. A national energy task force headed by Vice President Cheney was convened in April 2001. The task force's unnamed participants met in closed sessions, delivering the National Energy Policy Plan (NEP)[8] in May 2001.

The curious insistence on the Energy Task Force's secrecy brought a predictable barrage of criticism, including reports claiming the vice president had continuing financial ties to the Halliburton Corporation, a leading oilfield service organization. High-profile allegations of financial conflict of interest were found by a nonpartisan media watch organization to be distorted.[9] However, the obliviousness shown by the administration to blatant appearance of conflict of interest, both in the NEP and in a post-9/11, multibillion-dollar, sole-source contract to Brown and Root, a subsidiary of Halliburton, for work in Iraq, was unprecedented.

Why would a new leadership begin its administration with power politics separated from apparent wrongdoing only by legal distinctions? These actions invited cynicism. They promised to arouse not only Democratic Party anger but also more general public distrust. Action on the ensuing energy bill was stalled for several years and never gained its main objectives. I can only interpret such tactics as products of a longstanding partisan conflict that hardened people in leadership circles against expectations of cooperation in national policy. Under this hardball concept of politics, an uncompromising approach to goals would presumably be interpreted as purposefulness and integrity.

Campaigning for the 2004 election, Senator John Kerry, in turn, showed such strong animus toward the petroleum industry that he refused to use the term "oil industry" or "oil companies," referring to them and associated groups almost exclusively as "polluters."

> *John F. Kerry at the Democratic Nominating Convention, 2004*
>
> "I will have a vice president who will not conduct secret meetings with polluters to rewrite our environmental laws."

Even more intense expressions of the anger of many environmentalists appeared in a startling book, *Crimes against Nature* by Robert F. Kennedy, Jr. (Kennedy, 2004), Senior Legal Counsel to the Natural Resources Defense Council (NRDC) and son of the late Senator Robert F. Kennedy. The book is a litany of denunciations and accusations against the Bush administration and its perceived policies of blatant corruption and favoritism to the oil industry and associated interests. The term "startling" is used here because Robert Kennedy, Jr.'s status in a leading national environmental organization, the NRDC[10], meant that NRDC had essentially abandoned any thought of compromise or even communication with the Republican administration or industry representatives. Statements by Robert Redford (Redford, 2005), a board member of NRDC, and by an advocate for the oil industry (Lamb, 2004), offer an example of positions in the standoff.

Environmental advocate, Robert Redford, in statement for the NRDC Action Fund (Redford, 2005)

"The Bush Administration and Congressional leaders are shamelessly exploiting Hurricane Katrina as the latest excuse to hand over the Arctic National Wildlife Refuge to the oil industry.... Given the massive oil spills still devastating the Gulf Coast, it defies belief that our leaders are rushing headlong to hand over America's greatest wildlife sanctuary to the oil lobby.... Instead of making America more energy efficient – the fastest way to meet our energy needs and avoid oil supply shocks – they would sponsor yet another corporate raid on our natural heritage.... This cynical exploitation of a national tragedy has revealed, as nothing else could, the complete bankruptcy of President Bush's pro-polluter energy policies.... Five years of coddling the oil industry has given us higher gas prices and left us more vulnerable than ever to oil shortages – not to mention oil spills, air pollution, despoiled public lands, and catastrophic global warming.... You and I must not let the Arctic National Wildlife Refuge become the next preventable casualty of this president's failed policies.... Within the next few weeks, Congress will cast its make-or-break vote on a Budget Reconciliation Bill that would allow oil drilling in the Arctic Refuge. I urge you to pour your heart and soul into defeating that bill."

Industry advocate, Henry Lamb (Lamb, 2004)

"... ...As gasoline prices continue to climb.... Senate Democrats say they are 'outraged that the administration is not doing everything in its power to alleviate the strain on drivers, consumers and businesses'.... This same Ted Kennedy and Tom Daschle..... have done everything in their power to increase the strain on drivers, consumers and businesses by blocking every attempt to increase domestic oil production... ... Americans have every right to be angry.... But their anger should be directed toward the real cause of the unnecessary price increases: irresponsible reverence for the environment... .... Anger should be focused on the *League of Conservation Voters* and the senator they have endorsed for president. Anger should be focused on the *Sierra Club*, the National Wildlife Federation, Greenpeace, *Defenders of Wildlife* and the horde of environmental organizations that go ballistic whenever anyone proposes to drill a new oil well or build a new refinery.... Had these organizations and their well-funded congressional puppets not blocked exploration in the Arctic National Wildlife Refuge (ANWR) when it was first, oil from that abundant supply would soon be coming on line to relieve pressure that forces prices upward.... But no. In every Congress for a decade, efforts to open ANWR have been met by massive, misleading anti-oil campaigns."

President George W. Bush and Karl Rove, his principal advisor through August 2007, had presumably learned from the earlier successful style of Ronald Reagan not to reply in kind publicly to harsh attacks. Individual oil companies had historical reasons for opting out of open communication or warfare with environmentalists. But until the early 1980s, the American Petroleum Institute (API), the umbrella organization for major oil companies was willing to debate and discuss detailed policy issues.[11] After the furor of Interior Secretary, James Watt's administration from 1981 to 1983 (see Chapter 2), API joined individual oil companies in largely withdrawing from public communications.[12] Subsequently, API has limited itself generally to bland policy statements like the need to reduce America's foreign oil dependence.[13]

It requires digging to find the emotional counterpart to environmentalist expressions on the part of industry leaders and entrepreneurs. They may be reflected in comments of oil field columnist and blogger, Henry Lamb (Lamb, 2004). Lamb expresses the bafflement and anger of industry and supporting business groups over what they regard as irrational legislative, regulatory, and legal obstruction of the development of energy resources that are available and needed by the country. Regarding claims about the environmental risks of offshore oil production, the oil industry and its partners cannot understand why 50 years of technology advances and increasingly stringent regulatory standards should not permit opening government-controlled lands to expanded leasing for oil and gas.

1.3 Battles over Offshore Oil and ANWR

A practical consequence of the conflict over resource and regulatory policy during the first Reagan administration is that 80% of the US Exclusive Economic Zone (EEZ) in the lower 48 states, an area extending 200 miles seaward from the coast, remains under moratorium or is otherwise inaccessible for leasing for oil or gas (see Chapter 2). Leasing a coastal strip of the ANWR, estimated by the US Geological Survey (USGS) to hold from 4 to 12 billion barrels of oil and from 0 to 10 TCF (trillion cubic feet) of gas (Bird & Houseknecht, 1998), has been repeatedly opposed and successfully blocked by Democrats and a minority of Republicans in Congress for 25 years. ANWR evokes passionate and symbol-rich feelings in opponents, as a typical appeal indicates.[14]

Prudhoe Bay, part of the Beaufort Sea, is the source of most of the current oil production from Alaska. This oil is piped to the United States via the Trans Alaska Pipeline (Fig. 1.2). Large additional oil resources, up to 27 billion barrels, were identified in the Beaufort Sea area (Mouawad, 2007). These have taken on special significance with the rise in oil prices. Following lease sales and environmental impact statements, a lease was granted to Shell Offshore Inc. by the Minerals Management Service on February 15, 2007. However, on August 17, 2007 the Federal Appeals Court for the 9th Circuit granted a petition brought by the Alaska Wilderness League, Natural Resources Defense Council, and Pacific Environment and Resources Center with local residents (USC, 2007) to stay the Shell lease.

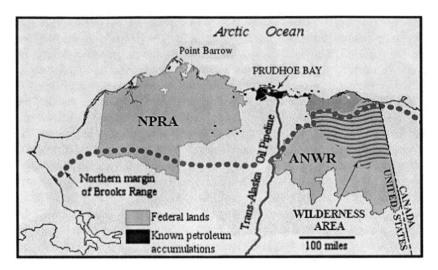

Fig. 1.2 Alaska North Slope, the National Petroleum Reserve Alaska (NPRA), and the Arctic
National Wildlife Refuge (ANWR)
Source: USGS

In 2006, a Congressional bill proposed modification of existing Atlantic offshore moratoria to allow southeastern Atlantic coastal states the option to authorize
drilling and leasing in federal waters off their coasts, as well as a greater share in
revenues. Virginia's small slice of the federal EEZ is in a region where natural gas
has been identified as the main potential resource. Gas is a commodity of special
value for its cleaner-burning quality compared with oil or coal, as well as its short
supply and increasing cost. However, Democratic Governor Tim Kaine blocked
action on a legislative study recommending opening the Virginia sector to offshore
exploration (Valtin, 2006).

These latest examples of thwarted expansion of leasing for oil and gas production are examples or results of the fact that US environmental organizations are now
unified in opposing any new development of domestic oil and gas resources, in spite
of oil exceeding $140 per barrel, and the fact that the US imports 60% of its oil. The
greatest impact of the oil shortage is on fuel for the transportation sector (auto, truck,
aircraft, and train), home heating, and petrochemicals. Even under the most favorable
scenarios, alternative (renewable) sources to meet the bulk of these needs will not be
available for decades, during which time shortfalls must be met by imports.

The success of the environmental movement in blocking energy recovery in
new oil and gas areas is not without costs, besides the issue of supply of transportation fuels. Attacked and opposed at every turn by environmentalists and their
political allies, American industry is embittered and cynical about what it regards
as irrational obstruction.

Both contenders in the standoff have taken on something of the attitude that
whatever the other side wants must be bad and should be opposed. Whereas environmentalists were able to stop expansion of oil and gas leasing and other industry

objectives, industry in turn used its influence to oppose (1) adoption of a national policy and timetable for energy conservation, (2) mandatory targets for reducing greenhouse gas emissions, (3) association with or serious negotiations by the United States to join major international agreements such as the Kyoto Protocol, and (4) accelerated replacement of conventional fossil fuel energy sources in the US economy by renewable energy sources. In short, both sides of the stalemate thwarted each other's goals, while not achieving their own (Manheim, 2004). This counter-productive status is opposite to conditions in the European Union (see Chapter 6). There, cooperation toward both environmental and economic goals (sustainable development) has gained universal acceptance, and is regarded as essential to trans-form energy use from fossil fuels without severe economic consequences.

1.4   Isolation of Information Systems Among Environmental Activists, Academic Analysts, and Producers

Six major groups produce communications about environmental policy and regulation. These are (1) environmentalists and their nongovernmental organizations, (2) industry and business and their trade organizations, (3) academic and professional research-ers, (4) popular media, (5) think tanks, and (6) government. Analysts and writers in think tanks may start as academic professionals, but usually have applied policy goals. Government, of course, is a huge source by itself, including federal agency publications and Congress, with its hearings, reports, and bills. The information systems used by environmentalists and industry are inherently asymmetric. A consequence of the long conflict over environmental policy has been their almost complete mutual isolation. Let us move back a step or two to look at information systems in perspective.

1.4.1   Environmentalist Communication

American naturalists in the nineteenth century, beginning with John J. Audubon shared their studies and love of nature in art or writing (see Chapter 3). Modern American environmentalists, with roots in the love of nature exemplified by John Muir, are noth-ing if not articulate and passionate in communicating. Beginning in the late 1960s, a new type of environmental organization with a dominant focus on political activism emerged (Environmental Defense Fund, 1967; NRDC, 1969; Greenpeace, 1971; Public Interest Research Groups (PIRG), 1972). After the "Watt wars" of the early 1980s older naturalist organizations, like the Sierra Club also moved toward political activism, their clout greatly enhanced after the Internet became widely accessible. Membership in environmental organizations is over 12 million nationally.

   In the past two decades, the goals and causes of the environmental movement have been widely publicized through a variety of media, including websites, mailings, newspaper and magazine articles, blogs, and effective, controversial TV specials such as those of Bill Moyers (Moyers, 2005), sponsored by the Corporation for Public Broadcasting and foundations. The environmental view is well represented in a majority of colleges and universities, where it may combine politically with PIRGs[15] and other environmental organizations.

1.4.2 Industry Communication

In contrast, industrial or business companies' focus is primarily on their products, services, and competitiveness. Within the corporation there is usually not the time, incentive to think about, or authorization to write about environmental issues beyond practical needs. The former practice of some corporate heads to speak candidly about environmental or policy issues was chilled by the political fate of candid comments by Charles Wilson, head of General Motors, in his Senate Confirmation hearings as Secretary of Defense in 1953[16] and by remarks of the President of Union Oil Co., Fred L. Hartley, after the Santa Barbara oil spill of 1969.[17] Both men's comments occasionally turn up in publicity campaigns taking critical aim at the business community.

Academic and research institutions that include focus on industrial perspectives are largely restricted to universities where regional petroleum activity is located, as in Alaska, Louisiana, Mississippi (Center for Marine Research and Environmental Technology, 2005), Oklahoma, and Texas. Much of the hiring for the petroleum industry is done from the pool of graduates who have become members or participate in meetings of local geological, engineering, or other professional associations in the oil industry areas of the country. This localization of hiring differs from patterns before the mid-1960s.

Natural resource or energy companies must plan more than 20 years into the future. In interfacing with regulations and society, such as preparing NEPA impact statements and presenting plans to public agencies, corporate leaders and their legal and policy staffs often coordinate with specialized contractors. However, main centers of policy development for business and industry have been omnibus associations like the US Chamber of Commerce, National Association of Manufacturers (NAM), and trade associations like the American Chemical Council, The Iron and Steel Institute, and the API.

In the early postwar decades, an open communication between the oil industry and the academic earth science community flowered. Oil companies hired many of the brightest earth scientists. The American Association of Petroleum Geologists' various publications media published key articles, treatises, and textbooks on geological and ocean research. Besides sponsoring workshops and supporting special publications, API also provided grants for geological and other scientific research.

API published an informative magazine, *Petroleum Today*, which reported on technical as well as policy notes beginning in 1959. One note from 1976 cited Thomas P. O'Neill, Lieutenant Governor of Massachusetts, who would later become the Democratic Speaker of the House of Representatives from 1977 to 1988.[18]

Ties between industry and academia dwindled after the late 1960s. *Petroleum Today* ceased publication in 1977, a reflection of growing polarization. The last major policy document released by API known to me appeared in 1980 (American Petroleum Institute, 1980). After 1989, API appears to have mainly kept communications within itself and its member industry leaders.

Becky Norton Dunlop, Vice President of External Relations for the Heritage Foundation, a leading conservative think tank, told me that conservatives were far behind liberal organizations in building an intellectual base for political policy analysis. She remarked, "Before the 1980s there were only a few books on public policy." The lean resources for analysis of political action received a major boost with the large-scale strategic planning and talent recruiting effort mobilized for the election of Ronald Reagan in 1980 by the Heritage Foundation (Heatherly, 1980). Buoyed by the Reagan administration and support of conservative foundations, conservative think tanks grew rapidly in the 1980s.

1.4.3 Industry Lobbying

Since the early twentieth century, lobbying in Congress, personal networking, and special advertising campaigns have been a significant part of industrial communication and politicking. An industry critic, Sharon Beder, is an Australian civil engineer who became a prolific environmental writer and passionate muckraker about Australian and US corporate activities. Beder recounts that by the early decades of the twentieth century, the Natural Association of Manufacturers (NAM) began systematic lobbying of Congress against union organizing, and sought to influence local elections (Beder, 2006a). With the capitalist system under attack during the Depression, business responded with a nationwide publications effort to promote free enterprise. Beder charges that business-allied groups in the United States have preached a free market ideology that corrupted American society and diverted the public and government from better controlling business activities. After a quiescent period following World War II, according to Beder, major campaigns resumed in the mid-1970s under leadership of the Advertising Council (Beder, 2006b). By the 1980s, the "anti-materialism of the hippie generation gave way to a more consumerist orientation," and corporate executives again returned to give courses and recruit on campuses.

Sharon Beder, an Australian civil engineer who became a prolific writer on environmental issues and a muckraker regarding corporate politics, has been raising alarm about the political mobilization of business as a class.... "The purpose of this [free market] propaganda onslaught has been to persuade a majority of people that it is in their interests to eschew their own power as workers and citizens, and forego their democratic right to restrain and regulate business activity (Beder, 2006a).

1.4.4 Academic Publication – The Separation of Theory from Practice

A query on "environmental policy" in Google Scholar browser yields 2,050,000 titles. These are not websites. These are scholarly articles and books! A search on the Amazon online bookseller list yields over 30,000 books (the Google book browser yields 38,700). Most of this mass of material is by university faculty and other scholars whose jobs are to teach and write about environmental policy, and who seek objective, verifiable research results. As a natural scientist who came to the social sciences as a "second language" in order to access knowledge about resource and environmental policy, I recognize the virtually hopeless task policymakers face if they were to try to extract practical guidance from this huge mass of information. Social scientists themselves have documented that policymakers do not try (Shulock, 1999).

Government agencies and other organizations use specific techniques developed by academic disciplines to quantify aspects of their work, but the fragmented literature on environmental policy in political science, sociology, economics, public administration, public policy, law, and other disciplines or approaches is essentially ignored for practical policymaking (Bogenschneider, 2002; Bromley, 1991; Sauri-Pujol, 1992).

Communication problems are not limited to the social sciences. In the case of Yucca Mountain as a high-level nuclear waste repository, a political scientist specializing in earth science noted that "the work of geologists and other scientists was rendered irrelevant in the policy debate (McCurdy, 2006)." Turning the acceptance of academic science for use in policymaking around, a memorial to the pioneering achievements in sustainability by the late geographer, Gilbert F. White[19], pointed out that whereas White was politically influential, among other things, questioning the usefulness of huge waterway diversion projects, the fact that his work focused on pragmatic results and did not respect disciplinary boundaries meant that "he was never fully embraced by the cloistered experts in the field" (Cohen, 2006). In other words, "normative" or applied assessments useful for decision makers have lower prestige (and professional promotion value) with many or most academic faculties than do empirical (quantitative) studies. Social scientists with whom I have talked usually recognize the dilemma, but feel powerless to do anything about it.

1.4.5 Popular Media, Blogs, and Government Publications

In addition to the "professional" literature, more than 450,000 blogs[20] are retrieved with the query, "environmental policy" by the Google Blog browser, along with almost numberless websites, and newspaper and other media references. (Federal science agencies and government publications will be discussed in Sections 3.4.2 and 3.4.3).

1.4.6 The Isolation of Information Systems Is Revealed in the US Global Climate Change Debate

In 1988, environmental modeler, James Hansen, gave his famous testimony to a US Senate panel.[21] Because top scientists are generally leery of making firm predictions, Hansen's blunt statement that global warming was occurring, and his detailed predictive maps of future temperature distributions throughout the world, made headlines in popular as well scientific media (Hansen, 1988; Hansen et al., 1988). Since then there has been a crescendo of news about global climate change, involving scientific articles, press releases about new research findings and environmental models, and conference and panel reports, capped by the Nobel Prize awarded to former Vice President Al Gore for his documentary film, *An Inconvenient Truth* (Gore, 2006), and to the Intergovernmental Panel on Climate Change for their definitive work (IPCC, 2007).

Feeding critical countercurrents was an earlier book criticizing environmental orthodoxies by a young Danish statistician and Greenpeace member, Bjørn Lomborg (Lomborg, 2001).[22] Displaying a virtuosic ability to chart disquieting anomalies in complex technical parameters, Lomborg unleashed a storm of controversy. Besides strong critiques in leading US scientific publications (*Scientific American, Science*), Danish environmentalists filed complaints charging scientific dishonesty by Lomborg to an ethics body under the Danish Ministry of Science, Technology, and Innovation. After several turbulent reversals, the charges were dropped. A center-right political administration in Denmark placed Lomborg in charge of a new Environmental Assessment Institute.[23]

In spite of the expanding volume of scientific articles, blue ribbon panel reports, and consensus among most European Union leaders that human emissions of CO_2 have already caused noticeable warming and posed dangers for the future, a remarkable level of skepticism in the United States about human effects on global warming remains. Leading US oil company chiefs startled many people concerned about global climate change by their noncommittal testimony in a Congressional hearing (Isidore, 2005). Skeptics cite dissident scientists and experts like Fred Singer and Dennis Avery (Singer & Avery, 2007, and references cited in their book). The idea that skeptics are limited to a few individuals is contradicted by the policy statements of the American Association of Petroleum Geologists as well as by a few hundred scientists who signed a petition sent to them by the Oregon Institute of Science and Medicine (OISM).[24,25] Another outpost of widespread skepticism was the *International Conference on Climate Change,* March 2–4, 2007 in New York City, among whose scheduled speakers was Czech Republic President Vaclav Klaus (Heartland, 2007). However, the IPCC 2007 report and periodic news reports of dramatic events, such as the rapid shrinking of the Arctic ice cover, have added momentum to the mainstream consensus about the urgency of action to curb greenhouse gas emissions.

Environmental groups monitoring industry activities like Exxon Mobil's support for conservative think tanks and skeptical scientists have denounced industry for obstructing political attempts to reduce greenhouse gas emissions; they cite the creation of front groups to confuse the public and oppose measures to prevent global warming (Greenpeace, 2007). A cottage industry has developed on both environmentalist and conservative sides to track funding sources and interrelationships between politically active organizations.[26]

Most scientists and environmentalists, bombarded with news from their media, feel that the growing evidence and clear urgency for taking action to control greenhouse gas emissions must be plain and convincing to every technically aware person. However, the assumption that US industry leaders are informed by comparable communication systems and have confidence in mainstream consensus may not be accurate. Although some leading companies – especially in industries that benefit from new energy developments, like General Electric Co. and DuPont – have embraced new paradigms, it is easy to underestimate the information barrier in the United States and the suspicion on the part of industry about the validity of scientific "consensus" position in which environmentalists lead or concur.[27]

> The assumption that US industry leaders are informed by communication systems comparable to environmentalists' and have confidence in mainstream consensus may not be accurate.

At least one influential Senator, James Inhofe (R) of Oklahoma, ranking member of the Environment and Public Works Committee, publicly pronounced the climate change movement a "hoax" as late as January 2005 (Inhofe, 2005).

1.4.7 Militancy of US Environmental Organizations

Environmental organizations since the 1980s have become more militant and ideological than their US counterparts in the 1970s and their current counterparts in Europe.[28] Most categorically reject any expansion of domestic fossil fuel production even where the net result of such production is an interim strategy. This is true even where the net result would reduce greenhouse gas emissions or have longer-term energy policy benefit. For example, successful offshore drilling off Virginia would likely recover gas, not oil. Use of cleaner gas would reduce greenhouse gas emissions in comparison with the current dominant Virginia energy supply from coal (see Chapter 8). Even if no commercial hydrocarbons were found, the revenues accruing to the state of Virginia could be used to subsidize environmentally beneficial technologies like carbon storage (Wagner, 2006). However, efforts to promote this option have encountered strong opposition.

The Sierra Club, the oldest environmental organization in the nation, prides itself on leading all other environmental organizations in the number of lawsuits that it files.[29] Its claim is qualitatively confirmed by a "litigation index" computed for a number of leading organizations (Table 1.2). The Sierra Club is followed by the NRDC and Environmental Defense Fund (EDF). The NRDC has the largest scientific and legal staff among environmental NGOs, and is famed for its aggressive litigation in environmental issues. Smaller organizations also show eagerness to do legal battle with business or governmental adversaries. For example, the motto of Friends of the Earth, displayed on its web site, is "Because the earth needs a good lawyer." An underevaluated issue is the effect of the current regulatory system, climate of litigiousness and suspicion, and the associated NIMBY (Not In My Back Yard) issue on development of renewable energies in the United States (see Chapter 4).

Table 1.2 Ranking of leading environmental organizations based on litigation activity

Nongovernmental activist environmental organization (NGO)	Members	Website	Founding Year	Legal action group[1]	Program outlay (latest, in million $)
Sierra Club	750,000	sierraclub.org	1892	1	53.7
Natural Resources Defense Council	1,200,000	nrdc.org	1969	1	48.6
The Nature Conservancy	1,000,000	nature.org	1951	2	532
World Wildlife Fund	1,200,000	worldwildlife.org	1961	2	126
National Wildlife Federation	4,000,000	nwf.org	1937	2	104
Environmental Defense	500,000	environmentaldefense.org	1967	2	41.5
Defenders of Wildlife	130,000	defenders.org	1947	2	24.6
National Audubon Society	600,000	audubon.org	1905	3	51.9
Earthjustice	70,000	earthjustice.org	1971	3	19.1
The Wilderness Society	300,000	wilderness.org	1935	3	17.9
World Resources Institute	([2])	wri.org	1982	3	17.0
Friends of the Earth[3]	1,000,000	foe.org	1969	3	2.7
League of Conservation Voters	([2])	lcv.org	1969	3	1.9
Greenpeace USA[4]	250,000	greenpeace.org/usa	1971	4	7.8
PIRG[5]	400,000	uspirg.org	1972	4	3.9

Source: NGOs' annual reports, ActivistCash.com, and DiscoverTheNetworks.org
[1] Measure of litigation activity based on number of Google hits for NGO's name plus "lawsuit," and name plus "litigation;" 1 is highest
[2] No members; staff performs research, information dissemination, political action
[3] Membership number is for Friends of the Earth International
[4] Greenpeace USA financial information is not available; Greenpeace Fund information appears in "Program outlay" column
[5] PIRG financial information is not available; PIRG Education Fund information appears in "Program outlay" column

Two organizations shown in Table 1.2 stand apart from the others. The World Resources Institute has no members and files few lawsuits. Instead, it has a large staff funded by grants and donations, who conduct research and topical reviews. World Resources Institute (WRI) actively seeks partnership with business and other entities. Nature Conservancy also, for the most part, eschews confrontation and litigation; its main activity and mission is to buy and preserve environmental properties. We note from the table the large aggregated budgets of the cited environmental organizations.

Environmental activist Non Governmental Organizations (NGOs) are funded largely by major foundations, many of which were created by corporations (Fig. 1.3). Curiously, the NGOs oppose actions of many corporations whose foundations originally funded them (Table 1.3).

LEADING ENVIRONMENTAL ACTIVIST NGOs	ED	NRDC	PIRG	NWF	WWF	SC	FOE	WS	EJ	GPU	NC	WRI	DW	LCV	NAS	NUMBER OF NGOs FUNDED
MAJOR FUNDERS																
Blue Moon	x	x	x	x	x	x	x	o	x	x	x	x	o	x	o	12
Bullitt	x	x	x	x	x	x	x	x	x	x	o	o	x	x	o	12
Beldon	x	x	x	x	o	x	x	o	x	x	o	o	x	x	x	11
MacArthur	x	x	o	o	x	x	x	x	x	x	x	x	o	o	o	10
Pew	x	x	x	o	x	x	x	x	x	x	x	o	o	o	o	10
RockefellerB	x	x	x	x	x	x	x	x	o	x	x	x	o	o	o	10
Hewlett	x	x	o	x	x	x	o	o	o	o	x	x	x	o	o	8
Tides	x	x	o	x	x	x	o	x	x	o	x	o	o	o	o	8
RockefellerFa	o	o	x	x	o	o	x	x	x	x	o	o	x	o	o	7
Schumann	o	x	x	x	o	x	x	x	x	o	o	o	o	o	o	7
Ford	x	o	o	x	x	o	x	o	o	o	o	x	o	o	o	5
Dodge	x	x	o	o	o	o	o	x	o	o	o	o	x	o	o	4
Goldman	x	o	o	o	x	o	o	x	o	o	o	o	o	o	o	3
Heinz	x	x	o	o	o	o	o	o	o	o	x	o	o	o	o	3
Moore	x	o	o	o	x	o	o	o	o	o	o	o	o	o	o	3
Packard	x	o	o	x	x	o	o	o	o	o	o	o	o	o	o	3
Brainerd	o	o	x	x	o	x	o	o	o	o	o	o	o	o	o	2
Carnegie	o	x	x	o	o	o	o	o	o	o	o	o	o	o	o	2
Johnson	o	x	o	o	o	o	o	o	o	x	o	o	o	o	o	2
Mellon	x	o	o	o	o	o	o	o	o	o	o	x	o	o	o	2
Merck	o	x	x	o	o	o	o	o	o	o	o	o	o	o	o	2
Duke	o	o	o	o	o	o	o	o	o	o	x	o	o	o	o	1
Mott	o	o	x	o	o	o	o	o	o	o	o	o	o	o	o	1
RockefellerFo	o	x	o	o	o	o	o	o	o	o	o	o	o	o	o	1
NUMBER OF MAJOR FUNDERS	15	15	11	11	11	10	9	9	8	8	8	6	5	3	1	

Fig. 1.3 Funding sources of leading environmental activist NGOs; Table 1.3 gives full names of NGOs and funding organizations
Source: Compiled by F.T. Manheim from online foundation and NGO documents, ActivistCash.com, and DiscoverTheNetworks.org

Table 1.3 Key for Fig. 1.3

Symbol	Organization
DW	Defenders of Wildlife
ED	Environmental Defense
EJ	Earthjustice
FOE	Friends of the Earth
GPU	Greenpeace USA
LCV	League of Conservation Voters
NAS	National Audubon Society
NC	The Nature Conservancy
NRDC	Natural Resources Defense Council
NWF	National Wildlife Federation
PIRG	Public Interest Research Group
SC	Sierra Club
WRI	World Resources Institute
WS	The Wilderness Society
WWF	World Wildlife Fund
Symbol	Major NGO funding sources
Beldon	Beldon Fund
Blue Moon	Blue Moon Fund
Brainerd	Brainerd Foundation
Bullitt	Bullitt Foundation
Carnegie	Carnegie Corporation of New York
Dodge	Geraldine R. Dodge Foundation
Duke	Doris Duke Charitable Foundation
Ford	Ford Foundation
Goldman	Goldman Fund
Heinz	Heinz Family Foundation
Hewlett	William and Flora Hewlett Foundation
Johnson	Robert Wood Johnson Foundation
MacArthur	John D. and Catherine T. MacArthur Foundation
Mellon	Andrew W. Mellon Foundation
Merck	Merck Family Foundation
Moore	Moore Foundation
Mott	Stewart R. Mott Charitable Trust
Packard	David and Lucile Packard Foundation
Pew	Pew Charitable Trusts
RockefellerB	Rockefeller Brothers Fund
RockefellerFa	Rockefeller Family Fund
RockefellerFo	Rockefeller Foundations
Schumann	Schumann Center
Tides	Tides Foundation and Center

Source: Compiled by F.T. Manheim from online foundation or
NGO documents, ActivistCash.com, and DiscoverTheNetworks.org

Chapter 2
Tracing the Roots of the Conflict

After World War II, the United States led the world in economic power and technology. The United States experienced boom conditions with only brief recessions through the 1960s, but in hindsight, we can discern trends that interacted to create stresses. These trends include suburbanization, with decline of public transportation systems in favor of the automobile; the growth of a new science paradigm, in which research scientists replaced engineers as leaders in national science management; the fragmentation and selective neglect of societal management and infrastructure; the rise of the modern offshore oil industry; the civil rights movement; a new environmental movement sparked by Rachel Carson's book, *Silent Spring*; and countercultural and antibusiness movements of the 1960s, aggravated by the Vietnam War.

Two of the movements, the offshore oil industry and the environmental movement, collided at the Santa Barbara oil spill of 1969. Preceded by rising pollution and other instabilities, the offshore oil spill created a sense of national crisis. It catalyzed passage of a series of groundbreaking laws that transformed US resource and environmental policy, and created new roles for Congress and the federal courts. The new system achieved rapid progress against pollution and other environmental problems. But it simultaneously opened up a rift in society. In the 1980s, the rift widened into political polarization (Fig. 2.1). Concern about global climate change has become an added source of potential conflict and malaise.

2.1 Engineers and Pre-World War II America

Prior to the 1950s, American engineers enjoyed leadership status in society. Respect for engineers was based on their reliability and their role in providing essential services to the growing society. Engineers designed and oversaw construction, power, and transportation systems in large cities; they designed and operated manufacturing facilities to produce products ranging from aircraft and automobiles to complex pharmaceutical chemicals and drugs. Engineers were in charge of public health systems like water supply and sewage treatment. Their status was enhanced by spectacular achievements like bridges and skyscrapers of unprecedented height like the Chrysler and Empire State buildings in Manhattan (Jackson, 1995).

F.T. Manheim, *The Conflict Over Environmental Regulation in the United States*, 21
DOI 10.1007/978-0-387-75877-0_2, © Springer Science+Business Media, LLC 2009

Engineers and pre-1950s America	A new science paradigm displaces engineers	1950s and 1960s; boom times with growing pollution and other stresses	Rise of the modern offshore oil industry	Rachel Carson and a new environmental movement	The collision at Santa Barbara and the 1970s environmental revolution	The 1970s and 1980s; a rift widens

Fig. 2.1 People, events, and trends leading up to the conflict
Source: F.T. Manheim

Civil engineers' training included business law and finance. It was designed to equip graduates to handle human and administrative factors as well as master requisite scientific and technological skills. Engineers often occupied administrative positions in local, state, and federal governmental agencies (see Chapter 3). A civil engineer, Herbert Hoover, was elected President, in part because of his reputation gained by organizing successful relief and rescue efforts during crises in both the post-World War I Soviet Union, and in the great 1927 flood in the United States (Barry, 1997). Engineers were proud of their status in society. They generally sought to maintain the standards of the profession and public trust (Baltimore, 2007), even where they had to work in conjunction with corrupt politicians.

Structurally and organizationally, the United States before the early 1950s was a far different place than now. Most major cities were centralized, coherently organized complexes where expansion mainly took place in preplanned ways at the edge of town. Planners of a more energy-conserving future may come to see useful insights in that history as reverse movement back to urban centers gains popularity (Lucy & Phillips, 2006).

All components of urban areas – business districts with major hotels, restaurants, entertainments such as night clubs, first-run movie theaters, performance centers, public parks, and sports and recreation facilities; residential areas with school facilities; and manufacturing and industrial districts – were connected by public transportation using street cars, trolleys, and buses. These conveyances were supplemented with subway or overhead rail systems for mass transit in larger cities.

Management of complex urban areas required centralized authority and professional skills. Before the advent of suburbanization and malls, city government and its ancillary public service and engineering departments, as well as state agencies, attracted qualified staff and leaders more easily than in today's highly fragmented, competing systems. Affluent people lived in the cities and had a stake in good services just as working class people did. Offices that managed public service and infrastructural responsibilities conducted regular reviews of transportation, school facilities, street lighting, water supply, and sewage treatment and public health facilities to plan for future needs. These were paid for by issuance of bonds (without federal funding) after review by city leaders, including the elected council, and in some cases, state bodies and the Governor (e.g., where regional waterways were involved, as in Boston).

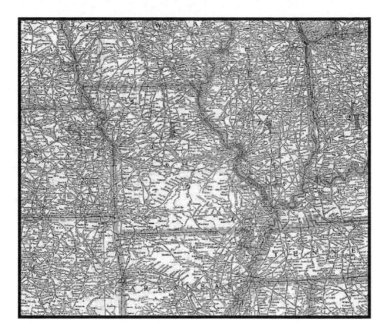

Fig. 2.2 Part of the US railway grid in 1918
Source: Library of Congress

The ability of urban managements from 1900 to the early 1950s to efficiently maintain attractive and efficiently functioning cities with low crime and good public services can seem striking today. Police in major metropolises like New York and Boston did not wear guns until after Prohibition and advent of urban mobs and bootleg liquor in the 1920s. Comments about the ability of women being able to walk at night in Harlem, NY are often heard. And for 44 years, from 1904 to 1948, a bus or subway ride in New York City cost a nickel (Markowitz, 2003).

In my home town of Kansas City, MO, scheduled bus service was maintained to outlying lakes and special recreational areas, as well as truck farms, dairy, and other intensive farming areas around the city. More distant farming areas were served by the dense rail network like that shown for the Midwest in 1918 (Fig. 2.2).[30] Such a rail grid would be a dream for those seeking to reduce greenhouse gas emissions from automobile traffic today.

2.2 1950s–1960s: Environmental and Other Stresses Begin to Erode the Boom

Postwar suburbanization and the accompanying surge in automobile use were stimulated by the availability of low-cost housing loans through the GI Bill and Federal Housing Authority (FHA) mortgages. Robert Moses, the great builder of parks and highways in New York City from the 1930s to 1964, became one of the most powerful men in the State of New York, with influence throughout the country.[31] In spite of his energy and achievements, his projects also created opposition. One of the reasons was his enthusiasm for the automobile and road building, which helped undermine public transportation in the greater New York City area (Caro, 1975). There remain unresolved claims that loss of streetcars and trolleys was accelerated by General Motors and other industries associated with auto use. But there is little doubt that the attraction of more open space, freedom of movement, mall development, and other factors were the prevailing forces in suburbanization.[32] The breakup of the urban system did not take place unnoticed. A classic book about this subject was written by Jane Jacobs, *The Death and Life of Great American Cities* (Jacobs, 1961).[33] I recall articles in magazines from the 1950s deploring "scatteration." The "scatteration" criticism was, in turn, criticized as an attempt by leftist academics to control society.

2.3 A New Academic Paradigm

Engineers do and serve; scientists discover and write

The most dramatic development of World War II, the atomic bomb, lent enormous stature to the renowned international physicists whose brains had conceived this horrific weapon. The most powerful conventional explosive device used during World War II was the blockbuster bomb, which weighed up to ten tons and contained the chemical explosive, TNT (trinitrotoluene). The atomic bombs that exploded over Hiroshima and Nagasaki occupied a space smaller than the blockbuster, but had explosive forces of more than 10 kilotons (thousands of tons of TNT).

Other scientific developments that contributed to the wartime effort included radar to detect aircraft and guide antiaircraft fire; the proximity fuze; geophysics used in antisubmarine warfare; rocketry; and antibiotics like penicillin and sulfa drugs – that greatly reduced deaths and incapacitation from infection in wounds – gave science and scientific research a new prominence. In 1944 Vannevar Bush,[34] science adviser to President Roosevelt and Director of the Office of Scientific Research and Development (OSRD), received a request from the President to develop a postwar science plan. This plan was to turn the awesome power of science, that been used to create weapons of war, to peaceful civilian development after the war.[36]

The influential book that emerged, *Science: The Endless The Frontier* (Bush, 1945), combined with competing ideas proposed by Kentucky Senator Harley Kilgore (Kleinman, 1995), and Vannevar Bush's energetic personal politicking, ultimately led to passage of the National Science Foundation Act, signed by President Truman in 1950.

The spindly National Science Foundation (NSF) Act of 1950 that President Truman finally signed had been stripped of much of what Bush wanted. He wanted military and medical research components. He wanted Department of Defense research to be limited largely to improvement of weaponry, and that NSF be the sole national leader in federal scientific research support. The new organization would be guided by an independent civilian board and have a robust budget (the initial authorization was only $250,000). However, Bush's insistence on the primacy of unfettered basic research stuck.

Bush's "linear concept" was that basic research discoveries by university scientists would automatically stimulate technical advances and vitalization of national life. Initially dwarfed by research support to academia by the Office of Naval Research and other federal agencies, NSF's unrestricted research grants, awarded to competing proposals by peer reviewers, nevertheless gave the new agency and its research grants high prestige. Minimally directed competitive research grants by the military and other agencies (e.g., Atomic Energy Commission and the National Institutes of Health) further enhanced the status of research science and scientists.

A final boost to the clout of academic research was the selection of leading scientists and scientific administrators as advisors to the President and to other key national advisory committees. After George Kistiakowsky, a Harvard professor of chemistry who had worked in the Manhattan Project and served as Dwight Eisenhower's science advisor, more than half of Presidential science advisors were physicists, in part because their scientific expertise had particular relevance to nuclear weaponry and energy during the Cold War. Engineers remained essential in putting scientific concepts into practice. For example, after President Roosevelt authorized development of the bomb, assigning to Vannevar Bush responsibility for organization of the Manhattan Project, Bush (himself an engineer) selected a colonel in that quintessential engineering organization, the US Army Corps of Engineers, Leslie Grove, to lead the effort. However, from the 1950s, scientists replaced engineers as the dominant leaders guiding federal policy in science and technology, and became the primary recipients of federal research funding.

An historian of science policy, B.L.R. Smith, notes that the humble status of academic science as it existed before the war would hardly have remained unchanged given the role of leading scientists in the war effort (B.L.R. Smith, 1990). According to the detailed biography of Vannevar Bush by Zachary (Zachary, 1999), before World War II, just a dozen universities in the entire United States spent more than $1 million for all on-campus research. In contrast, during the war, $3 billion was spent on radar equipment and research alone. The Massachusetts Institute of Technology's (MIT's) radiation (radar development) laboratory had 4,000 staff. The Manhattan Project was more costly than the entire US automobile industry, and at its height employed half a million people (Hughes, 2002). The ultimate outcome of Bush's efforts was that basic scientific research achieved a new prominence in American society (details presented in Chapter 8).

The university system grew rapidly in the 1940s and 1950s, influenced by the GI Bill of Rights (Servicemen's Readjustment Act of 1944), which was signed on June 22, 1944 by President Roosevelt. The bill's emphasis on education for veterans allowed 2,300,000 veterans to attend colleges and universities with most of their expenses paid. Research funding grew to more than $1 billion (in 2000 dollars) by the mid-1950s, dominated (80%) by outlays from the Department of Defense and the Atomic Energy Commission. NSF trailed far behind with 0.1% of the federal R&D budget.

By 1955, the Pentagon's enthusiasm for supporting discretionary scientific research had waned (Dickson, 1988, p. 27). But in 1957, the Soviet Union's Sputnik space satellite shocked the nation and its leadership circles. President Eisenhower moved quickly. In 1958, Congress passed and President Eisenhower signed the National Defense Education Act, which provided robust support to all levels of education, including scholarship aid to the universities – while prohibiting government controls on curricula, programs, and personnel of educational institutions. After 1961, the social sciences also got support and an impetus for growth. For many research scientists, the time from 1960 to 1968 became a Golden Age, during which research scientists consolidated their role as the new scientific elite of the nation (Greenberg, 2001). Federally funded scientific research was divided, in science historian, Derek de Solla Price's terms, into "Big science" and "Little science" (Price, 1963). Big science began as a special province of high-energy physics, promoting ever more powerful particle accelerators.

2.3.1 The Vision and the Reality

Passage of the NSF Act in 1950 gave Congress's and the President's stamp of approval to the *Endless Frontier* vision of basic research as a perpetual wellspring for technical and economic development. As one might expect, most subsequent NSF directors continued to articulate the vision, at least publicly, as did other scientific luminaries like George Kistiakowsky and Jerome Wiesner (science advisors to Presidents Eisenhower and Kennedy), Detlev Bronk, Lee DuBridge, I.I. Rabi, Glenn Seaborg, Michael Polany, and many others.

Doubts and contradictory evidence about the automatic link between basic research and technical and practical advances of value to the nation soon emerged. They increased further from the 1960s on, but the grip of the *Endless Frontier* concept remains strong in the United States. Scientific or intellectual concepts may linger on long after their validity is challenged, as has been noted in the famous book, *The Structure of Scientific Revolutions* (Kuhn, 1996) and by other authors (Hart, 1998).

By the 1960s, Vannevar Bush himself soured on growing trends in America, including scientific research outlays that he had helped initiate. He came to feel that "federal support for research had foundered. The profligacy of the 1960s – fiscal, military, and technological – ran counter to his belief in a limited charter for public enterprise" (Zachary, 1999). He carped that "there was much duplication and repetition of research. *We are being smothered in our own product* (Zachary, 1999)." Bush's complaints were not limited to academic research. He was critical of corporate advertising that was corrupting American tastes, and the US auto industry, that rejected innovation, including his own innovative design proposals for more fuel-efficient engines. Bush was ahead of his time with respect to energy conservation.

At the height of the scientific Golden Age of the 1960s, a thundercloud now remembered by few scientists created near panic in the American Association for the Advancement of Science (AAAS), the NSF, and among many scientific leaders. In 1966 and 1967, the Defense Department (DOD) released the results of Project Hindsight, a massive retrospective study of scientific and technological "events" leading up to 20 major advances in military technology (Isenson, 1969; Sherwin & Isenson, 1966, 1967). Among the conclusions from the work of 13 teams studying developments from 1946 (and earlier) to 1963 was that only 0.3% of contributions to the military breakthroughs (including satellite navigation) came from undirected university science.[36] The DOD study's implications that undirected basic research of the kind supported by DOD (and NSF) since World War II was largely irrelevant to technological applications aroused the AAAS and the NSF to crash projects that would justify the NSF and other grant programs.[37]

Pressure on NSF to seek increased practical yield from its programs mounted. President Lyndon Johnson signed a Congressional bill revising NSF's charter to authorize applied research in 1968 (NSF, 1968, pp. 83–84). A barrage of criticism from the engineering community complained about NSF's neglect of practical applications and engineering.[38] A National Technological Foundation (NTF) was proposed. There were warnings about the United States' export deficits. In 1969, the Nixon administration created the IRRPOS program (Interdisciplinary Research Relevant to Problems of Our Society) and in 1973 expanded it into the RANN program (Research Applied to National Needs) which was imposed on NSF "against its will (Green & Lepkowski, 2006)." RANN created successful innovations, including some 60 alternative energy systems that were subsequently spun off to what would later become the Department of Energy (Belanger, 1998). However, the program was controversial and unpopular among basic research advocates who felt that it had no place in NSF's mandate (Belanger, 1998; Green & Lepkowski, 2006). It was terminated in 1979.[39]

Though the RANN type of explicitly applied science projects did not become integrated into NSF's science mix, NSF did add engineering research support, and in later years cautiously moved toward "interdisciplinary, problem focused research."[40] The concern over global climate change is particularly discernible in the dramatic increase in funding for physics and related fields, noted in Fig. 2.3.

In short, Fig. 2.3 shows a downtrend in technological advances, contrasting with rise in research funding and publications. This relationship supports the criticisms of Belanger (1998) and others that fundamental assumptions of the basic research concept were flawed.

I do not criticize the quality of American research science. Since the early postwar years, American academic research funded by federal granting agencies has usually maintained internationally acknowledged standards of quality. But anyone seriously interested in potential yield to society found no answer to how the enormous mass of twigged disciplinary publications could ever find practical use. In fact, the academic trends were not limited to academia; they also affected federal science agencies[41] and, as noted in the later remarks of Charles Vest,[42] exercised influence on the training of engineers.

Scientific research activity measured in terms of numbers of research publications is poorly – perhaps even inversely – correlated with technical innovation

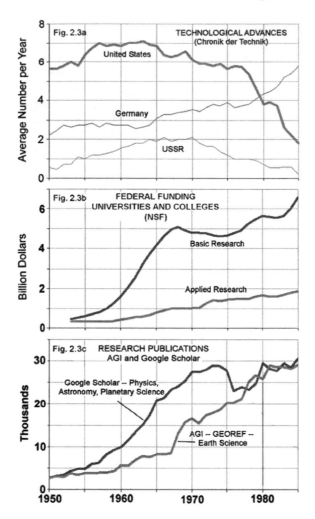

Fig. 2.3 Scientific research trends 1950 – 1985
Source: American Geophysical Union (AGI) (*Georef*, courtesy of Sharon Tahirkeli), *Chronik der Technik* (Paturi, 1989), NSF, and Google Scholar. Figure 2.3a, compiled from the leading chronological compendium of international technical advances, *Chronik der Technik* (Paturi, 1989) shows major advances per year on a five-year running average[*] for the U.S., along with Germany and the Soviet Union. Figure 2.3b shows federal funding for basic research (partly also applied research), from online databases maintained by the NSF (2008). Figure 2.3c provides the number of publications in selected research fields. Physics, Astronomy, and Planetary Science data are derived from queries for given years on Google Scholar. The query used the keywords "U.S.", "USA", and "United States", and excluded "patents". Earth science data are for U.S. publications in the leading English-language abstract journal, *Georef*
[*]This compilation differs from other compilations of advances in science and technology in its detail, use of consistent criteria (it was supported by research grants), and goal of providing data as free from national bias as possible. Its author's reference material included access to information from the Soviet Union through cooperation with the Soviet Union's Ambassador in Berlin

or other social value to the nation. Basic research communities, both natural and social science, tend to keep their work and interests inside "invisible colleges" (Price, 1963; Ziman, 1968).[43] They have justified public support for the research in terms of a powerful concept launched 60 years ago, whereby through undefined but automatic mechanisms, abundant benefits would be delivered to society from basic research – without active or special effort on the part of basic researchers. We can see and judge some of the relationships among these factors in Fig. 2.3, which compares average yearly productivity in technical innovation for the United States with numbers of research publications in specific scientific fields and overall basic research funding.

US technical innovation was high – by far the highest in the world – up to about 1965 (Fig. 2.3a). Thereafter it declined consistently. These were the years when the United States lost industries that it had founded or developed: radio, TV, rubber products, and household appliances of various kinds; and it declined in others areas: steel, automobiles, mining, furniture, footwear, and textiles. Patents, another index of national technologic productivity, slumped between 1965 and the early 1980s to a degree only exceeded during the great depression of the 1930s (data from the US Patent Office). During this time funding for scientific basic research rose dramatically, along with the number of scientific publications.

Studies of mechanisms for technical innovation in society during the past 25 years confirm that useful products do not spring automatically from isolated disciplinary science (with rare exceptions). They require the right combination of conditions, people, and incentives (Rycroft & Kash, 1999). By the 1990s, new insights and directions had emerged among US researchers of technical and economic policy like Ezra Vogel, Robert Reich, and Michael Porter. Analyzing experience of the progress and strategies of other nations brought out the remarkable growth of Japan, which had risen from having a quarter of United States' GDP per capita in 1964, until it temporarily passed the United States in GDP per capita in 1986 (Central Intelligence Agency, 1986). But Japan had a scientific strategy opposite to that of the United States. It invested heavily in applied science and technology, and little in basic research.

Harvard University technical policy researcher, Michael Porter noted in a milestone book, *The Competitive Advantage of Nations* (Porter, 1990) that since the 1960s the United States had experienced troubling losses in competitive advantage in sectors where it had traditional strengths. Its export gains were disproportionately in unprocessed raw materials like farm products, and its range of exports was low for a nation of the United States' size: "The US is being out-innovated."

Porter led a group that developed the concept of the "technology cluster" as an engine of technologic advances. Such clusters are illustrated by Silicon Valley in the San Francisco Bay area, and Route 128 around the Massachusetts Institute of Technology in Cambridge, where educational institutions, scientists, engineers, private companies, and government come together in cooperation toward advanced product development and market adaptations.

Just as close cooperation among research scientists, engineers, industry, and government achieved extraordinary feats during World War II (Zachary, 1999) in

England and the Soviet Union as well as the United States, so has this type of concerted cooperation been shown to be critical in recent years – whether it be within larger corporations[44] or with wider cooperation.

The productivity of close partnership between universities and industry was already demonstrated by German science in the nineteenth century – when Germany became the first nation to rank among leading nations in the world through the application of science and technology (see German science case study in Chapter 8).

A set of essays based on yearly addresses by Charles M. Vest, President of MIT from 1990 to 2004, provide a vignette of the transformation of thinking about research policy by a leader of one of America's premier technological universities (Vest, 2007).

In his inaugural address as MIT's president in 1991, Vest pointed with pride to the fact that MIT faculty in the 1950s and 1960s had spearheaded the infusion of basic science into engineering education and practice. Vest declared that:

"For four decades the American research universities had served the country exceedingly well. From virtually any perspective, they have paid enormous dividends in return for the public's trust and investment."

But by his annual address of 1997, 6 years later, Vest had a different tone. He turned away from the self-congratulatory attitude that was widespread among scientific leaders (though Vest always qualified his visionary rhetoric with realistic notes). His 1997 speech showed a new seriousness about "minding the store," remembering that economic activity paid the bills for the nation as well as for universities.

"It is time that we anchor ourselves more firmly with industry." Academic fragmentation now no longer got partial praise. "We have to deemphasize narrow disciplinary approaches, particularly in the structure of our curricula and in the way we help students learn to think. We need to pay more attention to the context in which engineering is practiced."

2.3.2 Diversion of US Scientific and Technical Talent?

The US policies for research support did not merely affect yield from tax dollars. They had a powerful influence on the way the nation used a scarce pool of scientific and technical talent. The proportion of the population with the ability and conceptual interests to pursue an advanced degree and frontier research is small – a few percent at most. Such people are exceptionally valuable to society when distributed in balanced ways.[45] In Chapter 4, we will examine the evidence that during the post-World War II period an arbitrary diversion of this talent pool took place. This diversion may have increased polarization and lost the opportunity for more balanced development in economic and infrastructural and environmental policy areas, as well as more creative and consensual resolution of societal problems.

2.4 The Modern Offshore Oil Industry

The modern offshore oil industry began in the Gulf of Mexico after World War II. Technically speaking, offshore oil wells had been drilled off California and in the Caspian Sea before the turn of the century. But these wells mainly tapped extensions of land oil fields under the sea. They involved vertical, shallow boreholes, and simple technology. The modern offshore oil industry emerged in about the same time frame as the space program. Landing a man on the moon in 1969 marked the space program's crowning achievement. The new offshore industry achieved technological breakthroughs rivaling those of the space program as it opened up major new oil and gas resources. But unlike the space program, the year 1969 brought not only oil discoveries in the Alaska North Shore (Prudhoe Bay). It also brought a crisis for the industry as well as for the federal government.

The policy development that had enabled the growth of the offshore oil industry in the United States was passage of the Outer Continental Shelf Lands Act (OCSLA) in 1953. Capping "tidelands" battles between the federal government and the states for ownership of offshore lands in the 1940s, OCSLA assigned responsibility for leasing and regulation of offshore energy and mineral resources outside state waters to the Department of the Interior (DOI). The act called for exploration and exploitation to be conducted by private industry through competitive bidding and leasing. Management responsibilities were delegated by DOI to the US Geological Survey (USGS) and the Bureau of Land Management (BLM).

The regulatory system was responsive to indications of interest in specific areas by industry, and yielded large revenues to the federal treasury from bonus bids for offshore lease tracts. It enjoyed bipartisan support in Congress. Points of debate mainly concerned the amount and disposal of revenues and the generally small size of lease tracts, which could break up natural geological reservoirs into leased subunits in inefficient ways. Some experts questioned the use of front-end bidding for leases, which, among other things, could place smaller companies at competitive disadvantages. Later, in the 1970s, the State of Massachusetts wanted to eliminate the front end bidding system for Georges Bank leasing because it precluded legal authority to stop operations if environmental hazards were encountered. Budgets allocated to the USGS for oversight were very limited, as will be discussed in more detail later.

Once ownership and a stable regulatory and management environment were established, advances in exploration, drilling, and production technology were rapid. The offshore energy industry first centered in the US Gulf Coast. By 1967, drilling had taken place off the coast of 74 nations with 20 nations producing petroleum. Major foreign oil companies such as Shell and BP, which, along with Atlantic Richfield Oil Co., developed major new oil fields in Alaska's North Shore in 1969, became active in US waters as well as other parts of the globe. The United States pioneered many technologies. Table 2.1 offers only samples of the astonishingly rapid development of major technical breakthroughs achieved during the 1960s.

Table 2.1 Selected
technological advances
in the offshore oil industry

Year	Technological advance
1961	Seafloor blowout preventer developed
1961	First moving seismic data collection method (Socony Mobil)
1961	First underwater pipeline trencher (Conoco/Phillips)
1961	First subsea well completed (Gulf of Mexico, Shell)
1962	Submersible drilling rig for 175 foot depths (Kerr-McGee)
1962	Second generation drillship (Glomar II)
1962	First commercial reel pipelaying vessel launched (U-303)
1963	Custom-built drilling fluid transport vessel developed (Baroid)
1964	Microwave data transmission begins in US Gulf of Mexico
1967	Computerized well data monitoring developed (Humble)
1968	US satellite navigation system set up
1968	Bright spot seismic technology (direct oil identification)
1970	Fixed platform water depth exceeds 1,000 feet

Source: MMS (Minerals Management Service, 2003)

The extraordinary rapidity and technical virtuosity of technical developments were achieved in part because of a highly competitive environment and ready availability of risk capital. I might take note here of the startling difference between the speed of development of the offshore oil and gas industry in the 1960s and the glacial pace of US technical and policy advances to reduce greenhouse gas emissions in the past decade. This difference involves policy barriers that will be examined in more detail in Chapters 5, 6, and 9.

In 1968, the American Phillips Oil Co., working under a cooperative agreement with the forerunner to the present Norwegian government oil company, Statoil (White, Kash, Chartock, Devine, & Lenard, 1973), discovered the huge Ekofisk field in the Norwegian sector of the North Sea. Since that time, exploration and leasing for oil and gas development has taken place along much of Norway's long coastline (Larstad & Gooderham, 2004). Norway is now not only a leading petroleum exporting nation, but also has become a world leader in petroleum engineering technology and in advanced environmental policy and technology development.

The achievements of the modern oil offshore oil industry include drilling ships capable of maintaining a fixed drilling position even in rough seas or currents like the Gulf Stream by "dynamic positioning," i.e., computer-guided propulsion units on four corners of the vessels. Such ships have been used since 1968 to conduct scientific drilling research in the oceans. Vessels can now drill to 30,000 feet depth.

In addition to the technologies noted above, multiple boreholes are initiated from a single platform, minimizing the drilling footprint. Drilling can be "bent" to penetrate a mile or more in the horizontal direction. Drilling bit assemblies can be equipped with sensitive "hydrocarbon sniffers" and other sensors so that the computer-steered drilling can follow potential oil and gas strata in any direction.

Major improvements in robustness of operations and safety have also been achieved. Since the disastrous Santa Barbara spill of 1969, which will be reviewed in detail in Section 2.7, a dramatic test took place in the north-central Gulf of Mexico during August 2005, when an area pincushioned with drilling platforms and producing oil and gas wells, and spiderwebbed with underwater pipelines to the land was hit in unprecedented manner by two class 5 hurricanes, Katrina and Rita.

As a result of the hurricanes the MMS reported that of the Gulf's 4,000 platforms, 115 were destroyed and 52 others were damaged. Of the 33,000 miles of Gulf pipelines, 535 pipeline segments were damaged (MMS, 2005). Although minor leakage occurred from the damaged platforms, automatic shutoff valves on the sea floor prevented any significant loss of oil or gas from all the producing oil or gas wells and undersea pipelines (Det Norske Veritas, 2007).[46]

Norwegian offshore petroleum operations in the North and Norwegian Seas achieved 2006 target dates for zero harmful discharge and subbottom completions of oil and gas wells and distribution pipelines. This means that no drilling hardware protrudes from the sea floor that could snag or impede fishing operations.

2.4.1 Regulatory Developments

Prior to 1969, both major political parties in the United States accepted development of oil and gas resources, onshore and offshore, as a normal and beneficial activity. With a few exceptions, political or regional disagreements focused mainly on bidding systems and allocation of leasing revenues. As late as 1968, the state of Maine issued exploration leases 80 miles offshore into Georges Bank. These claims were not settled until a decision by the US Supreme Court finally dismissed coastal state challenges and limited Atlantic state jurisdiction to 3 miles. For historical reasons, Texas and Florida have state jurisdictions to 3 leagues (about 9 miles) into the Gulf of Mexico.

In managing offshore leasing, USGS prepared preliminary offshore resource estimates, posted operational regulations, and was also responsible for monitoring exploration and drilling activities; it also maintained limited proprietary data on those activities. BLM posted lists of tracts for lease sale in the Federal Register and conducted the sales. Winners were determined by the highest bonus bids, i.e., "front end" bids, for the right to obtain and exploit petroleum found in a tract. Award of exploration and production rights carried stipulations regarding diligence, or minimum required levels of exploration and drilling activity. In addition to bonus bids, per acre fees and royalty payments for oil or gas production were collected by USGS. After awards of a small share of lease revenues to offset special costs or impacts on coastal states, the remainder of revenue was remitted to the US Treasury.

2.5 The Turbulent 1960s: Increasing Pollution, Environmental Problems, the Counterculture, and a Preoccupied Administration

Increases in motor traffic and transport, emissions from industrial plants, and space heating contributed to air pollution inversions in New York City that caused hundreds of deaths in 1963 and 1966 (ALA, 2002). In Southern California, population increased greatly during World War II leading to severe air quality problems (Dewey, 2000).[47]

Management structures became weakened through fragmented geographic and operational jurisdictions, diffused authority, and increased number of formal approval steps for site studies and operations. Increased bureaucracy reduced the number of firms equipped to cope with governmental bureaucracy, and increased contract costs. President John F. Kennedy took strong interest in pesticide and other pollution problems, and made a 5-day, 11-state conservation tour in September 1963. However, after Kennedy's assassination on November 22, 1963, the Johnson administration became preoccupied with the election campaign of 1964, Great Society programs, civil rights issues, and the Vietnam War. Countercultural movements, now largely forgotten, split society (Fig. 2.4). Congress became increasingly impatient with the lack of action to better mobilize and coordinate scattered federal agencies.

This impatience can be seen in both the increase in number of laws and in the sharper wording of environmental legislation during the mid-to-late 1960s.[48] The names of some statutes might sound similar to those of the 1970s, but their

Fig. 2.4 Michael Rossman before (a) and during (b) the countercultural revolution
Source: Dust cover from Michael Rossman's book, *On Learning and Social Change* (Rossman, 1969), with permission from Heyday Publishing Company; photographs by Ben Tarcher, Optic Nerve

scope was incremental compared to the revolutionary changes that would come later. The most influential law in this period was the Wilderness Act (1964), which opened the way to a flood of laws protecting environmentally attractive areas.

The increased Congressional activity in turn tended to cause states and municipalities to defer setting in motion major upgrading of systems. They anticipated new federal requirements, possible federal subsidies, or both (see history of Boston Harbor sewage treatment case study in Chapter 8). Lax controls of waste disposal in waterways by industries and municipal waste treatment plants added to pollution in inland waterways, harbors, and estuaries.

A history of pollution conditions in the nation's waterways can be reconstructed because bottom sediments in waterways have "memories" (J. Schubel, cited in Manheim, Buchholtz ten Brink, et al., 1999). When suspended matter settles to the bottom of a river or lake, it brings with it and retains a chemical composition that reflects the concentration of contaminants discharged into the waterways or absorbed from the air during the time the sediment was transported to its burial site. Closely spaced measurements of contaminant concentrations in age-dated sediment cores can therefore provide historical reconstruction of changes in environmental pollution with time (Valette-Silver, 1992).

An ingeniously chosen set of sediment cores taken by Texas A&M University marine scientists at the mouth of the Mississippi River (birdsfoot delta south of New Orleans) integrates sediments representing the entire Mississippi River, which in turn receives water from rivers draining 26 or more states, covering roughly 40% of the area of the lower 48 states.

The age of sediments in one 90-cm-long core was dated by ^{14}C and ^{210}Pb radiometric methods (Presley, Wade, Santschi, & Baskaran, 1998). Figure 2.5 shows the concentrations of three common polyaromatic hydrocarbons (PAHs), which make up the bulk of coal tar residues and petroleum, scaled against year. The PAH species show a rapid increase in concentration after 1945, probably due in large part to

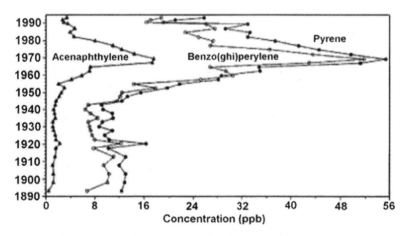

Fig. 2.5 Hydrocarbons (PAHs) from Mississippi River bottom cores
Source: Presley, Wade et al. (1998). The concentrations are given in parts per billion of dry sediment

transportation fuels. They peak in the early 1970s and decline rapidly thereafter, probably owing to measures taken in response to the 1970s environmental laws. Many other organic and inorganic pollutants (especially lead) show a similar rapid rise from the 1940s to a peak in the early 1970s.

Sediments have memories

Cores of bottom sediments at the mouth of the Mississippi River were used by scientists from Texas A&M University to recreate the degree of contamination in the River back to the 1890s. They show pollution rising sharply after World War II to reach a peak in the early 1970s.

2.5.1 Rachel Carson's Silent Spring and Rise of a New Environmental Movement

Rachel Carson was the first woman to pass civil service tests and become employed as a biologist with the US Bureau of Fisheries. She subsequently rose from a position as biologist with the US Fish and Wildlife Service to become chief editor of the Service's publications. She resigned in 1952 to write books. After the best-selling *The Sea Around Us* and *The Edge of the Sea*, Carson spent 5 years writing *Silent Spring*. She undertook her book on the effect of pesticides and other chemicals on the United States' natural environment with passion and intensity. Anticipating controversy, Carson researched her work exceptionally carefully for a book written in popular and accessible style. She listed more than 500 references and an extensive, carefully annotated index. A flavor of the book can be gained from a passage whose message she repeated with variations throughout the volume.

"These sprays, dusts, and aerosols are now applied almost universally to farms, gardens, forests, and homes – nonselective chemicals that have the power to kill every insect, the "good" and the "bad," to still the song of birds and the leaping of fish in the streams, to coat the leaves with a deadly film, and to linger on in soil – all this though the intended target may be only a few weeds or insects. Can anyone believe it is possible to lay down such a barrage of poisons on the surface of the earth without making it unfit for all life? They should not be called 'insecticides' but 'biocides'."

Saturday Review called *Silent Spring* "a devastating attack on human carelessness, greed, and irresponsibility...." Other publications described "terrifying revelations." The book was published in September 1962, and by the end of the year, it had sold 100,000 copies in bookstores and had been featured in the Book of the Month Club. In a 1962 speech, Carson observed "public reaction was reflected first in a tidal wave of letters: letters to congressmen, letters to government agencies, to newspapers, and to the author (Marco, Hollingsworth, & Durham, 1987)." Carson's impact can be gauged by the fact that by the following year, 63 bills curbing pesticides had been introduced by the states (Graham, 1970). Not all reviews were favorable, many national magazines asserted that Carson's treatments were polemic, and she exaggerated. Powerful in its own right, Carson's assault on chemicals and societal leadership and perceptions was heightened by additional serendipitous factors. The book emerged during a decade in which revolutionary social ideas attacking the status quo appeared.

Carson's most prominent target was a chemical that had been regarded as a miraculous and life-saving discovery – dichlorodiphenyltrichloroethane (DDT). To get the full impact of the conflict in values created by Carson's powerfully prepared appeal one must remember the history of DDT. DDT had been synthesized by a German chemist in 1874, but its useful properties were discovered by the Swiss chemist, Paul Muller, in 1939. DDT demonstrated powerful, broad-spectrum toxicity to insect pests, and was persistent and poorly soluble in water. Yet it had such low toxicity to mammals that soldiers in World War II carried delousing powder with 10% DDT content. It controlled a million-person postwar outbreak of typhus in an unprecedented 3 weeks. It became widely used for mosquito and crop-pest control in the United States and other nations, resulting in steep reductions in malaria and other diseases. It was credited, no doubt legitimately, with saving millions of lives. Muller received the Nobel Prize in Physiology or Medicine for his discovery in 1948 (Fischer, 1964).

2.5.1.1 The Second Order Biological Effect

Carson's data showed that minute concentrations of DDT in water could produce biomagnification through the food chain. For example, minute quantities of DDT or its breakdown product, DDE, could be taken up by phytoplankton, eaten in turn by zooplankton, minnows, larger fish, and ultimately, fish-eating birds. Modern analyses show, for example, that an osprey might have as much 25 ppm (parts per million) DDT in its tissue – over 600 times more than in zooplankton, which in turn are greatly enriched with respect to water and phytoplankton (Heinz & Hoffman, 1998).

Carson did not claim that all adult birds were necessarily sickened by their pesticide levels, but cited documentation to show that pesticides caused failure to reproduce in populations of our national symbol, the bald eagle, and other cherished birds. Many other chemical compounds that performed useful services in society were argued by Carson to have direct or indirect adverse environmental consequences to humans or their environment.

Complex concepts like biomagnification through the food chain were new even for professionals. They could not be understood by personal observation or common sense on the part of lay persons. This meant that effects unseen or ignored by society were having devastating effects on the environment – caused by the very agents that were accepted and praised by industry and government as supporting farm productivity and public health.

More erosion of confidence by persons with increased awareness came when the chemical industry and parts of the medical community combined to attack not only Carson's book but her as a person, according to Carson's biographer, Linda Lear (1997).

2.6 Remedial Action Falters[49]

During the 1960s, there was recognition that action on environmental pollution needed more coordinated federal management. As a Presidential candidate John F. Kennedy had shown detailed interest in national water use, river and wastewater pollution, and forests and parks. In a speech on natural resources in 1960, Kennedy emphasized concern about air pollution and wilderness preservation, and indicated plans to expand existing pollution control functions.

Kennedy read Rachel Carson's book in 1962 before its publication, and asked his science advisor, Jerome Wiesner, to conduct a panel investigation with recommendations for action. The results of the pesticide study by an advisory committee chaired by Colin McLeod were released as a formal White House Report on May 15, 1963 (MacLeod, 1963). A Federal Pest Control Review Board had already been established jointly by the Secretaries of Agriculture, Interior, Defense, and Health Education and Welfare. The pesticide report contained data and concrete recommendations, including a comprehensive data-gathering program involving pesticide residues in humans, air, water, and wildlife, expanded interdepartmental cooperation, funding to states to improve pesticide monitoring, and concentration of research efforts on nonchemical pest control and specific pesticides. Kennedy indicated in his preface to the pesticide report his plan to submit legislation to Congress to implement and fund the programs:

> "This report on the use of pesticides has been prepared for me by my Science Advisory Committee. I have already requested the responsible agencies to implement the recommendations in the report, including the preparation of legislative and technical proposals which I shall submit to the Congress. Because of its general public interest, I am releasing the report for publication (MacLeod, 1963)."

After the release of the report Kennedy took a 5-day pollution tour with Wisconsin Senator Gaylord Nelson. However, Kennedy's plans and potential influence were cut short by his assassination on November 22, 1963.[50]

Preoccupied by the Kennedy assassination, civil rights issues, his Great Society programs, and the Vietnam War, President Lyndon Johnson acknowledged pressure from senators on pollution issues. However, he did not act on the Kennedy pesticide report and appeared to rely primarily on the Senate and science advisor Donald F. Hornig for guidance. The President's Science Advisory Council (PSAC) pollution panel delivered a White House Advisory Report in 1965 (Hornig & Tukey,

1965). The PSAC pollution panel's status report had been directed to the ten scientific agencies, whose responses and recommendations (unedited) constituted the full report to the President. A quite blunt statement from one of the agencies, the Federal Water Quality Commission, commented that the entire report was more of a collection of ad hoc comments than a blueprint for action:

> "The Report, although thought-provoking, has major weaknesses and it is not a useful document from the standpoint of guiding, programming and planning for the water pollution control program."

It appeared as though the Pollution panel, which consisted of Hornig, 22 professors or academic administrators, and 2 participants from AT&T and IBM laboratories, was somewhat overwhelmed by the multiplicity of organizational choices potentially available.

A US Department of Natural Resources had already been proposed by former Presidents Herbert Hoover and Dwight Eisenhower. Such a cabinet-level department was recommended by the National Commission on Materials Policy authorized by the US Congress in 1970. According to Russell Train[51] and John C. Whitaker, after the Santa Barbara oil spill President Nixon weighed three proposals for organization of pollution control, two involving consolidation of federal agencies in a new department, and third, ultimately chosen, the EPA plan. However, discussions and planning for a Department of Natural Resources continued actively in the Nixon Administration among Walter Hickel, Rogers C.B. Morton, William Radlinski, and Robert White, Administrator of NOAA (National Oceanic and Atmospheric Administration), until Hickel's altercation with President Nixon.[52] A later initiative for a consolidated science department was primarily a device for drastic pruning of government, rather than for improving coordination.[53]

2.6.1 The Stratton Commission Report and President Richard Nixon

In 1966, Congress passed the Marine Resources and Engineering Development Act of 1966 (Congress, 1966). It created a Commission on Marine Sciences, Engineering, and Resources, which became known as the Stratton Commission for its chairman, Julius Stratton, former president of Massachusetts Institute of Technology and Chairman of the Board of the Ford Foundation. Over the next 3 years, the commission embarked on the most comprehensive review of the United States' ocean interface and activities ever undertaken. It conducted a broad series of hearings and prepared data-gathering reports in which virtually all affected sectors of the nation participated.

Through its balance and attention to detail the proposal attracted an unusual breadth of support from diverse sectors of society – including scientists, industry groups, labor unions, the US Navy, Congress, fishermen, and environmentally concerned organizations.[54] One can compare the development process (though not the outcome) of this proposal with the approach now standard in advanced European

Union nations like Austria, Germany, and Sweden, as will be mentioned in Chapter 6. The commission's final report, delivered in 1969, presented an integrated proposal for an independent US agency to deal with and coordinate all aspects of ocean affairs: *Our Nation and the Sea: A Plan for National Action.*

There was only one influential person who did not agree with the concept of an ocean agency independent of political entanglements. This individual, Richard Nixon, happened to be President. While Nixon accepted the need for an ocean agency, he wanted to maintain Executive Office control. Nixon issued an Executive Order, Reorganization Plan Number IV in 1970, placing the new NOAA under the Department of Commerce. Then as well as up to the present, Commerce was one of the more politicized Cabinet departments. In addition to this problem, which will be discussed in more detail later with respect to fisheries management, NOAA has never been covered by an organic act. This means that the agency as a whole is not funded on an annual basis. Its programs are supported by various laws or committees through annual appropriations processes. This has resulted in erratic and often unpredictable program development and support.

2.6.2 Contending Philosophies for Environmental Management

2.6.2.1 The Jackson Approach: Enhanced Federal Agency Focus and Coordination

Two Congressional policy streams with different approaches for reform of the federal government's management of environmental issues contended during the late 1960s. The first was led by Senator Henry A. (Scoop) Jackson, Democratic Senator from Washington. Jackson was aided by one of the most knowledgeable experts on environmental policy in the nation, Lynton Caldwell (Caldwell, 1963). Jackson favored a leadership role for federal science and regulatory agencies in environmental management. His approaches, as incorporated in Senate Bill S 1075,[55] submitted in February 1969, required increased consideration of environmental issues and coordination by all federal agencies. An environmental impact statement that would identify potential environmental risks would be required for all major federal initiatives. A new advisory body, the Council on Environmental Quality (CEQ), would be established in the executive office of the President. This council would be charged with reviewing environmental management in all federal agencies and in the country as a whole, proposing new policies, and preparing regular progress reports to Congress and the nation.

Jackson's concept was proposed at a time when "conservation" (environmental consciousness) was bipartisan. Partisan divisions that would affect appointments and the influence of such a council remained in the future. Today one might think of the Federal Reserve Board as an analogy. The council's members would be chosen from respected, qualified persons whose judgment would be professional in character.

2.6.2.2   The Movement for Tough National Environmental Laws

In the US Senate, reform movements concerned about air pollution and environmental problems were led by Senator Edmund Muskie (D-ME), Chairman of the
Committee on Air and Water Pollution. Also prominent among Congressional leaders concerned with environment in Congress was Senator Gaylord Nelson (D-WI),
who had been instrumental in passage of the Wilderness Act in 1964, and was the
prime mover in initiating Earth Day in 1970. According to Graham (1999), Senator
Muskie had favored step-by-step improvement in pollution control, retaining a primary role for states in creating plans for controlling air pollution. But new developments and Muskie's status as a potential Democratic candidate for President in the
election of 1972 caused a shift in Muskie's strategy. Ralph Nader attacked Muskie
as a pawn of the Maine paper companies, and President Nixon unveiled a surprising approach for air pollution, calling for national air quality standards (Landy and
Dell, 1998).[56]

According to descriptions by Richard Ayres and Leon Billings,[57] Muskie and a
group of five leading senators (two Democrats and three Republicans) worked out a
tough new strategy to be enforced by the newly created US Environmental Protection Agency (EPA). It included a number of provisions never before seen in federal
statutes: (1) national quality standards for contaminants (in this case, air), (2) rigorous,
scientifically supported procedures for pollution control, (3) specific objectives and
time frames for achieving them, (4) monitoring of pollutant levels, and (5) provisions
for empowering citizens to sue in federal court for noncompliance or laxity on the part
of the enforcing agency.[58] There were few precedents for such intensive analysis of the
problems by a bipartisan group of senators. Their careful work in framing the bill and
supporting its passage helps explain the Clean Air Act Amendment's (CAA) passage
by unanimous vote in the Senate. It was signed by President Nixon on December 31,
1970.

A major role for public interest litigation in support of tough laws was advocated
by Ralph Nader and his Center for the Study of Responsive Law (1969) and Public
Citizen and US Public Interest Research Group (PIRG) (1971). The law students
of "Nader's Raiders" even took on whole agencies and ultimately forced modest
to wholesale reforms (Time, 1970). Litigation was also advocated in an influential
book by Joseph L. Saxe (Saxe, 1970).

An important undercurrent of criticism in the 1960s involved the disproportionate
wealth and productivity of the United States.[59] In various shades of opinion,
voices argued that the United States should deliberately decrease its production
and consumption of material goods (Meadows, Meadows et al., 1972); more radical voices argued for tearing down nuclear power plants (Foreman & Haywood,
1993); and some groups formed agricultural communes (Miller, 1999). Some of
the ideas raised in books, newly started magazines and newspapers, pamphlets
and posters, made academic reformers like Barry Commoner sound like conservatives (Commoner, 1972). However, participants in legislative development did
not think that these countercultural movements influenced the framing of the new
laws.

2.7 The Collision: The Santa Barbara Oil Spill of 1969

On January 28, 1969, 4 days after the inauguration of Walter Hickel, President
Nixon's new Secretary of the Interior, Union Oil Company's Platform A blew
out in the highly faulted Santa Barbara channel. There were 12 drilling platforms
along the coast. Eight were in state waters; four were in federal waters, between
3 and 12 miles from shore. Three had encountered oil and were cased to their full
length, awaiting production. A fourth had been drilled to its maximum depth of
3,479 ft on Platform A by Union Oil Co., the operating company for a consortium
that included Gulf Oil, Mobil Oil, and Texaco. The group had paid a $61.4 million
front-end bonus for Federal Lease Tract 402. Besides annual rental of $16,200 for
the 5,400 acre ocean-floor tract, a royalty of 1/6 of market price per barrel of oil
produced would accrue to the US Treasury (Easton, 1972).

The crew on the platform, which stood in 188 ft of water, were removing drill
pipe in preparation for electrical logging. They were unscrewing the eighth 12-foot
stand of 12-inch pipe when gray, gassy mud shot out of the top of the pipe. As the
crew managed to disconnect the pipe and lay it in its rack, a column of mud and
gas projected 20 ft upward from the open pipe. The pressure prevented the mud-
drenched men from screwing a blowout preventer into the pipe. When mud turned
into a hissing gas cloud, they attempted to screw in the kelly (the kelly is the shaft
that fits over and turns the drill stem and pumps mud into the hole), but a further
accident destroyed this possibility. Now the only control left was closing the jaws
of the massive, rubber-jacketed steel ram below the rotary table. To clear the way
for it to operate, 2,759 ft of drill pipe hanging below the drill floor was dropped to
the bottom of the hole.

The ram had closed off the blowout at the platform, but large boils soon appeared
along a geological fault near Platform A's location. Discharge of what was later
variably estimated at between 3.4 and 4 million gallons of crude oil – the biggest
and most fateful offshore spill in US history not associated with a tanker accident –
had begun through fractures in the rock formation below the platform (Nash,
Mann, & Olsen, 1972). It turned out that the USGS regional oil and gas supervisor,
D.W. Solanas, in Los Angeles, had agreed to a request by Union Oil Co. to place casing to only 239 ft below the sea floor. Had he insisted on the 1,200 ft casing required
by California State regulations, the spill would have probably not occurred.

In 2 days, the spreading oil slick extended over 150 square miles from near
Santa Barbara to Ventura. The blowout was finally brought under control on
February 7. By that time, however, containment booms, floating pumps, and dis-
persants dumped from aircraft had not prevented the black tide from encroaching
on the shore. Grebes, ducks and other seabirds, seals, and sea lions were oiled
(Fig. 2.6).

Pictures of oiled beaches and wildlife on the nation's TV screens in nightly
news broadcasts and in newspapers and magazines, created a national sensation.
A local antioil movement, *GOO* (Get Oil Out) gained massive support. The new
crisis escalated doubts regarding the ability of the Federal government to protect

Fig. 2.6 Bird and straw coated with oil from the Santa Barbara oil spill, 1969
Source: William C. Jorgensen, Santa Barbara Community College

the environment from negative influences of powerful economic forces. It created a collision between the booming offshore energy industry and its Federal managers, and the new environmental movement.

Federal officials in whose areas of responsibility the Santa Barbara Oil spill fell were Interior Secretary Walter Hickel, and Director of the USGS, William Pecora. Hickel was the Republican Governor of Alaska when called by President Nixon to accept nomination as Secretary of the Interior (Hickel, 2003). Hickel's nomination was controversial. Senate confirmation hearings were so long and difficult that Hickel was not finally approved until after President Nixon's inauguration (Roberts, 2003; Wilkerson & Limerick, 2003).

Secretary Hickel surprised everyone with his decisive response to the Santa Barbara crisis. Before even moving into his office, he flew to California to inspect the spill and the public agitation. According to contemporary witnesses, while there he informed Fred Hartley, President of Union Oil, that he would shut the rigs down (which Hartley questioned) and then called the Attorney General's office to "find me authority" to declare a temporary moratorium on offshore drilling (W. Hickel, personal communication, 1990). He declared a temporary moratorium on all offshore operations, and subsequently supported strong environmental actions. On the other hand, he rejected excessive scapegoating of Union Oil (Marco, Hollingsworth, & Durham, 1987). Recognizing that Hickel's bold response to the offshore environmental issues calmed some of the furor, Nixon gave him broad discretion to act until Hickel overstepped his bounds in expressing sympathy with student protesters of the Vietnam War. Hickel's subsequent meeting with Nixon led to a departure for Hickel as dramatic as his entrance (Hickel, 1971).

USGS Director, William T. Pecora, was an accomplished earth scientist and a member of the National Academy of Sciences. Like Hickel, Pecora had people and political skills and a forthright approach to issues. He is regarded by agency historians and writers as one of its best administrators in history (Wallace, 1996). Pecora and Secretary Hickel quickly established rapport and a close relationship. It appears that Pecora advised Hickel regarding many of his decisions relating to offshore oil issues. Pecora's performance in the crisis led to his nomination and appointment as Undersecretary of the Interior in 1971, but Pecora would not serve long. He died prematurely of diverticulitis in 1972.

2.8 The 1970s and 1980s

2.8.1 The Environmental Revolution

The atmosphere of crisis triggered by the Santa Barbara oil spill sent the train of environmental reform steaming down the Congressional tracks. During the earlier 20th century, Congress observed self-imposed limits in legislation dealing with federal bureaus and regulatory agencies. It created new agencies, consolidated functions, specified missions, and authorized funding for executive bureaus and agencies. However, Congressional laws left operational discretion to plan for the future, create policy pursuant to Congressional mandates, deal with new problems as they emerged, and resolve disputes or conflicts, largely to agency leaders, or to the states and municipalities.

The first environmental law enacted after the Santa Barbara Oil Spill was the National Environmental Policy Act (NEPA). Only six pages long, NEPA's scope and boldness are breathtaking. It declares in Section 102 [42 USC § 4332] of Public Law 91–190 that:

> The Congress authorizes and directs that, to the fullest extent possible: (1) the policies, regulations, and public laws of the United States shall be interpreted and administered in accordance with the policies set forth in this Act, and (2) all agencies of the Federal Government shall....

Beginning on December 30, 1970, a series of groundbreaking laws that would transform the US environmental management system and the federal government itself were passed. These laws assumed irreducible conflict between environmental protection and economic interests.

- Clean Air Amendments 1970
- Federal Water Pollution Control Act Amendments (Clean Water Amendments) 1972
- Federal Environmental Pesticide Control Act (FIFRA amendments) 1972
- Marine Mammal Protection Act 1972[60]
- Coastal Zone Management Act (CZM) 1972
- Endangered Species Act (ESA) 1973
- Safe Drinking Water Act 1973

The new statutes departed dramatically from previous policies. They laid out rigorous, detailed operational guidelines for control of polluting activities as well as other potential harm to the environment. Economic and other issues were largely excluded from consideration. For example, the original Federal Water Pollution Control Act (FWPCA) of 1948, designed to address water pollution issues in the United States, consisted of only 13 sections and was 7 pages long. The FWPCA Law with amendments, or the Clean Water Act (CWA), as it was called after 1977, now includes 607 sections and about 360 pages (see Section 8.4.2). In short, Congress now assumed a leadership role that included detailed specification of operational guidelines as well as environmental regulatory policy.

The new laws also contained provisions, building on precedents set in the civil rights cases, for empowering citizen and their organizations to challenge proposed activities by both private and governmental organizations, as well as the operation of EPA, in federal courts. The laws specifically encouraged such action by allowing reasonable attorney and witness fees to be recovered for successful suits.

The new management system had teeth. It achieved rapid progress against pollution and other environmental problems. But it also had other effects, such as setting in motion a new surge of litigation (see Fig. 4.2), in which old and new environmental organizations like the Sierra Club and NRDC, the Environmental Defense Fund (EDF), and many other national and local organizations took active part. Another effect was to remove previous constraints on the passage of legislation. The proliferation of new environmental laws is shown in Fig. 2.7. Intrusive legislation also expanded into other societal areas. The new environmental management system

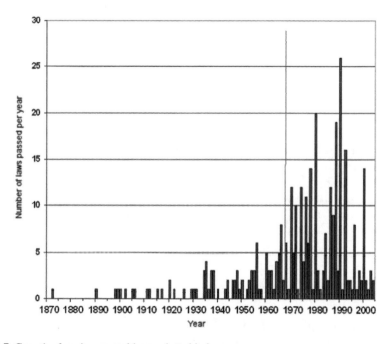

Fig. 2.7 Growth of environmental laws, plotted in laws per year
Source: Unpublished data compiled by F.T. Manheim assisted by G. Fuhs, George Mason University. *Vertical line* marks the beginning of the environmental revolution at about 1970

created a rift between environmentalists and the business community in US society (examined in more detail in Chapter 4). That rift would widen into partisan polarization in the later 1970s and 1980s.

2.8.2   The Nixon–Ford Years: 1969–1977

President Nixon was well aware of environmental issues and the growing challenge to public policy making that they posed. He ordered a study of major steps to overhaul the federal government's environmental management policies and system. In later speeches, Nixon was often able to articulate the key scientific concerns of the day in considerable detail, partly aided by Russell Train and other environmentalists on his staff. However, Nixon's actions were also influenced by his concern for tactical politics and his own political strength.[61]

In the aftermath of the first Arab Oil Embargo (1973), severe gasoline and home heating fuel shortages led to recognition of the problem of dependency on foreign oil imports. In response, President Nixon ordered a major expansion of offshore oil leasing; the first area selected for leasing was the MAFLA (Mississippi–Alabama–Florida) offshore area.

Gerald Ford, who succeeded Nixon as President, pursued a centrist policy on offshore leasing. His Secretary of the Interior, Rogers C.B. Morton, was a moderate, capable administrator. His proposed offshore oil and gas leasing in the MAFLA area in late 1973 faced challenges because lease sales for the MAFLA area were scheduled without formal input from the leaders and populations with the biggest stakes.

The oil spill off Santa Barbara and possibly other potential impacts due to drilling in the marine ecosystem were troubling to the Florida governor, Ruben Askew (D). Florida's image and economy were closely linked to sports fishing, tourism, and winter residents attracted by Florida's sunshine and beaches. Askew was in a quandary, for there seemed to be few legal ways to challenge offshore leasing scheduled for late 1973. But Republican Representative C.W. (Bill) Young came up with an idea. The State of Florida could deny pipeline permits for any oil found if the DOI did not agree to negotiations about the upcoming leasing.

Talks in Washington took place between Governor Askew and Interior Secretary Morton. An agreement was reached. Interior would sponsor a comprehensive environmental review of the MAFLA tract by the best scientific expertise available, and the State of Florida would let the lease sale to go forward.[62]

2.8.3   The Carter–Andrus Years: 1977–1981

Gerald Ford's defeat in the election of 1976 brought to the White House a president in whom an environmentalist commitment was one of the most powerful motivating forces. Jimmy Carter, the new President, chose an Interior Secretary equally dedicated to environmental issues. During his 1970 gubernatorial campaign in Idaho, Cecil B. Andrus stated his intention to become "the first environmental governor in the United States."[63]

Though Andrus began his tenure with strong environmentalist policies, he later modified them, and took personal initiatives in negotiation and working out compromise in oil and gas leasing agreements after the onset of the Iranian (oil) crisis in 1978.[64]

President Carter may be considered the first President to initiate overtly partisan tactics in land use policies that had heretofore been bipartisan. With Secretary Andrus' help, he used the obscure American Antiquities Act of 1906 to declare 56 million acres of northeastern Alaska "National Monuments."

As the National Parks Conservation Association describes it, "The Carter Administration responded in 1978 [to a deadline over decision regarding 80 million acres of Alaskan land designated in the 1971 Alaska Native Claims Act] by withdrawing over 100 million acres of federal lands from development: 40 million acres were withdrawn under the authority of the Secretary of the Interior and 56 million acres were designated as National Monuments with one swoop by President Carter's pen (NPCA, 2007)."

Prolonged negotiations led to a complex compromise, The Alaska National Interest Lands Conservation Act (ANILCA). It was passed in 1980, after Carter's defeat in the election of that year. The act designated nearly 80 million acres to National Park Service, National Forest Service, and US Fish and Wildlife Service jurisdiction, a third of which was set aside as wilderness areas. A coastal strip of the Arctic National Wildlife Refuge (ANWR), part of ANILCA, has been in intense controversy over potential oil and gas leasing for more than 25 years. Leasing has been blocked in the Senate by often narrow margins.

Another Carter action that advanced partisanship was Carter's removing USGS director Vincent McKelvey from a position previously treated as a professional rather than a partisan political appointment. He also suppressed a major Interior Department Task Force Report on the first comprehensive survey of offshore hard mineral resources that was completed in 1979. The report was not published until the ensuing Reagan administration (Hess et al., 1981). On the other hand, Carter was not antibusiness per se. He supported deregulation movements in the later 1970s, notably for the airline industry.

2.8.4 Reagan Administration: 1981–1989

As Governor of California (1967–1971), Ronald Reagan had taken moderate environmental positions. He had signed a bill creating the Air Resources Board that consolidated air pollution control agencies in 1967. During the stormy events surrounding the Santa Barbara offshore spill, Governor Reagan supported blocking further platform construction in the Santa Barbara Channel and buyback of channel leases. He also signed a law creating a 3-mile sanctuary around all channel islands (Easton, 1972).

During the later 1970s, however, Reagan became polarized over what he regarded as the encroachment of federal governmental regulation in the nation, and its effect on natural resource policy and business activity. Jimmy Carter (Carter, 2000) acknowledged in a reminiscence about Alaska that Reagan had been angered by his "land grab," i.e., use of Presidential authority to sequester a huge tract of Alaskan land (Carter, 2000).

Supported by a detailed pre-election plan led by the Heritage Foundation (Heatherly, 1980), incoming President Reagan launched a major conservative counterattack against government regulation. Ignoring the legislative reform elements in the Heritage Foundation blueprint, the President launched a four-pronged program. The Heritage Foundation had in advance identified offices that needed more business-friendly leaders and programs that needed reform. Vice President George H.W. Bush headed a Cabinet-level task force on Regulatory Relief that reviewed all new and existing regulations. Guided by Office of Management and Budget (OMB) director, David Stockman, Reagan issued an executive order that required all Executive Branch agencies to submit regulations to OMB for prepublication review. Those that involved costs over $100 million would be accompanied by a new concept, "regulatory impact analysis" (McGarity, 1991; Meads & Ballentine, 2002).

Battles over the Comprehensive Environmental Response Compensation and Liability Act (CERCLA, the "Superfund" Act) raged during the first Reagan administration. The Superfund Act's indiscriminate assignment of financial responsibility to any firm or person linked with contaminated sites had aroused much antagonism. Individual regulators, scientists, and politicians were amenable to modification of CERCLA. But powerful groups on both sides of the law became polarized. Environmental activists showed no interest or concern except for environmental protection. Some held that parsing out responsibility would complicate getting cleanup paid for, while others expressed open contempt for business. On the other side, Republicans David Stockman and Jack Kemp (cited by Greider, 1982) characterized the situation prior to Reagan's taking office as follows:

"In the 1970s Congress passed more than a dozen sweeping laws devoid of policy standards... and cost effectiveness. Subsequently McGovernite no-growth activists assumed control of most of the cabinet posts in the Carter Administration. They have spent the last four years tooling up for implementation, through a mind-boggling outpouring of rule-makings, ... guidelines, and major legislation, all heavily biased toward maximization of regulatory scope and burden."

2.8.4.1 Anne Gorsuch and the Environmental Protection Agency

Anne Gorsuch, President Reagan's first EPA Administrator, acted with partisan abandon.[65] Anne Gorsuch reduced EPA staff by 23%, slashed EPA enforcement budgets by almost 50%, referrals to the Justice Department by 50% (EPA, 1990), and undertook many other actions that inflamed environmentalists and created vigorous backlash in the US Congress. After an associate, Rita Lavelle, was indicted for perjury (related to misuse of EPA Superfund money), Anne Gorsuch refused to turn over subpoenaed documents under the instructions of President Reagan. The embattled Anne Gorsuch resigned after a turbulent 22 month tenure of office (Lash, Gilman, & Sheridan, 1984; Sullivan, 2004). President Reagan, in a pragmatic move, replaced her with the well-regarded first Administrator of EPA, William Ruckelshaus. Table 4.1 shows that in the second Reagan administration enforcement operations of the EPA regained or exceeded enforcement activity in the last year of the Carter administration.

The EPA assault is related in detail in a book whose senior author is now President of the World Resources Institute (Lash, Gillman, & Sheridan, 1984). (See further discussion of Burford (Gorsuch) in 7.3.3.)

2.8.4.2 James Watt's Interior Department

President Reagan's Interior Secretary, James Gaius Watt, had already aroused controversy before his official appointment on January 20, 1981. Watt was a slight, balding man who combined a rather self-effacing persona with implacable zeal and confidence in his mission. This mission was to restore access to federal lands for multiple use purposes (e.g., energy and minerals extraction), which he perceived had been arbitrarily and unwisely curtailed by misplaced liberal and environmentalist policies. Knowing that his quest would bring opposition, Watt predicted, not inaccurately, that he would double the Sierra Club's membership.

A longtime official in the Interior Department cited the opinion of knowledgeable experts that Watt began his tenure as one of Interior's all-time most effective administrators.[66] Many Presidential appointees have cited the extreme difficulty in effecting real change in entrenched bureaucracies, but Watt started off with high batting averages. He announced in his opening address to Interior employees that he was going to "clean house" not merely in terms of Presidential appointments. He would place changemakers in key positions throughout the Interior Department, which included half a dozen agencies with responsibilities in land use and resource policy.[67] Early administrative targets were met, and during his short tenure Watt oversaw several major national policy changes: the EEZ proclamation of 1983, which extended federal jurisdiction over nonliving offshore resources to 200 nautical miles, and transfer of responsibilities for offshore energy and minerals leasing from the USGS to a new agency, the US Minerals Management Service (MMS).

In contrast to the style of most high-ranking bureaucrats, Watt displayed the same breathtaking directness and candor (and certainty), whether he was speaking to a group of USGS geologists or senators on Capitol Hill. However, Secretary Watt was tone deaf to opposition, concerns, or even questions about his plans and objectives to a degree that can only be explained by his belief that he had a higher calling. Soon after his appointment, Watt sent California Governor, Edmund Brown, a brusque letter informing him that compromise leases off California, which had been laboriously negotiated by Cecil Andrus with local and legislative leaders and the Governor, would be restored to their original scope.[68] He proposed massive area-wide lease sales for the Gulf of Mexico, Alaska, and the Atlantic coast, while requesting reduced budgets to manage (including environmental investigations) lease sales up to 20 times the area of his predecessor's tracts. Area-wide sales were upheld by the courts in a multiorganization suit brought against the Alaska lease sale (Hopson, 1982; Office of Technology Assessment, 1984). Critics called the large tracts "fire sales" and "giveaways," but oil companies defended them and some independent observers felt the system allowed more efficient planning of the exploration/exploitation process, along the lines used by Canada and Norway (Office of Technology Assessment, 1984). A report to the Congress by the Office of Comptroller, General Accounting Office (General Accounting Office, 1981) found merit in the area-wide leasing concept but raised eyebrows over Watt's budget strategy.

Watt was equally aggressive regarding Western land policies for mining (especially coal) and grazing (High Country News, 1983). The Federal Land Policy and Management Act passed by Congress in 1976 had directed the BLM to conduct a wilderness review of its entire holdings within its land-use planning process. Soon after taking office, Watt announced an accelerated schedule for completion of wilderness review by 1984, combined with budget cuts of 42% (Baker, 1983). Watt suffered a serious setback when the Senate in 1983 rejected many of his land-use initiatives. An ill-advised remark in response to that defeat precipitated his resignation in September 1983 (Kumins, 1997).

2.8.4.3 Major Shift in the Structure of Offshore Leasing and EEZ Proclamation

Secretary Watt transformed the management of offshore leasing by moving leasing responsibilities from USGS and the Bureau of Land Management (BLM) to a new agency, the MMS in 1982.[69] Separated from USGS, which did not have a reputation for promptness and efficiency, MMS became purposeful and effective in carrying out its mandated functions. The tradeoff was that it lost extensive scientific expertise and with USGS's tradition and reputation for objective investigation and reporting. The latter capabilities could have been a source of independent judgment to the Congress and the public, especially needed for sensitive natural resource and environmental issues. MMS contracted for external scientific and technical expertise, which could be of a high order. However, on the principle that "he who pays the piper calls the tune," it was widely assumed, at least by skeptics or opponents, that in controversial matters MMS's planning and environmental impact statements would invariably reflect official policy. In other words, MMS's role became limited to that of a policy arm of the administration. That left Congress and the courts the only buffer accessible to the public.

Secretary Watt also aided in preparing President Reagan's Proclamation of an Exclusive Economic Zone of 200 miles from shore, which in addition to offshore oil and gas, opened exploration and leasing for hard mineral resources, including sand and gravel, and strategic metals like cobalt, manganese, and platinum that were found in ocean-floor ferromanganese crusts and nodules.

Before Secretary Watt's resignation and as a part of the backlash against Watt's policies, members of Congress had already begun creating specific offshore area moratoria
or restrictions on leasing and attaching them to Interior funding bills.[70] Liberal legisla-
tors and environmental organizations like the Sierra Club that earlier accepted lease
sales after close scrutiny and hard bargaining (Sierra Club, 1975), henceforth began to
categorically oppose all new offshore drilling. Michael Scott of the Wilderness Society
stated that up to the Watt administration, conservation had been a bipartisan policy
going back to the time of Theodore Roosevelt. Watt made it a partisan issue, and "The
Democrats see the advantage of it being a partisan issue (M. Scott, 1983)."

By the late 1980s, the individual moratorium areas had become so large that
President G.H.W. Bush consolidated them into a single offshore moratorium
declaration in 1990 (Fig. 2.8).[71]

> Cecil Andrus, Watt's environmentalist predecessor and former Idaho gover-
> nor, was quoted, regarding Watt's fall: "The astonishing thing about it was
> that his personal insensitive feelings brought about his eviction. It wasn't
> this administration's plunder of the natural resources that brought him down
> (Oakley, 1983)."

Watt's brief but stormy tenure (1981–1983) generated a level of animosity that
exacerbated an already tenuous relationship between the environmental community

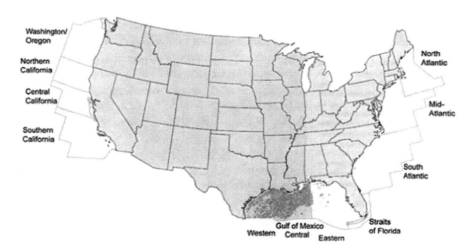

Fig. 2.8 Offshore oil and gas leasing areas for the lower 48 states
Source: Minerals Management Service (2003). Boundary lines refer to limits of U.S. EEZ
(Exclusive Economic Zone). *Lightest shading* indicates areas under moratorium or otherwise not
available for leasing. *Darker areas*, in the Gulf of Mexico, are available for oil and gas exploration
and production

and extractive industries. In the past controversial issues like civil rights and conservation had crossed party lines. The EPA and revolutionary environmental bills had become law during the term of President Nixon. Henceforth, polarized issues and political party lines would tend to coincide.

2.9 Back to the Present

More moderate Republican leaders took the controls in Interior after Watt but little changed in geopolitical conditions during the remainder of the 1980s. The final chapter for Atlantic exploration and leasing was written with the Manteo Block off North Carolina. Leases purchased in 1981–1983 were estimated by Mobil Oil Co. to hold gas resources upward of 5 trillion cubic feet (tcf) (1 billion barrels of oil equivalent (bboe)). In spite of unusually comprehensive environmental research preceding the EIS, including submersible investigation of bottom fauna and flora, the State of North Carolina waged a seesaw battle against leasing that ended in its favor in 1990 (Minerals Management Service, 2000). Throwing in the towel, the first President Bush decided to proclaim a consolidated moratorium on OCS (Outer Continental Shelf) areas outside those areas where local support and historic activities were active (Kumins, 1997).[72]

A new environmental disaster occurred when the *Exxon Valdez*, a large tanker, ran aground in Prince William Sound, Alaska, in 1989, spilling an estimated 81 million gallons of heavy crude oil (EVOSTC, 2005). Exxon paid out more than $2 billion in remediation costs. Though vastly larger and more damaging than the earlier Santa Barbara spill, the policy consequences were limited compared to the earlier event.

President Clinton extended the OCS exploration moratorium to 2112. Early in the George.W. Bush Administration (2001), MMS announced plans to update inventory estimates for OCS energy resources in areas including moratoria areas (e.g., the Atlantic OCS). Vociferous opposition from environmental organizations erupted. Senator Jon Corzine (D) of New Jersey warned that if the action plans were not withdrawn, he would offer legislation to permanently ban all drilling on the Atlantic OCS ("COAST Anti-Drilling Act", 2005).

The administrations of G.H.W. Bush and Bill Clinton have been referred to in environmental law textbooks as a time for "refinement, recoil, and rejuvenation of environmental regulations" (Percival, Miller, et al., 2000). This time includes the last major revision of the Clean Air Act (1990) in a massive 800-page statute during the Bush administration, and a dual approach to environmental policy during Clinton's administration. On the one hand, Clinton appointed tough and initially uncompromising environmentalists like Carol Browner (EPA Director) and Peter Kostmayer (EPA Region III chief). It was also the time of a multipronged "Reinventing Government" program under Vice President Al Gore. A change in administration policy away from confrontation toward listening and cooperating took place after the Republican majority took over the House of Representatives in 1994

(Dunlop, 2000; Nakamura & Church, 2003). However, in his final days in office, President Clinton outraged conservatives – mainly Republicans – and won praise from environmentalists – mainly Democrats – for again using presidential authority to sequester nearly 60 million acres of forest land under the 1906 American Antiquities Act, from road building and logging.

The George W. Bush administration began without expectations of escalation of conflict over environmental policy, since Bush had had good relations with Democrats as Governor of Texas. He also committed his administration to a reduction of CO_2 emissions and had pushed for renewable energy development. While still in his former incarnation, Bush is said to have favored former Senator William Danforth, a moderate Republican and ordained Episcopal priest, as a Vice Presidential running mate. Once taking office, however, the Bush administration became marked by an astonishing escalation of conflict over environmental policy (alluded to in Chapter 1). The administration showed evidence of extreme partisan desensitization and radicalization widely attributed to the influence of Vice President Richard Cheney. Cheney had served as CEO of Halliburton, a leading oilfield service firm. His attitude may be considered an example of the product of the 35-year conflict between environmentalists and the oil industry (Devine, 2004).

In September 2008, following a mortgage crisis and a dramatic rise in gasoline prices, President George W. Bush rescinded the offshore oil drilling moratorium.

Chapter 3
Why History Is Important for Environmental Decision Making Today and Tomorrow

Science and technology have made enormous advances since the American Revolution. But we need only think of the Nazi nightmare that overtook Germany, the nation of Goethe, Schiller, Beethoven, and of major medical and scientific advances, to recognize that human behavior and political developments do not follow the same laws as science. Nor has the United States just marched forward since its inception, taking advantage of new knowledge and ideas to systematically improve the quality of its public affairs. This is demonstrated by the longstanding internal turmoil that is the subject of this book and which has paralyzed the nation's ability to achieve solutions to major problems. Recent public polls show that less than 30% of voters express confidence in the ability of government to "do the right thing," contrasted with over 60% prior to the 1960s.

In contrast, the US Constitution continues to be relied upon and revered by Americans with very different political views. Except for flaws such as the slavery compromise (see Section 3.2), the Constitution not only enabled rapid and largely harmonious growth of the new nation, but also provided a robust political framework that has worked under conditions the framers could not have remotely predicted. It is regarded by foreign scholars as one of the important political achievements in history (Auduc, Meissel, et al, 1998).[73]

The Constitution may be the preeminent example of the use of political history to build new institutions.[74] The participants in the Constitutional Convention of 1787 faced an enormous range of challenges in crafting the Constitution. Later in this chapter (Section 3.4) we will look at enviromental use of political history in Marsh's *Man and Nature*, written in 1864; historical analyses in the book launched the environmental movement in the United States and other nations.[75]

Two recurring themes in US political history are especially important to examine in the context of this book. The first theme is the constructive role that mediators can play in defusing conflict. Mediators appreciate the concerns of people or groups in conflicts, but see beyond those issues to seek solutions. The second theme is "bad governance," which the UN Economic and Social Commission points out as being old as human civilization. Good governance is a goal difficult to achieve. But as global population has grown and the climate challenge has emerged, good governance is now regarded as not just desirable

F.T. Manheim, *The Conflict Over Environmental Regulation in the United States*,
DOI 10.1007/978-0-387-75877-0_3, © Springer Science+Business Media, LLC 2009

but indispensable if sustainable development is to be achieved (UNESCAP, 2008).

This year (2008) the United States experienced major economic shocks triggered by the collapse of speculative ("subprime") mortgages, suggesting serious lapses in the judgment of financial managers and others. Commentators and historians have noted that such problems have their root in the political, business, and societal leaders' focus on short-term preoccupations and self-interest, pushing important policy issues and longer-term consequences into the background. The intense focus on the "now" has become prominent enough in the United States to warrant the label, "presentism" (Hunt, 2002).

3.1 Communications and the Importance of Mediators

3.1.1 George Washington as a Mediator

George Washington's importance in the early history of the United States is too well known to need comment. However, his role in containing the deep philosophical disagreements between Thomas Jefferson (Secretary of State) and Alexander Hamilton (Secretary of the Treasury) during Washington's first administration offers a good example of mediation relevant to today's conflicts. The rift between the two cabinet members shows how, even with brilliant thinkers, ideas that have a core of validity can grow into exaggerated assumptions or fears when communication breaks down. Jefferson had long disapproved of Alexander Hamilton for his morals (drinking, supposed corruption, and womanizing) and authoritarian preferences. By 1792, Jefferson became convinced that Hamilton was scheming to convert the US republican system into a monarchy. He wrote Washington a long and passionate letter detailing his fears. To Jefferson's concerns about Hamilton's monarchist schemes, Washington replied, "While there may be *desires*, I find no *designs*" (American National Archives, cited in Koch & Peden, 1944).

Washington validated Jefferson's suspicions that Hamilton harbored royalist ideas, demonstrating that he was not unaware of the source of Jefferson's concerns. But he was also aware of Hamilton's important contributions to building a sound fiscal framework for the government. Washington made a clear distinction between feelings ("desires") and potentially subversive actions ("designs"). The importance of recognition of sources of conflict and mediation or injection of reality at early stages, before situations get out of control, are now among the basics in contemporary conflict resolution (Cheldelin, Druckman, & Fast, 2003). The availability of a respected mediator made it possible for the new nation to retain the skills of both Jefferson and Hamilton even though the two did not get along personally and clashed on basic principles.

Mediation played a critical role in the Revolutionary War period. Had Lincoln lived to pursue his goals as a mediator, the history of Reconstruction after the Civil War may have been different. The progression of environmentalists and industry into isolation and radicalization in the past 35 years took place in part because of the absence of mediation or other intervention at high levels,[76] so that polarization progressed to breakdown in communication between political parties and other contending groups after the 1980s.

3.1.2 Abraham Lincoln as a Mediator

Lincoln's revulsion against slavery was the driving force in his leadership in forming the new, antislavery Republican Party in 1854 (Donald, 1996). As President, after the outbreak of war with the breakaway slave states, he had shown frustration in not being able to find generals sufficiently aggressive in taking the offensive against the Confederate forces. Yet, in his speech at the bloody Gettysburg battlefield in 1863, Lincoln did not issue a call for the North to crush the South. This could have been justified by the fact that the Confederacy was fighting to retain an institution that had earlier been recognized as inhuman even in Charleston South Carolina.[77] Instead, Lincoln urged reconciliation and generosity.[78] In other words, Lincoln distinguished between military policy and human policy, keeping long-term unity for the nation an ever-present priority.

According to a comprehensive treatise on Reconstruction (Foner, 2005), Lincoln, while resolute and astute in promoting Negro voting rights, had an "amazingly generous" plan to provide recompense to the South for former slave property. He was cognizant of the risks and potential opposition, but felt that this gesture of generosity could assuage the bitterness and rancor of defeat and destruction of much of the Confederacy's heartland, and help restore the nation to harmony.

Assassination deprived the nation of Lincoln's skills and determination to achieve his dual Reconstruction goals: establishing full rights for the African American population and reuniting the nation. He was replaced as President by Vice President Andrew Johnson, a Tennessean with "opposite" character traits. According to Foner, though courageous and opposed to slavery, Johnson was prejudiced against Negroes, driven by emotions, impulsive, and insensitive. He was regarded as a turncoat by many southerners, and distrusted by northerners. Southern observers have stated that Johnson "held up before us the hope of a white man's government, and it was natural that we should yield to our old prejudices (Reagan, 1906)." By the late 1870s, whites in the south systematically suppressed and disenfranchised the African American population. As if the failure of Reconstruction in the South were not enough, Ken Burns brought further poisonous legacies of that failure to public attention in his television documentary, *The Civil War* (1990). President Woodrow Wilson, influenced by the provocative and powerful film, *Birth of a Nation* (1915), set in motion resegregation of the federal government's personnel system, which had remained desegregated since the early Reconstruction period.

3.1.3 Theodore Roosevelt as a Mediator

An example of intervention to reduce antagonisms and forestall social instability is provided in President Theodore Roosevelt's response to inflammatory attacks on a group of wealthy US Senators.[79] When *Cosmopolitan* ran a series of articles in 1906 accusing 23 senators of "treasonous behaviors" (Phillips, 1964), Roosevelt[80] made what would later be called his "Treason of the Senate" speech, named after the title of Cosmopolitan's series. In it he coined the term "muckraker," denouncing those who were trying to discredit public figures with exaggerated or unfair charges merely because they were rich. Though Roosevelt himself came from privileged family circumstances, he gained popular respect for his renunciation of the rapacious behavior of monopoly-seeking tycoons whom he called "malefactors of great wealth" (Roosevelt & Davidson, 2003).

Roosevelt also negotiated an end to the Russo-Japanese War in 1905, for which he gained the Nobel Peace Prize; mediated international disputes over Venezuela, the Dominican Republic, and Morocco; and was the first head of state to propose what would become known as "the Second Hague Peace" in which he obtained for Latin American nations equal status with the rest of the world.

3.2 Bad Governance Produces Bad Consequences for Society

Since World War II the definition of good governance[81] has become more than just a matter of intellectual debate. The United Nations and the Organisation for Economic Cooperation and Development (OECD) have acquired massive background information on the characteristics of successful and failed states, and formulated a pragmatic model (UNESCAP, 2008).

UNESCAP points out that bad governance is as old as human history. Frequently employed to describe problems in developing nations, bad governance also affects advanced societies with representative systems. Because of the high stakes and complexities involved in the evolution of the European Union, now encompassing 27 diverse nations, effective governance continues to be a prime goal of regular negotiations. (Practical experience has yielded insights and changes that will be discussed in more detail in Sections 3.3, 3.4, and 7.4.3.)

For the purposes of this book bad governance encompasses political actions, laws, or court decisions that are inconsistent or not accountable, and actions that fail to follow principles or the rule of law. It includes actions that deny participation in decision making to given groups in society, and those that consider mainly short-term goals and fail to deal with critical issues when they involve painful choices. Bad governance promotes militancy and tends to create cumulative or increasing stresses on society. These stresses may lead to crises (Osborne, 2004; UNESCAP, 2008).

A fateful early example of bad governance was failure to resolve the slavery issue in the Constitution. Ten of the 13 original states (including the slave-holding states of Maryland, Delaware, and Virginia) wanted to outlaw slavery in the Constitution. Georgia and the two Carolinas held out against abolition.[82] Faced with the breakup of the original colonies, the Constitutional Convention delegates accepted the "3/5 rule" and other accommodations to slavery. In Virginia, George Mason railed against "this infernal traffic" and opposed Virginia's ratification of the Constitution because of it, and Edmund Randolph supported him (Davis, 1964). But many rank and file landowners, including people of Scotch–Irish descent from Pennsylvania who settled in the Shenandoah Valley, favored the economic advantages of slavery.[83]

> Bad governance produces bad results: examples are retention of slavery in the Constitution, and the spoils system, introduced in the administration of Andrew Jackson (1829–1837).

Once ratified, the addition of 3/5 of the slave population gave the slave states 47 seats in the House of Representatives instead of 33 that they would have otherwise held. This advantage and later drift of Maryland and Virginia into slave-state politics caused southerners to dominate the Presidency, the Speakership of the House, and the Supreme Court, and permitted passage of proslavery laws (Wills, 2003). It allowed new slave states to enter the union. The invention of the cotton gin in 1793 increased the commercial value of slaves in cotton cultivation. In the same year, Congress passed the Fugitive Slave Law, which authorized judges or magistrates to decide the status of fugitives. Angered by what amounted to validation of slave status in free states, northern states enacted various provisions to hinder the application of the new law. The seesaw battle over slavery continued to escalate to its inevitable outcome in 1861 – the Civil War.[84]

Another development that promoted bad governance was the spoils system, inaugurated in the administration of Andrew Jackson (1829–1837). The first six presidents of the United States appointed most federal employees themselves. The expectation and convention was that these employees should be well qualified and discharge their responsibilities efficiently and fairly, avoiding favoritism for political or personal advantage. As pointed out in a Ph.D. dissertation that became a famous book by Arthur Schlesinger, Jr. (Schlesinger, 1946), this system had achieved efficiency. However, it favored an educated stratum of American society. President Jackson saw this as contrary to egalitarian principles of the republic. He thought ordinary Americans should be able to handle the duties of government offices, and that winning parties were entitled to install their supporters in them.

The new system fostered incompetence and rampant corruption in government, though author Schlesinger found redeeming qualities in Jackson's reasons for adopting it.[85] Among the later consequences of lack of responsible officials and institutions to guide management were arbitrary and poorly controlled land use policies – as in the 1849 gold rush in California and other developments in the laissez-faire era (to be discussed in Section 3.3.2).

3.3 Environmental and Public Health Management

3.3.1 Pre-revolutionary War Period to the 1830s

In the course of research on contaminated sediments in Boston Harbor in the early 1990s (Manheim, Buchholtz ten Brink, & Mecray, 1998), I became interested in the history of the harbor, making use of the fact that the city of Boston has the oldest continuous records of city management in the United States. An historical summary (Dolin, 1990), town ordinances, and historical documents show that Boston, Massachusetts was remarkably strict in the degree to which it managed public health issues, land use, and labor law (Tichi, 1979). Though the germ origin of disease was yet unknown, communicable diseases were assumed to be associated with "filth." Therefore, sewage discharge into Boston's harbor was forbidden from the early 1700s until the early 1830s.[86]

This and other measures help explain the fact that Boston escaped the worldwide cholera epidemic of 1832, which claimed thousands of lives in other American cities, especially New York and New Orleans, where sanitation was poorer or minimal. The environmental policy history of Boston Harbor will be taken up again in discussing the rise of engineers in society later in this chapter.

Thomas Jefferson, with his limitless interest in every aspect of natural science [87] and society, made useful observations about societal management issues in his Notes on the State of Virginia. He commented on well-organized societal management (Jefferson, 1782), taking considerable pride in what we would now call "community-based services." For example, Jefferson referred to community subsidy for placement of sick, handicapped, or indigent persons with farm families, where they got far better care than would be the case in institutions in European countries. Jefferson indicated that this kind of system was typical throughout the colonies [though perhaps mainly in the white population] and indicated that, unlike the teeming beggars in cities of Europe, the only beggars to be found throughout the colonies were recent immigrants from abroad.

The colonists did not invent careful land management. The countries from which they emigrated, especially England and Holland, generally maintained such oversight, especially on Crown and church land. King George III appointed John Bartram, an American-born Quaker, Botanist for the North American Colonies (Norton, 1991). In Europe before the industrial revolution riparian owners of land could use water on their property but were required to leave rivers in unpolluted condition (MacDonald, 1990).[88] In Sweden the first forestry law was passed in 1600, and in Austria and Germany forest management had even older roots. In 1752, the Swedish crown undertook a general survey and reallotment of all land, acquiring most of the forest now owned by the state. A state college of forestry was founded in Sweden in 1828, providing training for forest managers for state, industrial companies, and church forest reserves.[89]

3.3.2 The Laissez-Faire Era

President Jackson's spoils system for federal government employees became a key influence, directly or indirectly, on environmental policies during the nineteenth century. Whatever the sympathies for the egalitarian principles that brought the spoils system into being, it encouraged a short-term, ideological focus that prevented more thoughtful and longer-range development and management of federal policies – putting it clearly in the category of "bad governance." It especially influenced land use in the territories.

Under the pressure of immigration, growth of manufacturing, and influence of national political policies, local governmental control on health gave way to short-range preoccupations. In 1849 and 1865, epidemics of Asiatic cholera, a bacterial disease spread through diarrheal discharge dispersed in water, took place in urban centers in the United States.[90] In addition to cholera, typhoid fever, which is also communicated by contaminated water and sewage, erupted in the years immediately following 1865, and yellow fever (mosquito-borne) recurred.

Before the Civil War the federal government had been on a lean fiscal diet. Laws to permit homesteading had been inhibited by southern political forces in order to forestall the growth of free-state areas.[91] However, the gold rush of 1849 began the exploitation of rich mineral deposits in the Western territories of the United States. The absence of federal water policies resulted in gold miners sequestering for themselves scarce water supplies via dams and impoundments. After California entered the Union in 1850, state law adopted the *prior appropriation* ("first-come-first-served") principle, allocating prior claimants water for "beneficial use," the remainder of the water going to subsequent claimants (Gillilan & Brown, 1997). The other Western states adopted mixtures of prior appropriation and riparian law.[92]

During the Civil War, new taxes were levied on a variety of commodities and services, along with tariffs on imports to partly offset costs of the war. After the war these sources of federal income were not immediately discontinued or applied solely to reduction of the national debt. These funding sources, combined with the existing spoils system, burgeoned after the Civil War into the "gilded age"[93] riddled with governmental corruption and scandal.

Land distributions for homesteading and resource exploitation expanded rapidly. This in turn increased the importance of connection of the nation's railway network between East and West. New subsidies provided by the federal government to encourage the expansion of railroads offered tempting targets for under-the-table profits.[94] Massive growth in manufacturing and trade, railroad transportation, and land distributions offered opportunities for profits through the formation of monopolies, cartels, corners on markets, or sweetheart deals obtained through the intercession of Congressmen.

The lack of responsible federal government organizations, competent officials, and policies for land use control in the federal government caused the federal government to largely ignore land use policies in the far west (Andrews, 1999). Operating essentially illegally, new areas were opened up to mining under "the most generous conditions in the world" (Wright & Czelusta, 2002). Future pollution problems were left behind at old mining sites. Examples include mercury residues in California gold mining sites, where mercury amalgams were used to process gold in placer deposits. Huge volumes of tailings from mining and smelting residues, such as those left behind at the Coeur d'Alene silver, lead, and zinc mining region of Idaho and other areas. These can contain lead, copper, zinc, cadmium, arsenic, and other potentially toxic metals that contaminate sediments and water in streams, and leach into ground waters (National Research Council, 2005). Coal mining had impacts on both land and water and caused severe loss of life in mining accidents (see Fig.3.2).

Relicts from poor land use and mining practices (Andrews, 1999) have done double damage. Besides local impacts, the bad image of old mining practices and working conditions[95] became part of the psychological background for the 1970s US environmental laws. The stigma on mining affected the US minerals industry more than any other economic sector (see Fig 4.7).

3.4 People and Milestones in American Environmental History

Historical review of pioneering naturalists, environmental thinkers, and events in US history must begin with Thomas Jefferson. Jefferson showed theoretical and practical interest in every facet of natural science that he encountered. A sophisticated horticulturist and viniculturist, he maintained systematic records of temperature from early times at Monticello. He hoped to initiate systematic temperature recording throughout the United States, and was particularly interested in the effects of large-scale forest cutting on temperature as well as on water retention and erosion. He was interested in health and diseases. Like Washington, he had all persons on his estate, white and black, inoculated against smallpox.[96]

Jefferson was an archeologist and anthropologist before those disciplines were formalized. One incentive to write his *Notes on the State of Virginia* (Bedini, 2002; Jefferson, 1782) is said to have been to refute a claim of the great French naturalist, Comte de Buffon, author of the 36-volume encyclopedia, *Histoire naturelle, générale et particulière* (1749–1778, 36 volumes). Buffon had claimed that animals were smaller in the New World and implied that human development was also degenerate there. Jefferson proved that, in fact, animals were larger in the North America than in Europe by shipping mastodon (mammoth) bones that he had collected to Paris (J. McLaughlin, 1988). Following Jefferson's instruction, the Lewis and Clark Expedition of 1906 collected 178 species of plants and 122 of animals (Burroughs, 1961).

Finally, Jefferson was keenly interested in Indian cultures and languages. In Notes on the State of Virginia, he speculated that Indians arrived in North America via the Bering Straits from Siberia (or possibly Siberians were derived from Indians traveling westward). He compiled a number of dictionaries of Indian languages, most of which have not survived (American Philosophical Society, 2006).

According to environmental historian, William Cronon, (William Cronon, in his forward to Lowenthal, 2003), the three most influential books on the relationship between humans and the environment in America are George Perkins Marsh's *Man and Nature* (1864), Aldo Leopold's *A Sand County Almanac* (1949), and Rachel Carson's *Silent Spring* (1962), each of these books following its own distinctive approach to the subject.

Lewis Mumford called Marsh's book the "fountainhead of the Conservation movement" (Mumford, 1971). Marsh's biographer, David Lowenthal, cites the book as the most influential in its time, next to Darwin's *Origin of the Species*, linking nature, science, and society (Lowenthal, 2003). "Although Thoreau's Walden and John Muir's writings on the High Sierra are now far better known, in fact neither had anything like the political impact of [*Man and Nature*] (William Cronon, in his forward to Lowenthal, 2003)."

Marsh was the Jefferson of his day in breadth of interests. He was a leading philologist, speaking 20 languages besides enthusiastically advocating for sign language. He published the first English-language Icelandic grammar in 1838 and translated classic works from Old Norse, Danish, and Swedish into English. First trained in law, he had experience in farming, manufacturing, was a commissioner in Vermont in charge of restoring fish populations, served four terms in Congress as a Representative from Vermont, and helped found and foster the Smithsonian Institution. *Man and Nature* drew on more than two decades as a US diplomat in Turkey and Italy. Marsh combined observations during long trips to the countryside in Europe and the Middle East with information from books and maps in the great libraries of Europe.[97]

Marsh charged – and documented in detail – that "ancient civilizations around the Mediterranean had brought about their own collapse by their abuse of the environment" while observing contemporary destruction of birds, buffalo, elephant, walrus, and narwhale. He demonstrated the damage due to irrigation by salination and other effects, calculating from historic sources that one fifth of the volume of Nile waters was used for irrigation in ancient times.

Marsh was as comfortable quoting from the classic Greek of Herodotus about the deforested country of the Scythians between the Danube and the Don rivers, as describing the work of leading French, German, and Italian scientists, or bog ecology in Sweden and Denmark from Scandinavian authors. The impact of his book derived in part from his warning that what civilization had done to the environment in Europe could happen in the New World if Americans did not behave in more responsible ways. His massive footnotes documented his conclusions in the kind of detail used by Rachel Carson, nearly 100 years later, to drive home her message about pesticide misuse. The breadth of Marsh's interests and his scientific intuition is indicated by the fact that, in discussing the Gulf Stream's importance for European climate he noted that if a breach in the western boundary of the stream occurred, and the Gulf Stream flowed into the Pacific Ocean, the result would have a marked chilling effect on Europe – perhaps even an ice age (Marsh, 1864, p. 442).[98]

Individuals and milestones involved in the further development of natural history and environment from the Revolutionary War period to 1970 are listed in Table 3.1.

Table 3.1 Milestones in US environmental history: individuals and events to 1970

Entity	Date	Historical note
Thomas Jefferson	1743–1826	President, horticulturist, meteorologist, paleontologist, compiler of Indian languages; commissioned Lewis and Clark Expedition with instructions to observe and collect fauna and flora
Coast Survey	1807	Coast Survey, designed and first led by Austrian immigrant, Ferdinand Hassler (1777–1843) who "invented the modern science agency": CS used scientific measurements, formal publications, emphasized merit appointments, and integrity
John J. Audubon	1785–1851	Ornithologist, naturalist, hunter, renowned painter of birds
George Perkins Marsh	1801–1882	Called father of the environmental movement in United States and influential in Europe; book, *Man and Nature*, described effect of man on nature; philologist, Fish Commissioner for Vermont, Congressman, envoy to Turkey and Italy
Henry D. Thoreau	1817–1862	Abolitionist, irrepressible individualist, naturalist, and writer
Asa Gray	1810–1888	Leading nineteenth century American botanist, author of definitive manuals on flora of North America; Professor of Natural Science at Harvard; provided information for Darwin's *Origin of the Species*
Spencer F. Baird	1823–1887	Professor of Natural History, ornithologist, natural history author and compiler, assistant Secretary of Smithsonian, US Commissioner of Fish and Fisheries
Frederick Law Olmsted	1822–1903	Most prominent landscape architect and park designer
Yellowstone Park	1872	First National Park, set aside by President US Grant
US Geological Survey	1879	Under first three directors (Clarence King, John Wesley Powell, and Charles Walcott), USGS became world's foremost earth science organization and conducted first forest and soil surveys. Powell created the first ethnological maps of North American Indian cultures
John Muir	1838–1914	Scottish-born champion of Western environment and wilderness, campaigned for Yosemite Valley, wrote influential articles and influenced Presidents Benjamin Harrison, Grover Cleveland, and Theodore Roosevelt; founded Sierra Club in 1892
Forest Reserve Act	1891	Act to temporarily protect forest areas of high value
Theodore Roosevelt	1858–1919	As President, established forest and wildlife conservation as a key principle for US government; added many parks, established US Forest Service and increased National Forest area from 46 to 148 million acres
Gifford Pinchot	1865–1946	Forester, prominent conservationist and first Director of US Forest Service; Governor of Pennsylvania
US Forest Service	1905	Created by Theodore Roosevelt as bureau in Department of Agriculture
Forest reserves	1907	Included limited lumbering, mining, grazing, and recreation; also National Wildlife refuges; overseen by FWS (Department of the Interior)

Table 3.1 (continued)

Entity	Date	Historical note
National Park System	1916	Bureau in Department of the Interior; now oversees 83 million acres in 51 areas; including memorials, wild and scenic rivers, battlefields
Aldo Leopold	1887–1848	Forester in US Forest Service, pioneer in game management and predator preservation, Professor at University of Wisconsin; gifted writer, author of *A Sand County Almanac*, a bible for wildlife conservationists
Rachel Carson	1907–1964	Federal biologist and chief editor for US Fish and Wildlife Service; best selling author of coastal natural history; landmark book *Silent Spring* (1962) launched new environmental movement
National Wilderness Preservation System (Act)	1964	Created major wilderness preservation areas, and opened way for subsequent designations
National Environmental Policy Act	1969	This and succeeding environmental laws in the 1970s have dominated environmental management to the present

Source: Compiled by F.T. Manheim from various sources cited in this chapter

Modern texts and books that include earlier environmental history usually focus on John Muir, Henry Thoreau, and William Wordsworth as representatives of the preservationist movement (along with John Burroughs, Olaus and Margaret Murie, and the founders of the Wilderness Society, including Aldo Leopold). The preservationist movement emphasizes the beauty and spiritual importance of unspoiled nature as a prime good. Conservationist icon Gifford Pinchot, first forester for the Department of Agriculture (1897) and first Chief of the US Forest Service (appointed by President Theodore Roosevelt), espoused wise management of resources to provide value for the present while maintaining the resources for future generations (Andrews, 1999).

The prominent environmental think tank, Resources for the Future (RFF), commissioned a philosopher, Byron Norton (Norton, 1991), to seek a strategy that could bring together divergent factions of the environmental movement. Norton showed that Muir and Pinchot shared similar reverence for forests and the national environment as well as appreciation for efficient and sustainable production from trees. Their difference was in degree rather than absolute, and the two initially cooperated. But the proposal to construct the Hetch Hetchy dam, designed to create a reservoir for San Francisco by impounding and inundating the beautiful Yosemite Valley, put them on opposite sides of a controversy. Muir eventually lost the long battle (1901–1913), and the differences between the two renowned leaders intensified. Norton offered a conceptual model of how both their skills could have been utilized in national government, avoiding acrimony and benefiting the country. Unfortunately, the Muir–Pinchot conflict became a microcosm of a larger conflict in the 1970s.

The high watermark of conservation philosophy was reached in the presidency of Theodore Roosevelt, who came into office as a result of the assassination of

President McKinley in 1902. Under Roosevelt, who was strongly influenced by Marsh and Muir, a "progressive conservation ethic became embedded in federal policy."[99] He signed the American Antiquities Act of 1906, established the Forest Service in the Department of Agriculture, and was instrumental in creating 150 national forests, 51 bird reservations, 4 game preserves, 5 national parks, and 24 reclamation projects (Harbaugh, 1975; Theodore_Roosevelt_Association, 2005).

> "To waste, to destroy, our natural resources, to skin and exhaust land instead of using it so as to increase its usefulness, will result in undermining in the days of our children the very prosperity which we ought by right to hand down to them amplified and developed (Theodore Roosevelt, State of the Union speech, December 3, 1907)."

Aldo Leopold (1887–1948) joined the Forest Service a year after gaining his Masters degree in Forestry from Yale University. It was 1909, at the height of the Hetch Hetchy controversy. Over time Leopold deepened his interest in the total environment, becoming a pioneer in game management and preservation, including then despised wolves. In 1921, he teamed with fellow forester Arthur Carhart to successfully convince his superiors to preserve a beautiful part of the Colorado White River Forest. A skilled writer and editor, Leopold's articles advocating wildlife preservation began to spread his reputation. On his recommendation, the Forest Service declared a half-million acre tract in the Gila National Forest of New Mexico as the nation's first wilderness area within a National Forest (Byrnes, 1998). Resigning from the Forest Service, Leopold conducted the first multistate inventory of game for the Sporting Arms and Ammunitions Manufacturers Institute, and ultimately accepted a position created for him as the nation's first professor of game management at the University of Wisconsin in 1934. In 1935 he founded the Wilderness Society with seven others. At his family farm, Leopold wrote the book that made him famous, *A Sand County Almanac,* finally finding a publisher in 1948, 7 days before suffering a fatal heart attack while helping put out a fire in a neighbor's property (Meine, 1991).

In her nuanced account of environmental pioneers, Patricia Byrnes selects a passage from Leopold's book that articulates his ethic (Byrnes, 1998).

> "All ethics so far evolved are set upon a single premise: that the individual is a member of a community of interdependent parts. [Man's] instincts prompt him to compete for his place in that community, but his ethics prompt him also to cooperate.... The land ethic simply enlarges the boundaries of the community to include soils, waters, plants, and animals, or collectively: the land."

New insights about the North American environment prior to European influence became popularized through Jared Diamond's best-selling book, *Guns, Germs, and Steel* (Diamond, 1997). The real North American Indian population was shown to be up to an order of magnitude greater than the few millions earlier assumed (Dobyns, 1983). Diseases like measles and chicken pox, developed over countless generations of Europeans living with domesticated animals, ravaged American Indian populations like wildfire, in many cases before they ever saw a white man. Thus, estimates of the American Indian population observed by European settlers were far too small.

Adding an array of evidence to expanded estimates of American Indian populations (as well as the Central American and Andean cultures), the recent book, *1491* (Mann, 2006), makes the case that:

"The Americas were a far more urban, more populated, and more technologically advanced region than generally assumed; and Indians, rather than living in static harmony with nature, radically engineered the landscape across the continents."[100]

Mann suggests that the towering forests that greeted settlers in western parts of the 13 colonies had, in many cases developed in the interval since the decimation of Indian tribes. In like fashion, the huge herds of buffalo that reached as far east as Ohio, may have likewise expanded after Indian hunters' numbers were greatly reduced.

> Recent studies show that North American Indian populations before European invasions were much larger and had greater influence on the environment than has been assumed.

US environmental movements have taken a direction different from the conservationist, multiple-use philosophies in advanced European nations. The more activist organizations have pushed to totally preserve as large a proportion of forests and other natural environments as possible, excluding roads, tourism, or natural resource recovery. Wilderness areas from which all access by motorized transport is excluded now totals more than 107 million acres. The antipathy of many environmental organizations to multiple use or "conservation" goals differs from the conservationist traditions of Europe, which have cultivated respect for land and nature in the larger population.

3.4.1 The Rise of Civil Engineers and Civil Engineering Management in America

The US Army Corps of Engineers (USACE) was established in 1802, formalizing temporary engineer units within the Continental Army during the Revolutionary War and again in 1794 (Layton, 1971). The first formal engineer training program in America was established at West Point in 1818, based on the model of the French Polytechnic University. Engineering carried high responsibilities and prestige. Only top cadets were admitted into the Corps of Engineers or the Corps of Topographical Engineers. There was intense competition for engineers who left the military (Barry, 1997).

American engineers followed more practical directions than their counterparts in Europe. Most gained their skills through apprenticeship, which continued even after initiation of engineering degrees at schools like Rensselaer (1835) and, after

1850, at Michigan, Harvard, Yale, Union, and Dartmouth. The American Society of Civil Engineering (ASCE) was founded in 1852 and the Massachusetts Institute of Technology engineering program was initiated in 1864.

By 1860 engineers had taken on important roles in society. In 1861, Andrew Atkinson Humphrey, a West Point graduate, completed a monumental, 11-year, Congressionally authorized study of the hydrology of the Mississippi River. The report was hailed throughout Europe as a major breakthrough in hydrographic science, and every major scientific society in America and many in Europe elected Humphrey to membership (Barry, 1997).[101] The study and his service as a general in the Civil War earned Humphrey the post of Chief Engineer of the Corps of Engineers.

Engineers' image was heightened by "impossible" achievements like the world's first steel bridge across the Mississippi River, designed by James Buchanan Eads, and completed in 1874.[102] Britain's Royal Society of the Arts awarded him the Albert Medal (previously received by Napoleon III, Louis Pasteur, and Lord Kelvin), and "in 1932 deans of American colleges of engineering named him one of the five greatest engineers of all time (Barry, 1997)." The influence of Eads' achievement went beyond bridge building. In 1867 America produced 22,000 tons of steel. The year the bridge was finished (1874), production was 242,000 tons.

Civil engineers could not consider major construction projects in isolation. Engineers' training included administration and finance so that they were prepared to estimate the costs of construction and activities, as well their subsequent operation and maintenance. In the early twentieth century, the construction of towering skyscrapers in the middle of major cities had to be conceived as part of a complex development involving many aspects of civic design and governance. Beside power grids, water, and sanitation, the design of such tall structures had to take into consideration the frequency of storms and even hurricanes. They might require expansion of subways and elevated rail systems to provide reliable and efficient transportation to people who worked in the buildings. Funding of construction, tax policies, and local politics all had to be part of the development concepts.

Officials' and the public's confidence in the professional competence and integrity of engineers was a key factor in willingness to entrust them with costly and often potentially dangerous constructions. The engineering profession prided itself on its high standing in society. An example of an engineer who maintained his integrity even while working in corrupt conditions of the "gilded age" was Peter Anthony Dey. Chief Engineer of the Union Pacific Railroad, Dey resigned from his position on becoming aware that Union Pacific leaders had inflated cost estimates he had provided for westward expansion of the railroad line from Iowa. His resignation was a part of revelations of the *Crédit Mobilier* scandal, which would become one of the nation's most notorious frauds involving government corruption.[103]

"An honest engineer triggers discovery of the giant railroad scandal (Johnson, 1939)."

Besides the important role of engineers in the nation's manufacturing industries, civil and chemical engineers were responsible for designing and maintaining public health facilities in cities. Cholera epidemics spurred action in building sewer systems for large cities in the mid-nineteenth century, both in Europe and America. In the

1850s, rapidly growing Chicago experienced devastating cholera outbreaks, which led to the initiation of successive innovative sewage collection and transport systems from 1863 until 1889. In the latter year, Chicago engineers completed a system that gained worldwide attention and awards. They had reversed the Chicago River, and ultimately directed the flow of Chicago's sewers through an enlarged canal system to the Mississippi River. However, this did not solve all problems, for the State of Missouri promptly sued in court to stop sewage discharge into the Mississippi River – on behalf of the downstream city of St. Louis. Ultimately, treatment of water had to be instituted.[104]

Breakdown of city prohibition against discharge of sewage into Boston Harbor in the mid-nineteenth century likewise led to cholera outbreaks in 1849–1854 and 1868–1873. The last series of epidemics triggered a doctor-led public protest movement that led to the city's commissioning of a comprehensive wastewater plan. The first designs were drawn up in 1875. The system was built in stages and was completed in the early 1900s. Underground collectors, interceptors, and tunnels with integrated gates constructed under Boston Harbor led to the major entrance to the harbor. The ingenious system discharged sewage at depth in the main channel connecting Boston Harbor to Cape Cod Bay on outgoing tides. Of importance from the policy and city management point of view, the plans included periodic engineering reviews that would propose expanded capabilities to keep pace with population growth and new conditions. The detailed plans received expeditious approval by the City Council and the Massachusetts legislature. Design drawings of the subharbor structures are striking enough so that many continue to be displayed as framed original drawings on walls of the Massachusetts Water Resources Authority in Boston.

In 1893, the Massachusetts state assembly ("General Court") passed an act directing the State Board of Public Health to assess and recommend a solution for a consolidated plan to solve multiple conflicting requests for water supply to growing Boston and suburbs around Boston Harbor. In 1895, the Board's Chief Engineer, Frederick P. Stearns recommended an extraordinarily ambitious plan involving construction of the Wachusett Reservoir in a watershed up to 45 miles west of Boston center. The new plan was acclaimed and accepted, in spite of its great cost, in large part based on Stearns' previous achievements and the detail and care with which the plan was developed.[105] The plans gained a gold medal at the Paris Exposition in 1900.

Stearns oversaw completion of the then largest reservoir in the world in 1906. He was elected President of the ASCE and a Fellow of the American Academy of Arts and Sciences, and retired from the Metropolitan Water Board in 1908 to become a consultant. In 1920, Stearns former protégé, Walter Goodenough, who had been named Chief Engineer after Stearns, met projected future water needs by a new, even more ambitious plan. The Quabbin watershed, which extended to nearly 70 miles from Boston, could supply Boston water until 1970. After much controversy, Goodenough's proposal, which required moving four villages from the watershed, was approved in 1926 and completed in 1930, 20% under budget. The water supply system continues to supply Boston with some of the nation's best water.

Fern Nesson's detailed history of the Boston water supply systems points out the "uncharacteristic willingness" of the Massachusetts legislature "to choose long-range, grandscale, expensive plans, when both in 1895 and 1926, cheaper short-term solutions were available (Nesson, 1983)." Nesson cites as a primary reason for acceptance the fact that the long-range (extendable) project was proposed by Frederic Stearns, a highly respected engineer known for his objectivity and skill. An important factor in approval for the far more controversial Quabbin extension was the provision for pure, unfiltered water from an uncontaminated watershed. This was perhaps overemphasized by the proposing engineers, since competing plans involved filtration, which by this time was a safe, standard procedure. A further positive factor was presentation of the plans in a dispassionate, thorough, and reasoned way by engineers who were well educated, literate, thoughtful, polite, and had long terms of public service behind them. In the long run, going for the Quabbin option not only secured high quality, low cost water for Boston, but also preserved a large area for wildlife and recreation. The engineers, where feasible, recognized and anticipated objections or conflicts and compensated for them.[106]

The highlights of their achievements offered here may seem to idealize engineers (especially civil engineers). The brief treatment obviously cannot offer a fully balanced account of engineering before 1945. However, key points relevant to this book are as follows: (1) engineering in the nineteenth century began as a well-esteemed profession, and continued to attract a good share of "the best and brightest" talent until after World War II; (2) prior to the 1950s engineers were generally given broad, hands-on training that enabled them to interface effectively with industry and society; and (3) bold, innovative, and expensive projects gained acceptance by officials and the public when they had trust in the responsible authorities.

"Normal" management with administrative safeguards and quality controls might have achieved low rates of human error. However, even "low" rates of mistakes could be sufficient to allow spectacular failures that would have eroded public trust for potentially high-risk structures including underwater tunnels, bridges, and overhead rail systems. Likewise, poor financial planning or management would affect the trust-based system that gained authorization, financing, and public acceptance of bold innovations like skyscrapers. In other words, where risk was high, it was understood that only exceptionally high levels of motivation and no-compromise performance and advance planning were tolerable. Through World War II, engineering attracted the kind of individuals who welcomed the challenges and risks, in a profession that maintained unusually high quality and performance standards through internalized, mutually reinforcing motivation.

Because of their broad-based training, engineers were often appointed to public service leadership or management positions in city, state, and federal government agencies. It is a commentary on the effectiveness in designing and maintaining operation of urban transportation systems that the cost of a subway or bus ride in New York City from 1904 to 1944 remained a nickel (Markowitz, 2003).

American engineering training retained breadth and solutions to the practical problems of society even after formal training in colleges and universities replaced apprenticeship in the twentieth century. It did so in spite of continuous employment of more quantitative methods until after World War II (Layton, 1971).

After World War II, the new prominence for basic research in the sciences promoted by Vannevar Bush and his book *Science the Endless Frontier* not only enhanced the status of scientists but also influenced university development. It also affected the training of engineering students. In his book, aptly entitled, *Pursuing the Endless Frontier*, Charles M. Vest, President of MIT (1989–2003) pointed out that:

> "... the engineering curriculum in this country was largely developed by MIT faculty in the 1950s and 1960s. They spearheaded the infusion of basic science into engineering education and practice (Vest, 2005, p. 10)."

Bruce Seely[107] offers a detailed account of the transformation of American engineering curricula to the research model (Seely, 1999):

> "Between 1940 and 1960, calculus and introductory engineering mechanics were slowly moved forward in the curricula. While not shown in the tables, calculus moved to the freshman year, while mechanics moved to the sophomore year. Second, engineering science courses displaced several traditional courses. Classes in specific subjects, such as bridge structures and roads and pavements, started disappearing, as did practical laboratory classes. Engineering Mechanics became a three- and four-hour course, not two three-hour classes. Thermodynamics replaced Steam and Gas Power; Electric Circuits and Machinery had appeared by 1960. Soil Mechanics and Foundations replaced Masonry and Foundations. Theory of Structures and Hydraulics of Drainage Structures also appeared. Even when content remained similar, the course name now reflected the emphasis on engineering science. By 1970, the shift was completed. Upper-level drafting courses were gone and fluid dynamics, differential equations, numerical methods, materials science, indeterminate structures, and systems design were in the curriculum."

These changes were driven by both prestige and money. They brought sophistication and advanced concepts into the training of engineers.

> "Only schools that embraced scientific engineering received large federal projects, and only schools with federal funding developed large research programs. Engineers at Texas A&M, for example, who still conducted practical studies, performed very little federally sponsored research. The same pattern held at many similar land-grant schools."

> "The most valued faculty were not designers but theoretically oriented researchers, who published papers in academically oriented research journals. Eventually, academic and industrial engineers developed different conceptions of engineering, and they almost stopped talking to each other. By the 1960s, practicing engineers routinely complained to professional societies about the limited utility of much academic research. As a result of changes in research funding and practice, two subcultures had appeared in engineering."

The downside was that engineering schools were basically training engineers to become professors at engineering schools – not to work in industry.[108] The lure of theory for bright minds had already manifested itself in European training for engineers.[109]

Recent failures of major structures, like the collapse of a tunnel in part of Boston's "Big Dig," a steam tunnel explosion in Manhattan, and the collapse of the I-35 bridge over the Mississippi River in Minneapolis, focused negative attention on the engineering profession and America's infrastructure. If the main federal research support, prestige, excitement, and hence attraction to the brightest talents in the engineering field are concentrated in more advanced research and theory, it follows that this talent will not be available to initiate and oversee activities that play a direct role in society. Newspaper reporters are reminded of the ASCE estimate that it would take 5 years and $1.6 trillion to bring the entirety of the United States' infrastructure back into a structurally sound state (McLaughlin, 2007). The United States may have to relearn the value of practical engineering, and the lesson this time will be extremely costly.

3.4.2 The Role of Federal Science Agencies Prior to 1969

The first two United States science and technology agencies formally established to deal with resource or "land" issues were the USACE (1802) and the Coast Survey (1807) (Fig. 3.1). New formal federal science agencies did not come until after the Civil War, although state geological surveys grew after the 1830s. Before the 1960s, states and larger metropolitan areas had primary responsibility for regulatory functions in their geographical areas. These included water supply and quality, health, wastewater collection and treatment, pollution hazards and solid waste disposal, and building, highway, transportation, and public safety ordinances. Federal agencies like the USACE, the US Department of Agriculture, the USGS, and the US Fish and Wildlife Service maintained specialized professional services, products, and expertise beyond those available in states and smaller units. The federal agencies conducted research; compiled data, maps, and other products; established guidelines based on authoritative data and experience; and investigated and regulated matters of national or interstate scope. The Public Health Service (PHS) was authorized by Congress to conduct national drinking water surveys in 1912, and issued its first national drinking water standards in 1914. As in the case of the PHS law, Congressional legislation on environmental or health issues often originated in and was guided by experts from federal agencies.

Prior to 1970, Congress created new agencies, consolidated functions, specified missions, and authorized funding for executive bureaus and agencies. But Congress also maintained self-imposed limits in legislation dealing with federal bureaus and regulatory agencies. It rarely extended the scope of laws to operational detail. Agency leaders were generally granted broad latitude to develop methods and policies and plan for the future and deal with new problems, disputes, or conflicts as they emerged (Table 3.2).

The first federal agency to have major regulatory responsibilities was the Corps of Engineers, being directed by Congress to prevent dumping and filling in harbors

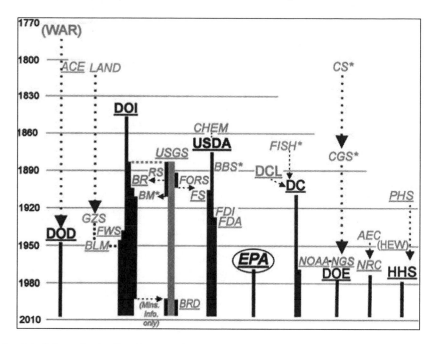

Fig. 3.1 Schematic timeline for evolution of U.S. federal scientific agencies associated with environmental policy

Source: Compiled by F.T. Manheim from online history of each agency and written accounts cited in this chapter. *Solid vertical line* segments represent the timeframes for given agencies, which are added to or removed from host agencies. Thus, for example, Reclamation Service (RS) and mining functions were moved from USGS to become separate bureaus (Bureau of Reclamation, BR; and Bureau of Mines, BM) in the Department of the Interior in 1905 and 1910, respectively. The Bureau of Mines was abolished in 1996, except for one function, which was returned to the USGS in 1996. The USGS is itself an Interior bureau but is shown as a separate *gray column* for this discussion. *Dotted lines* connect the founding of independent agencies to a later host organization. Many changes are too complex to be shown on the graph. For example, in 1996 much of the Department of Agriculture's Bureau of Biological Survey, along with some biologists from the Fish and Wildlife Service were moved to USGS and combined to form the Biological Resources Division. EPA is *circled* for emphasis because of its importance in environmental regulatory enforcement.* indicates an agency that has been part of agencies other than as shown

in the 1880s and 1890s (USACE, 2007). The Rivers and Harbors Act of 1899 gave the Corps responsibility to manage navigable waterways, including waste discharge. A Corps of Engineers website acknowledges that Corps officers came to assume that pollution was in the hands of the state and did not intervene unless there was a threat to navigation. However, the Corps' historical review points out that the Corps' early efforts to stop sewage discharge from New York were blocked by courts that ruled pollution was under state jurisdiction. The evidence suggests that where the Corps has been given clear jurisdiction it has tended to be vigorous in enforcement – often more rigorous than EPA, consistent with its identification with

Table 3.2 Symbols of US federal scientific agencies associated with environmental policy Asterisk as in Fig 3.1

Symbol	Organization	Symbol	Organization	Symbol	Organization
ACE	Army Corps of Engineers	DOD	Dept. of Defense	HEW	Health Education and Welfare
AEC	Atomic Energy Commission	DOE	Dept. of Energy	HHS	Health and Human Services
BBS*	Bureau of Biological Survey	DOI	Dept. of the Interior	LAND	Land Office
BLM	Bur. Land Management	EPA	Environ. Protection Agency	NGS	National Geodetic Survey
BM*	Bur. Mines	FDA	Food and Drug Administration	NOAA	National Ocean Atmos. Administration
BR	Bur. Reclamation	FDI	Food Drug Insecticide	NRC	Nuclear Regulatory Commission
BRD	Biol. Resources Div.	FISH*	Fish Survey, Commercial	PHS	Public Health Service
CS*	Survey of the Coast	FORS	Forestry Survey	RS	Reclamation Survey
CHEM	Bureau of Chemistry	FS	Forest Service	USDA	Dept. Agriculture
CGS*	Coast Geodetic Survey	FWS	Fish and Wildlife Service	USGS	Geological Survey
DC	Dept. of Commerce	GZS	Grazing Service	WAR	War Department
DCL	Dept. of Commerce and Labor				

Source: Compiled by F.T. Manheim from online history of each agency and written accounts cited in this chapter

a military organization. Books by Shallatt (1994) and Barry (1997) indicate that the Corps' work involved both monumental constructions and fiascoes.[110]

Before the 1970s, many federal agencies had broad mandates, that is, research, supporting, and regulatory functions. Thus, the Corps was charged with aiding commerce and issuing permits for waste discharge into rivers. The Department of Agriculture supported farmers and improvements in agricultural methods, controlled animal diseases, and use of pesticides, and, with an Act in 1890, inspected meat. Besides its wide-ranging earth science research activities, the US Geological Survey after 1953 also defined offshore oil leases, collected front-end bid money and royalties, and oversaw safety and proper practices on offshore oil rigs.

Other Department of the Interior bureaus, such as the BLM and FWS had both supporting and regulatory functions. The Atomic Energy Act Amendments of 1954 assigned to the Atomic Energy Commission the task of overseeing the initiation of commercial use of atomic energy, and regulating its safety.[111] The PHS had a key role in water quality, disease, and other human health issues. However, its mission did not include the health of biological species in the natural environment.

3.4.3   Science Agencies Before World War II: Professional, Apolitical, but Buffeted by Politics

Federal science agencies in the nineteenth century came into being because of specific needs for professional services – fish, chemistry, biology and entomology, public health, earth science, engineering, and land management, mapping, etc. With jobs to be done, leaders of the agencies were chosen for their professional qualifications. The leaders' commitments tended to be primarily with their professional identification, not to political or ideological loyalties.

Creation of the Civil Service system by the Pendleton Act of 1883, which required prospective employees to pass qualifying examinations, was indispensable in strengthening the professionalism of Federal agencies as a whole. Federal science and natural resource agencies attracted well-qualified and motivated staff. Some of the nation's noted environmental leaders gained experience in or became known in the course of their employment in federal agencies. Examples include Gifford Pinchot, first head of the US Forest Service (1906). Aldo Leopold who trained in forestry science at Yale and immediately joined the Forest Service, and Rachel Carson, who rose in the Fish and Wildlife Service to become chief editor of all its publications before retiring to write books. Nine of the first 30 scientists hired by the first director of the USGS, Clarence King (1879–1981), were (or became) members of the National Academy of Sciences. Although different in kind from the service-oriented agencies, one should not leave out the Smithsonian Institution – part library, part museum, part laboratory, and research center, whose founding director was the United States' greatest physicist, Joseph Henry (Dupree, 1957).

The professional independence of science agency leaders, especially those with highly independent personalities, could lead to conflicts with the President, or equally independent members of the Congress. European science agencies, not subject to the highly unpredictable quirks of the US Congress, which controlled the purse strings for federal agencies, tended to have less turbulent histories.

> "Thus for example, the great scientist-founder of the Coast Survey, Friedrich Hassler, had to leave his agency in 1818, returning 12 years later to reassume leadership under a friendlier Congress. Gifford Pinchot ran afoul of disagreements with the increasingly conservative President Grover Cleveland, who fired him as director of the Forest Service in 1910. First USGS director, Clarence King, resigned after only two years, in 1881, in part because he would lose the support of Interior Secretary Carl Schurz (under President Rutherford B. Hayes), who would not serve under incoming President Garfield. Eleven years later, his successor, John Wesley Powell, had aroused so much ire in Congress that it slashed budget, leading to a disastrous RIF. Two years later Powell was finally forced to resign (Rabbitt, 1989)."

The USGS's science and policy history is provided as a detailed case study of federal science in Section 8.3. The third Director of USGS, Charles Walcott demonstrated leadership qualities that can be considered a model for successful operation of American federal science agencies. Beyond requiring standards of quality, good science, and prompt performance, Walcott was responsive to Congressional missions and mandates, and strove to maintain open and courteous relationships with USGS constituents and all others. He also took independent initiative to fulfill emerging needs and take advantage of opportunities. Once "bread and butter" service was accomplished, employees were free to extend their work to the frontiers of then-known science – which in turn expanded the Survey's abilities to undertake new tasks.

The USGS long enjoyed organizational and operational stability, in part because it provided important services in every state – for example, topographic mapping and water resource research. This gave it a broad political constituency and support.

In other cases, however, rapid and inadequately planned organizational changes decided by the Congress posed problems for the ability of science agencies to plan and operate effectively. The Department of Agriculture's Bureau of Biological Survey and the Department of Commerce's Bureau of Fisheries were moved to the Department of the Interior in 1939 and combined to become the US Fish and Wildlife Service in 1940.[112] Further changes took place in 1956. Many FWS biologists were abruptly moved into the National Biological Survey[113] in 1993, and in 1996 the NBS became part of the USGS; there the group was renamed the Biological Resources Division.

The US Bureau of Mines was created in 1910 in response to a series of major mine accidents, including the explosion in the Monongah No. 6 and 8 Mine, West Virginia, in 1907. That event claimed the lives of 362 miners – the worst mining catastrophe in US history. Three other major disasters in Pennsylvania, Illinois, and Alabama together added more than 450 deaths (Fig. 3.2).

Although the bureau had qualified scientists and engineers who compiled and analyzed mining performance, developed new technologies, and safety recommendations,

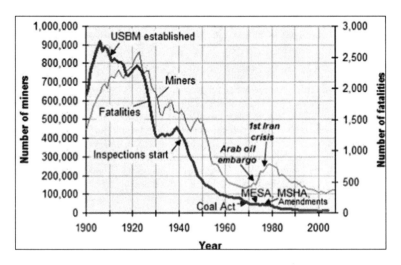

Fig. 3.2 Mining fatalities in U.S. mines (mainly coal mines) 1900–2007
Source: MSHA (2008). During the Arab oil embargo and the first Iran crisis, coal was mined to supply some electric power plants in order to decrease the need for a portion of imported oil

interest group pressures dissuaded Congress from granting the agency inspection authority until 1941. Although the level of mining fatalities declined with time, another explosion costing the lives of 125 miners in 1972 and the new mood of Congress in the 1970s brought passage of a detailed, proscriptive law (Federal Mine Safety and Health Act of 1977, Public Law 91–173) and transfer of responsibility for mining safety to a new agency, the Mine Safety and Health Administration (MSHA) in the Department of Labor in 1977.[114] This separation ultimately led to dismantling of the Bureau of Mines in 1996.

Recognition of the problem of coordination of scientific activities and expertise dates back to at least the Theodore Roosevelt administration. Consideration of a Department of Natural Resources, comparable to that in many states, was taken up at some level during the Hoover, Roosevelt, and subsequent administrations. It reached advanced stages of planning during the Nixon administration and continued after the formation of EPA through cooperation between the NOAA Administrator Robert White and USGS Director William Pecora, actively supported by Interior Secretary Walter Hickel. However, the effort collapsed as a consequence of the firing of Hickel, in 1970.[115] The Penny–Kasich amendment for consolidation of selected federal science agencies came close to approval in the House of Representatives in 1993, but this was more of a drastic retrenchment measure than a plan for coordination (for detail see USGS history, Section 8.3).

The decade of the 1950s saw burgeoning industrial production, suburban population growth, and automobile traffic – all of which increased pollution. Federal science agencies generally maintained high levels of skills and commitment to their tasks. However, achievements like the moon landing, a nuclear submarine fleet, mechanized agriculture, and many significant US medical advances were not matched by corresponding attention to environmental management and infrastructure, nor to the organization and coordination of federal science capabilities to meet new pollution and environmental challenges that crossed state borders.

By 1970, the turmoil in society in the 1960s had seeped into many of the federal agencies. Sheila Jasanoff, a political scientist specializing in science policy, has noted the corrosive effect of partisanship and conflict over regulatory policy on the quality of science in regulatory agencies in the 1970s:

"It was a common complaint of industry and members of the scientific community during the 1970s that the Environmental Protection Agency (EPA) and the Occupational and Safety Administration had systematically distorted their assessments of cancer risk so as to build the case for more regulation." ... " By contrast in the early years of the Reagan administration environmentalists accused EPA of a pro-industry and antiregulatory bias into its principles for assessing carcinogenic risk (Jasanoff, 1998)."

Similar observations were made about the operations of the US Forest Service by Randal O'Toole, President of the Thoreau Institute of Bandon, OR. According to O'Toole, up to 1952 the US Forest Service mainly used selective cutting of mature trees:

"When done right, a selection cut forest is hard to distinguish from wilderness." ... "Clearcutting, in which all trees on 20 to 100 acres or more are cut regardless of maturity, imposed higher reforestation and rehabilitation costs. But since agency managers got to keep those costs out of timber receipts, they had an incentive to clearcut even when other cutting techniques were more compatible with recreation and other uses." ... "By 1970 the Forest Service was cutting almost four times as much timber as it did in 1952, almost all of it clearcut. The resulting controversies over both clearcutting and the volume of timber cut rocked the agency and led to lawsuits, congressional hearings, and tree-sitting protesters." " ....and by the 1980s the Forest Service was losing billions of dollars a year on the timber and other programs (O'Toole, 1988)."

Summing up the history of federal science agencies, an overall impression emerges that the creation of professional agencies in response to perceived needs by Presidents yielded positive results. Changes initiated by Congress, as well as funding irregularity, have been more capricious and often interfered with effective operation. Special damage was done to national programs and problems by abolition of the Office of Technology Assessment through the influence of House Speaker Newt Gingrich. This action removed an important, unbiased source of evaluation of the performance of federal agencies as well as assessment of national problems. One of the recommendations to help restore trust and reduce conflict and gridlock is to "reprofessionalize" agency management, removing or reducing the partisan influences of the past 30 years (see Chapter 7).

Chapter 4
The Environmental Revolution of the 1970s and Its Outcomes

4.1 Problems Prior to the 1970s

Stresses emerged and grew under the surface of the expansive mood in America after World War II. Many engineers and other men who had key positions in maintaining operating facilities and infrastructure went into the military during the war. This led to deferred maintenance or modernization of some health, sanitation, and waste disposal systems (Reid, 2008).

African American populations from the southern states migrated north and to the coasts to work in the war industries. But black members of the armed services who came home to their communities faced discrimination and isolated atrocities after having served their country in war. They experienced a new awareness of injustice, fueling the Civil Rights movement of the generation of Martin Luther King (1930–1968). After several years of debate, President Harry S. Truman issued an Executive Order on July 26, 1948, banning discrimination in the armed services, and thereby desegregating it for the first time since the Civil War and Revolutionary wars.

Movement of rural populations to the cities increased pressures that promoted water pollution and other problems. Suburbanization and motorization gained momentum. The integrated management systems that characterized major cities before the war began to fragment. All states established water pollution agencies by 1948, but their levels of effectiveness differed widely. The upper midwestern states, sites of the Progressive movements in the 1920s, were among states that most conscientiously sought to maintain public health services.

Boston, which had experienced large population influxes, lagged severely behind prewar maintenance of its wastewater treatment system. New waste treatment facilities on Deer Island, near the entrance to Boston Harbor, were already inadequate to meet the needs of Greater Boston and its suburbs upon their opening in 1968. When holding capacities were exceeded, raw sewage was discharged to Boston Harbor. In the later 1960s, willingness to make investments in upgraded facilities was held back by anticipation that increasingly restive Congress would provide subsidies for the expansion and modernization of municipal wastewater treatment facilities, especially those that discharged their wastes to the ocean or navigable waterways (Metropolitan District Commission, 1968).

F.T. Manheim, *The Conflict Over Environmental Regulation in the United States*,
DOI 10.1007/978-0-387-75877-0_4, © Springer Science+Business Media, LLC 2009

New York City's air pollution was severe, and the Greater Metropolitan area had major waste disposal problems and serious coastal pollution. The industrial and steel manufacturing belt south of the Great Lakes belched smoke and combustion products. Severe local pollution was partly "solved" by raising the height of stacks. Now, however, the smoke, soot, and sulfur dioxide (which oxidizes in the air to form sulfuric acid) was carried by the air currents at higher elevations and affected former clean areas hundreds of miles away.

Pittsburgh had long been one of America's most polluted cities and industrial areas. Joel Tarr cited Herbert Spencer, English philosopher and social theorist, who visited Pittsburgh in 1882: "Six months residence would justify suicide." Tarr continued:

> "By 1945.....sewage pollution was only one of the several serious issues facing Pittsburgh. Its populatio n was stagnating, its industries were exhausted by the war effort and badly in need of renewal, its physical plant was deteriorating, and its air and land, as well as its water, were badly polluted (Tarr, 2004)."

Most federal agencies had well-qualified staff and professionally experienced leaders. But, as mentioned in the previous chapter, scientific agencies could be subjected to the equivalent of musical chairs and be abruptly shifted from one department to another at the initiative of the President or in response to serendipitous concerns or political alignments or pressures in Congress. Through the 1960s and 1970s, political influences intensified on both federal science and regulatory agencies, and also science advisory bodies (Jasanoff, 1998).

Seen from the perspective of the environmentally concerned citizens, Congressional leaders, and environmental organizations like the NRDC, Sierra Club, and Environmental Defense Fund (EDF), tough measures were needed to halt pollution.

The enormous dynamism of US industrial performance in World War II – followed by the ability to overcome staggering difficulties to land humans on the moon lent US industry and technological capability an aura of near invincibility. The perception had a twofold influence on the framers of the new environmental laws:

[The power of industry] reinforced the need for a framework of protections for the environment that would be robust, comprehensive, and not subject to the discretion of enforcement agencies which, it was feared, might "do nothing" to enforce laws when faced with powerful economic interests.

The framers felt reassured that the "inexhaustible vitality of the US industrial system" would be able to maintain economic viability in spite of limitations and requirements imposed to protect the health of the environment.

4.1.1 The New Environmental Management System

4.1.1.1 The National Environmental Policy Act (NEPA)

The first of the new environmental laws passed after the Santa Barbara oil spill was NEPA.[116] After precursor developments, Senate bill S 1075 was submitted by Senator

Henry M. "Scoop" Jackson in February 1969. It was enacted by Congress late in December 1969 and signed by President Nixon on January 1, 1970 (Congressional Quarterly Service, 1973), after Jackson acceded to demands by Senator Muskie[117] that its provisions be strengthened.[118] Assisting Senator Jackson in crafting the Act was Lynton Keith Caldwell, a political scientist and leading expert on environmental policy, who had made a comprehensive review of all available references and considerations relating to environmental policy development (Caldwell, 1998).

NEPA required every federal official and agency to consider environmental aspects of every action, submit detailed impact statements, and consult and obtain written comments of "any Federal agency which has jurisdiction by law or special expertise with respect to any environmental impact involved." The law itself does not specify enforcement means, but court interpretations soon held that NEPA's environmental impact statement provisions could be used as an administrative tool for citizen challenge of proposed actions in court (Percival, Miller, et al., 2000).

In a book describing the Act and 30 years of work in environmental policy, Lynton Caldwell pointed out that:

"By mid-20th century the quality-of-life objective of material growth was being frustrated by excess and misapplication in economic activities and technology."

He noted that at the White House Conference on Natural Beauty in 1965, President Johnson declared:

"Our conservation must be....a creative conservation of restoration and innovation. Its concern is not with nature alone, but with the total relation between man and the world around him."

But there was a disconnect between Johnson's philosophic recognition and his actions as a politician (as noted in Section 2.6).

I regret passing quickly over Caldwell's book into which a lifetime of careful experience and reflection was obviously poured. It contains no statements that are not nuanced. His book is deeply idealistic. He affirmed that NEPA was not designed as a regulatory statute, but as a policy act; it does not provide specific enforcement measures or refer to court action.

"The National Environmental Policy Act (NEPA) offers a set of goals that could guide the nation toward an economically and environmentally tolerable, sustainable future....Reconciliation of environmental values and economic interests is a major objective of NEPA (Caldwell, 1998, preface and p. 4)."

He indicated that it sought not to dictate but to broaden insights and thinking, and to create greater harmony between man, society, and nature.

Caldwell states (as do Lindstrom & Smith, 2001) that NEPA has been widely misinterpreted, and not enough attention has been given to its deeper messages and goals.

"Although mainstream environmental nongovernmental organizations tacitly supported Senator Jackson's bill at the time of enactment, they subsequently showed interest in NEPA primarily for the environmental impact requirement that enabled them to stop or delay specific government programs or projects to which they objected."

The environmental impact assessment (EIA) and statement (EIS) concept gained wide acceptance not only in the United States but among virtually all other advanced nations.

It has become the single most widely emulated feature of American environmental law. The EIA concept in advanced European nations has tended to facilitate the fact-finding model espoused by NEPA's founders. However, US adoption of NEPA took more legalistic and adversarial pathways. The use of the EIS as a basis for legal challenge was already anticipated in the *Scenic Hudson Preservation Conference v. Federal Power* and *Sierra Club v. Morton* court rulings – and became firmly established by Circuit Judge J. Skelly Wright's decision in *Calvert Cliffs v. US Atomic Energy Commission* (449 F.2d 1109). Moreover, though NEPA formally dealt only with federal government activities, its application as a legal tool also became extended, directly or indirectly to state and private initiatives.

The burgeoning of litigation (see Section 4.3.1) that invoked NEPA along with the *Administrative Procedures Act* of 1946 as a device to stop activities that various groups disliked deeply disappointed Senator Jackson, especially when it was used to challenge bills that he supported, like the Alaska Pipeline and the Boeing Supersonic Transport. He in turn became the object of criticism by environmental activists (Ognibene, 1975).[119]

4.1.1.2 The Revolutionary Environmental Laws[120]

In December 1970, the Clean Air Act Amendments (CAA) became the first of a series of groundbreaking new laws and developments that would transform the US environmental management system.[121] The pollution-related laws were designed to be enforced by the new Environmental Protection Agency (EPA) that had been created by President Nixon under Reorganization Plan 9, sent to Congress in July 1970, and approved in December.

More laws followed in rapid succession. The laws listed below are only the most important of the more than 116 environmentally related statutes created from January 1, 1970 through 1979. Of the 116, about 40 were wilderness areas authorized through the Wilderness Act of 1964 and other land designations.

- Wilderness Act of 1964
- Clean Air Act Amendments 1970 (CAA)
- Occupational Safety and Health Act 1970 (OSHA)
- Federal Water Pollution Control Act Amendments (Clean Water Act) 1972 (CWA)
- Federal Environmental Pesticide Control Act (amendments) 1972 (FIFRA)
- Marine Mammal Protection Act 1972 (MMPA)
- Endangered Species Act 1973 (ESA)
- Safe Drinking Water Act 1974 (SDWA)
- Fishery Conservation and Management Act 1976 (FCMA)
- Federal Land Policy and Management Act 1976 (FLPMA)
- Resource Conservation and Recovery Act 1976 (RCRA)
- Federal Mine Safety and Health Act Amendments (1977) (FMSHA)

CERCLA (1980). The most controversial and expensive of all environmental laws, the *Comprehensive Environmental Response, Compensation and Liability*

Act (CERCLA, or Superfund Act) of 1980 (Nakamura & Church, 2003), was triggered by national attention focused on a "chemical soup" containing toxic and carcinogenic compounds that seeped into schools and homes built on a site known as Love Canal in 1978. The tract had been transferred to the Niagara Falls, NY, Board of Education by Hooker Chemical and Plastics Corporation. Although later analyses revealed that dangers of the site had been exaggerated (Wildavsky, 1995), exposure of the event in newspapers and other media drew public attention to many other sites, lakes, and lagoons that had been dumping grounds of toxic wastes.

CERCLA contained absolute liability provisions that had their origin in the Water Quality Act of 1970, designed to deal with oil spills like that of Santa Barbara (Leon Billings, personal communication, June 2008).[122] Later criticized as an unwise idea, CERCLA's cleanup standards also became linked to standards established for the SDWA.[123] In effect, CERCLA peered into the history of Superfund sites and could assess costs for cleanup to any organization found to have linkages with them and deep enough pockets to pay. Herein lay many problems (see Section 8.4.3).

Political scientists and Brookings Institution scholars, Robert Nakamura and Thomas Church (Nakamura & Church, 2003) explored options for reform of the Superfund Act after observing its operations for some 15 years. They stated that Superfund represented "regulation in its most unpalatable and politically vulnerable form":

> "Its liability scheme, grounded in the broadest possible notion of the "polluter pays", offends many elemental principles of fairness. It uses these controversial liability principles to extract – under threat of treble damage vast sums of money from entities ranging from corporate titans to cash-poor municipalities to mom-and-pop dry-cleaning establishments. These extracts typically have nothing to do with the current operations of these businesses or governmental units. Rather, compliance is commanded in a highly adversarial and quasi-prosecutorial forum in which the program's targets … are often treated more like criminals than responsible citizens and businesses (Nakamura & Church, 2003, p. 9)."

According to the Rand Corporation, from one third to a half of all money spent on the CERCLA program went to legal and administrative expenses (Gates & Leuschner, 2007). Several front-page articles in The New York Times in 1992 revealed huge costs to remove and dispose of soils and other materials posing low risks while other more serious problems were neglected. In spite of the known defects, Nakamura and Church concluded that legislative reform of CERCLA was unlikely, and that "administrative reforms" (more listening and cooperative attitudes) of the kind developed in that second Clinton administration could soften Superfund's harshness and were the most promising avenue for improvement (Nakamura & Church, 2003).

The pollution-related laws had common features:

1. *Specificity of application.* Unlike NEPA, which promoted many-sided evaluation of environmental issues, each of these laws had specific focus on air, water pollution, toxic wastes, etc.
2. *Unprecedented preemptive detail.* Laws prior to the 1970s generally gave discretion to regulatory agencies to plan, create and execute policies, and solve problems. The new laws essentially eliminated this discretion. The directives

extended to operational procedures, kinds and requirements of permits, types and frequency of reports, the training and experience of agency employees, details of research grants, the makeup of boards, review and coordinating panels, yearly limits on spending for given operations, technologies mandated to be applied to given functions, details of enforcement and penalties, and much more. The environmental objectives were preemptive; that is, they took precedence over economic and all other competing values except where explicit (and usually difficult) procedures to seek waivers were provided.

3. *Uniform standards.* EPA or other enforcing agencies were to establish regulations and rules to implement the various provisions of the laws, including standards for maximum permissible pollutant levels (e.g., stack emissions or pipeline discharges). These would be uniform and apply nationally or as otherwise specified. The objective was to systematically eliminate pollution or reduce it to safe or acceptable levels using the best available technical means. Plans would be required for all producers of pollutants who exceeded standards to move toward compliance with the norms within given time frames. The standards were designed to be *technology forcing.*

4. *Monitoring and publication of results.* Emitters and dischargers of pollution had to systematically monitor their emission products and report the results to EPA (independent checking might be employed). The enforcing agency in turn would make data available to the public in accessible form.[124]

5. *Strong enforcement and punitive measures for violations.* Various penalties were defined for violators, who could be taken to court either by agency legal staff or through attorneys for the Justice Department.

6. *Delegation of monitoring and implementation of regulations to states.* States were to be given authority to oversee application of the given laws. If they did not agree or if their plan for compliance was found unacceptable, the federal government could take over and enforce the laws.

7. *Citizen monitoring and enforcement.* Citizens and citizen groups (e.g., environmental organizations like the Sierra Club, EDF, Friends of the Earth, NRDC, as well as local ad hoc groups) were empowered by most of these and later laws to challenge activities that were not in compliance with standards, or to sue regulatory agencies including EPA in federal court for inadequate enforcement of laws and regulations (Table 4.1).[125] Citizen action was further encouraged by provisions allowing reimbursement of reasonable legal and witness fees in successful suits. In some cases even reimbursement for unsuccessful suits could be authorized (Kramer, 1982). (Historical examples involving other laws, such as the Endangered Species Act of 1973 and the Marine Fisheries Act – later named the Magnuson–Stevens Fisheries Act of 1976 – are included in Section 8.4)

Another law that aroused widespread consternation was the ESA. ESA's unequivocal championing of the lowliest species, such as the 2-in. snail darter fish, the Furbish lousewort, and the Prebbles meadow jumping mouse against mega-economic development drew lines in the sand between environmental supporters and a large part of the general public (see Section 8.4.4).

Table 4.1 Judicial action by EPA during the Reagan administration

Law or regulation	Judicial cases filed in court, FY 1981–1987						
	1981	1982	1983	1984	1985	1986	1987
Air-stationary	56	29	77	55	66	82	74
CWA	30	11	56	81	60	103	68
SDWA	2	3	20	6	9	5	12
RCRA	13	2	2	9	6	23	43
CERCLA	6	5	30	31	32	30	54
TSCA	0	0	4	5	7	4	8
FIFRA	0	0	2	5	8	6	7
Air-mobile sources	8	1	13	17	24	6	19
Total	115	51	204	209	212	260	285

Source: www.epa/gov/history. CWA Clean Water Act, SDWA Safe Drinking Water Act, RCRA Resource Recovery Act, CERCLA Superfund Act, TSCA Toxic Substances Control Act, FIFRA Environmental and Pesticides Act

4.1.1.3 Impact of the Major Pollution Laws

The main features of the pollution-related laws were designed to optimally mesh with and support citizen litigation. Their specificity fit with development of clear-cut, uniform standards. Monitoring and regular reporting procedures with provisions for full public disclosure provided citizens access to data, while fixed standards offered performance indices that nonspecialists could check on. Such features as training programs and research grants, which might seem to mark unnecessary interference in agency operations, had specific purpose. The laws were complex. Training was therefore essential to make sure enforcement agency employees understood and could implement detailed provisions. Research was likewise an essential component for standard setting. The operating structure of EPA mirrors the structure of the laws, that is, EPA's operating divisions are devoted to water, air, solid waste, pesticides and toxic substances, enforcement and compliance, etc. Lawyers are trained to pay close attention to real-world detail. These perspectives help us understand why the 1970s laws became so detailed.

Standardized operating procedures and minimal discretion on the part of EPA limited the ability of regulated organizations to evade the intent of the laws. Hawkeyed citizens and environmental organizations like the NRDC were empowered and encouraged to monitor performance and sue in federal court if they found laxity. The availability of reimbursement for legal fees obviously reduced citizen organizations' litigation costs and helped them recruit capable lawyers to file lawsuits.

Such extreme emphasis on enforcement, punitive action, and aggressive legal enforcement, directed not only against polluters, but also against the enforcing agencies, testifies to the cynicism and skepticism of the framers of the legislation. We can see why the system quickly became effective. But when we look at criteria for good lawmaking, it will be clear that a price was paid to erect this powerful adversarial system.

The highly motivated and sophisticated reformer-framers of the laws (among which a committed bipartisan group of senators was identified in Chapter 2) were a legislative *tour de force*. The new laws were so detailed, so knowledgeably written, and contained so many creative features that they overwhelmed the reservations of potential opponents who might have been concerned about their complexity and rigidity. Something had to be done, and the new laws offered an in-depth, professional approach. There was "no other game in town" that could compete. Both the Clean Air Act and Clean Water Act won unanimous votes in the Senate. Many of the later laws were likewise spearheaded by the Muskie Air and Water Pollution Committee.

President Nixon had appointed Russell Train, an environmentally motivated Republican and former Chairman of the World Wildlife Fund, as Chairman of an environmental advisory panel that developed concepts for the EPA. According to Caldwell (1998), Nixon had initially opposed the provision for a Council on Economic Quality (CEQ) in the Office of the President, as called for in NEPA.[126] But when the law was passed he acceded to it and appointed well-qualified individuals. Train became Director of the Council and later, the second Administrator of the EPA. President Nixon signed the Clean Air Act Amendments but vetoed the Clean Water Act Amendments. His veto was overridden with a wide bipartisan margin (1972).

4.2 Results of the New System

It is a daunting task to try to do justice to the results of environmental regulation, based on the huge and often controversial literature (21,000 hits for books on a query for environmental regulation in the Amazon.com bookseller web site and 1,810,000 titles in the Google Scholar search engine). (I describe in Chapter 9 the approach I have taken to arrive at the conclusions in Sections 4.2.1 and 4.2.2). Many of them can be documented by multiedition textbooks and reviews of environmental policy such as those by Vig and Kraft (2005), Percival, Miller, et al., (2000), Rosenbaum (2005), Easton & Goldfarb (2003), and in references cited in these works (explored in more detail in Chapter 7). Chertow and Esty (1997), Fiorino (2006), and others have called for regulatory reform.

4.2.1 Positive Outcomes

- Enactment of the rigorous, uncompromising laws of the 1970s (with subsequent amendments) was greeted with relief and enthusiasm by environmentalists and reduced a major source of political dissatisfaction. Most environmentalists continue to regard the laws as the United States' principal bulwark for the environment. Many other people – initially even businessmen – initially welcomed resolution of the confusion and chaos of the past.
- The comprehensiveness, specificity, and level of detail in the new laws established the United States as an international environmental leader. Specific innovative features like the EIS were widely copied abroad.

- Workable systems for delegation of authority to states to enforce statutes became established for air and water quality.
- The United States gained a central agency, EPA, which generated regulations from the laws, enforced laws, provided expertise on a wide range of pollution and environmental quality issues, and sponsored research. Since the mid-1990s, EPA has pursued outreach and coordination of user group activities such as the Annual Environmental Partnership Summit Conferences. Emphasis on documentation accessible to the public through EPA and other scientific/regulatory agencies made the United States the world leader in availability of detailed environmental data. Much research has been sponsored and undertaken.
- The new laws provided groups and local citizens with incentive and means to maintain vigilance against environmental abuses or lax regulation, through appeal in the courts.
- The powerful teeth of the laws soon began to bite, making rapid inroads on pollution, especially industrial pollutants like lead, polyaromatic hydrocarbons (PAHs), and polychlorinated biphenyls (PCBs) (Fig. 2.5). Between 1970 and 2002, particulate matter emissions fell 34%, carbon monoxide decreased 48%, volatile organic compounds (which contribute to ozone pollution) dropped 45%, sulfur dioxide declined 52%, and nitrogen oxide emissions fell 17%. These reductions decreased incidence of adverse health effects attributable to air pollution, including lung damage, asthma, bronchitis, reduced oxygen in blood, eye irritation, and reduced resistance to infection (Schroeder & Steinzor, 2004).
- The National Pollution Discharge Elimination System (NPDES) has been remarkably successful in achieving documentation of waste dischargers, private and public, throughout the nation. All pollution monitoring data, including that from NPDES sites as well as most toxic release sites, are enumerated in inventories accessible on the web through the EPA web site (www.epa.gov), or through an interagency database (Landview, U.S. Census Bureau, 2007) that allows users to plot the location and nature of NPDES, TRI, and Superfund sites on a built-in mapping system developed by NOAA. The data on the Landview DVDs are combined with hydrographic features from the USGS and demographic data from the Census Bureau, showing model cooperation among scientific agencies.
- The EPA *Brownfields and Land Revitalization* web site lists examples of former waste sites that have been successfully restored to use.[127]
- Although a number of threatened or endangered species remain on cautionary lists, the populations of prized endangered species like the bald eagle have rebounded.

4.2.2 Negative Outcomes and Criticisms

- The system has led to burgeoning litigation and litigiousness in society.
- Industry and related business groups complained about the command and control character of the system soon after its inception. Observers have pointed

out that the imposition of detailed, inflexible regulatory controls by the federal government is fundamentally inconsistent with the liberty-loving traditions of American society.

- The system's complexity, cumbersomeness, inflexibility, and cost, along with delays or obstructions imposed on economic and other functions in society, have placed significant burdens on entrepreneurs or businesses and organizations.
- The combination of the impacts and the public litigation that the system was designed to encourage, has created a rift between environmentalists and industry. (Chapters 1 and 2 detail how this rift widened into political polarization resulting in gridlock, and inability to amend or reauthorize major laws since 1991.)
- The concept of the supremacy of environment over all other societal interests, backed by legal teeth, has given rise to the NIMBY phenomenon. This has impeded not only economic development and lost opportunity costs[128] (Kagan, 2003), but also infrastructural maintenance and environmental management. The United States has successively fallen farther behind other nations in international ratings for environmental management and potential future development.
- The prolonged conflict has fostered radicalization on both sides. On the environmentalists' side there has been growth in movements not only opposed to extraction of natural resources – especially oil and gas, but also to any kind of human use of attractive land involving the general public (the wilderness movement). On industry's side antagonisms have fostered attempts by political supporters to evade, or undercut the environmental laws, and weaken or influence the actions of EPA (Kennedy, 2004), and to foster in the business community simplistic reliance on the free market system as a solution of all societal problems (Beder, 2006b). (I argue in Chapter 9 that the conflict has affected the quality of industrial leadership.)
- The antagonism between environmentalists and industry has fostered skepticism in the business community about global environmental change and promoted opposition to accelerated energy conservation or other measures. Internal conflict has impeded the United States' ability to achieve consensual policies about energy use.

The fragmented system has frustrated attempts to achieve integrated environmental management or ecosystem management, and has led to overloading, erratic or underfunding of regulatory agencies. Other complaints by environmentalists include environmental justice issues (Schlosberg, 1999).

4.3 Underexamined Problems

The preceding issues (Section 4.2) have received considerable attention in literature. Now we turn to fundamental issues about which there has been surprisingly little discussion.

4.3.1 Is Congress an Appropriate Environmental Manager?

The problems and circumstances of the late 1960s offered ample reasons to reform US environmental management. The question is whether the nation's prime political debating forum was suited to design and maintain detailed regulatory functions in the complex and changing scientific field of environmental management.

Congress lacks built-in scientific expertise. Only a few senators and representatives have scientific background. Expert scientific witnesses can be called – but the political orientations, ideologies, and instincts of the leaders determine who will be called and who will actually write legislation. The current system encourages partisan, reactive, serendipitous legislation serving short-range goals. This was dramatically revealed in the impact subsidies for corn-based production of ethanol have had on world hunger (Brown, 2006, 2007).

As a result of the chaotic lawmaking process, Congress has a huge annual burden of legislation, estimated at around 10,000 bills. It is continually barraged with ongoing business, while in the House of Representatives preoccupation with reelection begins soon after election. Competition among committees, the influence of powerful individuals and industry lobbying campaigns, economic woes, arguments of academic and scientific leaders, and the blizzards of letters and messages generated by environmental organizations all complicate good lawmaking.

A minutely detailed law is virtually obsolete on the day it is passed. Yet, the legislative gridlock of past decades means that most of the prime environmental laws used by EPA could not be reauthorized or amended since 1990. An example is shown for the CWA in Fig. 4.1. The Act has not been significantly amended since 1987.

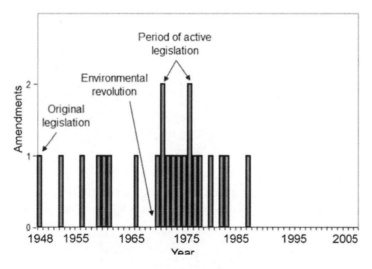

Fig. 4.1 The Federal Water Pollution Control Act (including Clean Water Act Amendments) Source: Compiled by F. T. Manheim assisted by Greg Fuhs, George Mason University. Each bar unit after the original FWPCA statute (1948) represents an amendment. Note clustering of amendments in the 1970s and absence of amendments after 1987

Lifting of pre-1970s constraints on regulatory legislation opened the floodgates to virtually unconstrained legislation.

Since the 1970s, members or groups of members of Congress who perceived any unmet environmental need have felt free to propose a law, even if the law dealt with detailed societal functions about which Congress has little expertise, and the law intruded into what clearly should be within the operational duties of agencies. Micromanagement that can include the stipulation of lettering on rechargeable batteries lends the whole legal and regulatory system an air of dysfunction and arbitrariness.[129] It complicates and demeans the work of agencies, and fosters continuing use of lawmaking as a kind of merit badge system for Congressional members to gain credit for productivity and environmental concern. A current database lists more than 430 environmentally related laws passed since January 1970.[130]

The massive scope and complexity of the current system makes it almost impossible to reform.

Thousands of bills to significantly modify the laws have been rejected in the past, many by opponents seeking to weaken the laws or ameliorate specific impacts. Small or incremental modifications may offer only ad hoc patches on more systemic problems. The industry–business community and its supporters naturally have the greatest incentive to propose reforms. But without real communication and holistic approaches such "reforms" are simply taken as partisan tactics and are fought by the environmental community and their supporters.

The laws are anachronistic with respect to US norms for good law, and the principles accepted for effective environmental law in the European Union.

An often cited set of criteria for efficient laws are those suggested by Thomas McGarity (McGarity, 1983; Percival, Miller, et al., 2000):

1. administrative feasibility;
2. survivability (under existing conditions of judicial and political review);
3. enforceability;
4. efficiency;
5. fairness and equity;
6. ability to encourage technological advances.

The environmental laws have certainly demonstrated administrative feasibility and survivability, thanks to the vigilance of the environmental community. Enforceability is more of a question – though on the surface it would appear that this quality is especially guaranteed by empowering citizens to use the courts to support enforcement. However, Yaffee (1982) was the first to carefully document the realities behind the enforcement of tough and rigid statutes.

Yaffee (whose main focus was the ESA) said that the 1970 CAA was such a strong statement because Muskie wanted to be more committed to the environment than President Nixon going into the 1972 elections. Because the deadlines and goals of the CAA were unrealistic and impossible to meet, a series of delays and weakening of standards were subsequently created by the EPA administrators or by Congress (Costle, 1985). The problem of having to follow formal regulatory and enforcement procedures regardless of economic consequences occurs in most of the statutes. It occurs because the laws and environmental nongovernmental organization (NGO) watchdogs do not allow consideration of economics or innovative technological or policy development that would create deviations from the law. Therefore, even the most conscientious EPA or cooperating state agency staff are forced to hide noncompliance and delays behind paragraphs and in various administrative niches and procedures, lest "going by the book" create such dislocations and bad publicity as to bring on public and legislative backlash. Noncompliance rates with the NPDES requirements of the Clean Water Act of over 50% (Table 4.2) are tolerated. Agencies will get tough in given cases if they find bad performance getting out of hand or beginning to jeopardize their reputation.[131] It is always more powerful to assert, as in the case of the ESA that "no species will consciously be allowed to go extinct" than to say "we will do our best to conserve them but we may have to consider trade-offs."

The necessity of hiding unrealistic standards and deadlines behind procedures and paragraphs forces civil service administrators who would like to operate openly and honestly into uneasy compromises. The system perpetuates suspicions or fears about agencies being coopted by their clients or lack of commitment to enforcement because one can always document failure to live up to the law. It routinely gives rise to misguided litigation and inflammatory newspaper articles that frustrate and can demoralize conscientious operational staff. Municipal and state agencies with good records are largely powerless to counter the massive confusion resulting from barrages of newspaper articles by reporters who have insufficient background to offer more than typical adversarial interest coverage (NGO A accuses Agency B of violations; Agency B denies or gives pro forma response); issues are rarely resolved. The Milwaukee case study details the public misinformation and sensationalization that citizen litigation can introduce into media accounts (see Chapter 8). In the Milwaukee case, the decision of a three-judge Appeals Court was acknowledged to have been formed mainly on the basis of the superficial and distorted newspaper stories.

The current system forces both public and private agencies to prioritize the safest and most legally defensible, rather than the best and most innovative strategies. The system inevitably creates overwhelmed agencies and complaints about inadequate support for enforcement, when, in fact, the chronic lags and failures are an inevitable consequence of the laws themselves. The European Union (EU), as we will see, prefers reality and cooperation to such charades. Advanced EU nations as well as more recent members have outstripped the United States in

Table 4.2 Lapsed National Pollutant Discharge Elimination System facility permits

Rank	State	Major facilities	Lapsed permits	Percent lapsed
1	District of Columbia	4	4	100%
2	Nevada	10	7	70
3	Rhode Island	25	17	68
4	Oregon	76	51	67
5	Nebraska	60	40	67
6	New Mexico	34	20	59
7	Colorado	102	51	50
8	Massachusetts	148	74	50
9	Louisiana	247	116	47
10	Indiana	175	81	46
11	Washington	89	41	46
12	New Hampshire	62	27	44
13	Minnesota	85	37	44
14	Connecticut	117	45	38
15	Hawaii	27	10	37
16	Idaho	66	24	36
17	California	235	85	36
18	Ohio	268	93	35
19	Delaware	24	7	29
20	New Jersey	168	49	29
21	Virginia	145	40	28
22	Iowa	123	31	25
23	Oklahoma	93	23	25
24	Michigan	181	44	24
25	Missouri	146	35	24
26	Arizona	46	11	24
27	Maine	93	22	24
28	Alaska	7	18	23
29	Texas	582	135	23
30	North Carolina	216	49	23
31	Tennessee	152	34	22
32	Maryland	100	21	21
33	Montana	44	9	20
34	South Carolina	191	37	19
35	Kansas	58	11	19
36	Wisconsin	132	23	17
37	Illinois	268	45	17
38	Florida	253	42	17
39	South Dakota	31	5	16
40	Alabama	211	28	13
41	Arkansas	108	13	12
42	Vermont	34	4	12
43	Mississippi	86	10	12
44	West Virginia	93	10	11
45	Pennsylvania	387	41	11
46	Utah	34	2	6
47	New York	361	14	4
48	Kentucky	130	2	2
49	Georgia	172	2	1
50	North Dakota	26	0	0
51	Wyoming	26	0	0

Source: EPA (U.S. Environmental Protection Agency, 2007a)

environmental performance according to recent environmental performance ratings (Esty, Levy, et al., 2008).

Of the last three criteria, efficiency, fairness, and ability to encourage technical innovation, efficiency was obviously sacrificed when uniform national standards and procedures were built into the laws. It is widely observed that maximum efficiency is obtained by decentralized and adaptive management.[132] Fairness can hardly be claimed when most of the apparatus of environmental protection is aligned against industry and designed to defeat its influence. Proponents for placing environmental values above all else can argue that this is necessary or desirable, but it certainly is not "fair."

Evidence of the inefficiency of the system is present at every level. Significant policy innovation would require major reform or modification of old laws. Such reforms are not only stalled by gridlock. Effective reforms would defeat the inherent design and intent of the original laws and amendments, which are to prevent "innovation" that could undermine and circumvent their fundamental goals.

An example of the present system's tendency to limit technological advances to the "safest" (technologically as well as politically) method is illustrated by the status of waste incineration in the United States. There is only one approved incinerator in the United States certified for burning PCBs; it is operated by EPA and it is underused. In Germany, which over the years developed the highest efficiency incinerators in the world, residents can go to certified industrial plants and get potentially dioxin-producing wastes (like PCBs) burned with minimum impact from dioxins (Ludwig & Schmid, 2007). Instead we landfill most toxic wastes.

4.3.2 Litigation and Litigiousness

Adversarial legalism[133] is a peculiarly American institution. No other advanced country has this curious and damaging form of civil combat. It arrived only in the late 1960s. Political scientist and legal scholar, Robert Kagan, somewhat cryptically described the growth of adversarial legalism in the United States as having:

> "….accelerated and persisted in the last third of the 20th Century because American… interest groups wanted their governments to wield power (Kagan, 2003)."

Kagan went on to say that the process ended with growth of mistrust of governmental power, many groups in society feeling compelled to "arm themselves with lawyers and legal claims."

Litigation rose from around 60,000 civil cases per year in federal courts in 1960 to more than 300,000 per year in the early mid-1980s (Fig. 4.2). The time of the rise correlates closely with escalating tensions over environment in the late 1960s and with the inclusion of new civil suit provisions in environmental laws. As the number of cases increased, so did the number of lawyers (Fig. 4.3). The number of judges did *not* keep pace, however, so today the wait for some cases ready for trial is 4 years.

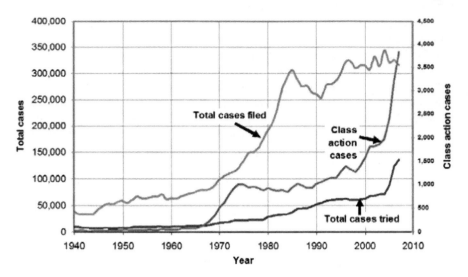

Fig. 4.2 Federal court cases, 1940–2006
Source: Compiled by F. T. Manheim

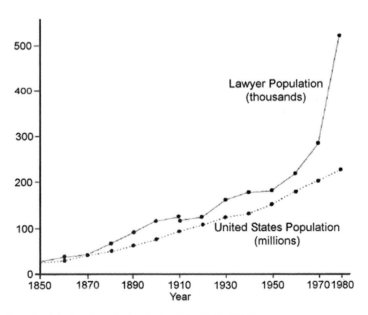

Fig. 4.3 Growth of the legal profession in the U.S., 1850–1980
Source: Modified from Halliday (Halliday, 1986). The number of lawyers increased dramatically in the U.S. in the 1960s

The nation's first landmark environmental law case was *Scenic Hudson Preservation Conference v. Federal Power.* Consolidated Edison proposed an innovative plan to the Federal Power Commission to generate hydroelectric power by cutting a reservoir into the top of Storm King Mountain, a scenic hill adjoining the Hudson River. Initial approvals were reversed by an Appeals Court ruling that remanded the proposal back for further environmental consideration. Legal disputes continued until 1980, when the effort was withdrawn.[134]

Earlier civil rights cases (e.g., *Brown v. Board of Education*, 1954) had opened up two new precedents and avenues for legal claims, the ability to sue governments, and class action suits, that is, bringing together a number of plaintiffs who did not necessarily know each other but who shared common grievances.[135] Activists like Ralph Nader and others understood how powerful the civil rights precedents could be. Undergirded by influential theoretical justification in a book by Joseph Saxe (1970), *Defending the Environment: A Strategy for Citizen Action,* the principle of citizen suit was incorporated into the laws of the 1970s and later (described in Chapter 2).

Modest rises in court cases occurred from the mid-1960s, before the explosion took place in the 1970s. How did this happen? The historic plot of the lawyer population (Fig. 4.3) (Halliday, 1986) shows that until the late 1960s, the number of lawyers in the United States roughly correlated with the rise in population. The 1970s explosion in the number of lawyers parallels the 1970s regulatory revolution in Congress. Halliday points out that the sudden rise in the lawyer population in the 1970s was unprecedented in the history of the US legal profession. Other criteria for changes in legal activity (criminal prosecution, corporate, finance, welfare, etc.) did not account for the changes. Nor could normal institutional response (law schools) have supported such a rapid increase. The post-World War II background of events and movements make it clear that new Congressional legislative and legal changes were "in the wind." The signals grew stronger and promoted expansionary pressures on the law schools (always sensitive to the nuances of implications of political development) until explosive rises took place in conjunction with passage of the revolutionary environmental laws. Because the scale of the court activity far exceeds the total number of specific environmental cases, it is clear that there was recognition that expansion in litigation would not be limited to precedent-setting environmental statutes but would affect wider national affairs.[136] In short, the current litigiousness that pervades the United States had its origin in the tumultuous period just prior to and during the 1970s era of legislative and legal changes.

In short, the current litigiousness that pervades the US had its origin in the tumultuous period just prior to and during the 1970s era of legislative and legal changes.

4.3.3 Economic Effects

The disparity in wealth and technology between the United States and other nations was great in immediate post-World War II period. The economy continued to look robust in the later 1960s even though under the surface evidence of structural imbalances could be discerned (Fig. 4.4). US industry had been able to achieve unprecedented production feats in World War II. It had overcome almost impossibly complex technical problems in order to send men to the moon. It had tapped oil miles under the sea floor, and pioneered new breakthroughs in agriculture, communications technology, and medicine. Its economic vitality could have plausibly been assumed to be virtually inexhaustible and in the period leading up to 1970s arguments like that of Barry Commoner that the United States produced and consumed too much and should reduce its affluence were frequently heard.[137] These factors offer a partial explanation for the 1970s environmental laws' explicit prohibition of "weighing benefits against costs in setting environmental standards (Cropper & Oates, 1992)," by reducing concern that curbs on industry would affect the economy.

The image of a vigorous and indestructible US economy has been shaken by the subprime housing crisis, accompanied by the unprecedented slide in the value of the US dollar, moving from par to 1.59 in terms of the euro by April 2008. The loss of industry by outsourcing to other nations is a painful reality that has impacted many areas in America, especially its industrial heartland. Economists (Reich, 2008) point out that that apparent positive indicators like GDP and stock market performance have obscured net loss of income for many Americans during the past 20 or more years.[138]

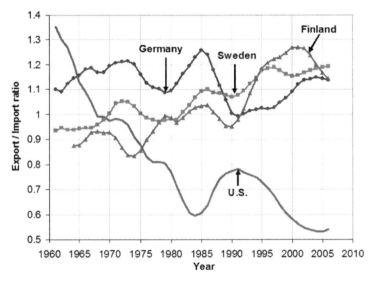

Fig. 4.4 Export/import ratios for durable goods for the U.S., Finland, Germany, and Sweden 1960–2006; moving 5-year average
Source: Data from OECD

Table 4.3 Steel production
in selected countries (million
metric tons)

Country	1980	1990	2000	2005
Finland	1.5	2.9	4.1	4.7
Sweden	4.2	4.5	5.2	5.7
Germany	43.8	38.4	46.4	44.5
United States	101.5	89.7	101.8	94.9

Source: Data from UNCTAD (UNCTAD, 2008),
quoted in OECD (Organisation for Economic
Cooperation and Development, 2007)

The statement has been widely heard that deindustrialization is a normal phenomenon for advanced nations and that the service sector will take up the slack for any shortfall (Mankiw & Swagel, 2006; Rowthorn & Ramaswamy, 1997).[139] Comparison of US export/import ratios with those of three leading European nations offers no support for a link between deindustrialization and advanced, competitive economic development. Figure 4.4 shows that the United States starts strong in the 1960s but except for a brief reversal in the late 1980s suffers ever greater trade imbalances. In contrast, the three northern European nations maintain or increase positive export–import ratios to the present.

The conflicting trends cannot be dismissed by the argument that "war-ravaged countries built new, modern factories," for Sweden was neutral in World War II and had a brisk demand for its products, like ball bearings.

Moreover, Table 4.3 shows that between 1980 and 2005, Finland, despite its lack of significant iron ore and coal resources, tripled its production of steel – especially alloys of specialty steel for which it opened the most modern underground plant in the world in 2003. This nation of only five million has produced a remarkable number of breakthroughs, from the invention and development of flash smelting of copper and other metals (Fig. 4.5) to world leadership in cell phone production in the last decade. In contrast, US steel production over the same period stagnated or declined.[140]

Finland built on earlier breakthroughs in flash smelting technology to achieve new levels of efficiency in the processing of copper ores. Smelting within a closed environment has made it possible both to use the energy value of sulfur in copper ore, and to essentially eliminate all emissions. That process is now being used by Kennecott Copper Co. for US copper production, and is being extended to other ores.

The expectation or claim that the United States could replace export of manufactured goods by export of services is "cruelly" contradicted by Fig. 4.6. US service exports were always too small to have any chance to compensate for loss of manufacturing exports. The trade deficit between exports of services and manufactured goods widened rather than narrowing with time, and imports of services now approach levels of US service exports.

4.3.4 US Industrial and Manufacturing Losses

How can one explain why the United States, with traditional strength in manufacturing, engineering, entrepreneurship and marketing skills, abundance of raw materials,

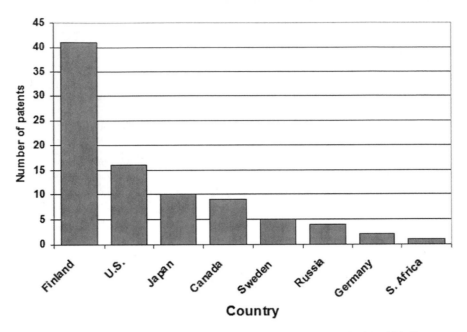

Fig. 4.5 Patents for flash smelting, the premier metallurgical breakthrough of the 20th Century
Source: Data from IBM (2008)**

**Neither Finland nor Sweden have the prejudices and problems associated with mining industries that the U.S. has. Sweden is the leading mining nation of Europe, but among the 14 priorities or areas of concern earlier listed on a web site of Sweden's largest environmental organization, *Naturskyddsforeningen*, mining was not included. The current home page of the organization *http://www.naturskyddsforeningen.se/om-oss/policydokument/* carries a policy section that argues that geological features deserve the same concern that other natural resources do. The treatment is sophisticated and complex. It points out, for example, that many extraordinary geological phenomena are only known and visible because of exploration and exploitation (e.g.. creating a fresh rock face where rock layers are preserved or fossils can be studied). In other cases older mines that reveal rare minerals or rock types are afforded preservation. This attitude is possible partly because of a historically positive association with mining in the Scandinavian nations, but also because mining is expected to be conducted with care and sensitivity to local environments and populations

excellent transportation, fine research universities, a huge internal market, and advanced communications skills, could have lost or experienced disproportionate attrition in so many of the industries that it once created or led?

The United States has long since abandoned production of radios, televisions, and most home appliances; it has only remnants of a former robust textile and shoe industries, no longer produces thread, and has stopped mining or refining many essential mineral products like lithium for the burgeoning battery market, rare earths, molybdenum, or bauxite, the ore of aluminum (Fig. 4.7). Large supermarket chains have been absorbed by other nations. Major US oil companies have been bought by foreign companies (Cities Service oil company was bought by the Venezuelan national oil company, and Amoco and Atlantic Richfield (ARCO) have been absorbed by British Petroleum). The national Russian oil company, Lukoil, is

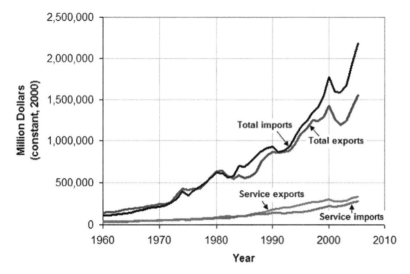

Fig. 4.6 U.S. exports and imports 1960–2005
Source: U.S. Department of Commerce, Bureau of Economic Analysis (2008)

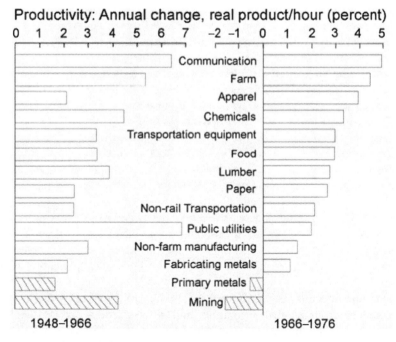

Fig. 4.7 Annual change in U.S. productivity before and after enactment of major environmental laws
Source: Kendrick & Grossman (1980)

expanding rapidly in the United States. The United States imports most of its stainless steel. It is increasingly reliant on foreign makers of advanced wastewater treatment technology, as in the case of the new German sewage treatment facilities for the Massachusetts Water Resources Authority (Greater Boston).

A furor arose over a proposal for management of major US ports by Dubai Ports World (DP) in 2006; DP is headquartered in Dubai but has subsidiaries in the United States, United Kingdom, and elsewhere. Ironically, DP had just bought the British firm Peninsular and Orient, the largest port-handling company in the world, which handles significant port operations in New Jersey, New York, Baltimore, New Orleans, Miami, and Philadelphia and some dockside activities at 16 other ports in the United States (Bridis, 2006). Apparently, US companies no longer have the expertise to carry out these operations as they did in the past.

A pioneer in space travel, the United States increasingly relies on Russia for basic components of space development. Competition in key US industry sectors has tended to shrink to a few giant companies, as in the case of Boeing Aircraft (short-range commercial aircraft are now dominated by manufacturers in Canada, Brazil, and Europe), General Electric, Texas Instruments (cell phones), and chemicals (DuPont and Monsanto).

The once-greatest shipbuilding nation in the world, with one of the longest coastlines and the greatest sea trade, has no nonsubsidized or nonmilitary-related shipbuilding industry, nor many American flag vessels in its fleet, beyond those mandated by law to be required to carry specific goods. Because of failed domestic fishery management and marine aquaculture, a significant proportion of the US's seafood is now imported. (Historical vignette, Whatever Happened to the Blue Revolution, in Section 8.2, contrasts the bright post-World War II expectations for ocean developments in the United States with the outcomes.)

An increasing proportion of fruits and vegetables are imported by a nation that was once a cornucopia of diverse food products made possible by its wide climate range, favorable growing areas, and innovative farming skills. Despite its huge forested area and surplus of wood production up to the 1970s, the United States imports an ever-increasing proportion of its wood products (Knudson, 2004; Western_Wood_Products_Association, 2008 – 38% in 2005).

Comparisons between the United States and European nations suggest that the US loss in industrial and economic productivity since 1960 is not paralleled by experience in any other advanced industrial nation except perhaps in Great Britain after World War II. The UK's economic slump has been linked partly to loss of its colonies, and policies that included nationalization of the coal and steel industries. The Lord Rothschild Report of 1971 also pointed to adverse effects of science and research policies that had features in common with US science policies (discussed in Chapter 2).

Taking into account heedless growth in the United States after World War II, widespread environmental deterioration, and unwise federal policies including clear-cutting forestry management initiated by the Forest Service in scenic and visually sensitive areas of the West,[141] it was understandable that there would be a reaction to restore environmental and philosophic balance. The United States has always tended to have more exaggerated swings of opinion or popular movements than in Europe or other advanced nations. But the pendulum swing toward environmental protection in the 1970s brought about measures that prevented future corrections and adjustments. The environmental reform movement built into law a system of inflexible, top-down environmental regulation that focused on environmental protection to the virtual exclusion of all other public interests, and backed it up by incentivizing private enforcement. The new laws opened up sources of internal conflict and problems that affected many areas of national life beyond environmental policy. These areas include partisan politicization of federal agencies and the court system, and setting the precedent for Congress to involve itself in attempting to micromanage all manner of societal problems through often rapidly generated and poorly considered laws. The total number of environmentally related federal laws is now over 620[142] – in contrast with a single law for Sweden, two laws for Finland, and three laws for Canada. The growing national energy problem alone has resulted in generation of some 500 bills on energy efficiency and renewable energy (Sissine, 2008), and in 2005 there were some 34 million lawsuits in state courts. Lawsuits concentrate the energy and ingenuity of contestants on defeating opponents – not on advancing productivity and creative solutions to problems.

4.3.5 The US Environmental Management System: Additional Implications and Comparisons

The United States has long been a nation where "movements," trends, and ambient psychologies of various kinds have been strong and experience rapid growth.[143] Without forgetting or minimizing the serious problems that brought the current system into being, let us step back and review some of the features of the new environmental management system of the 1970s and effects on entrepreneurial or administrative decision making that the system fostered.

1. The environmental laws establish tens of thousands of regulations and require setting uniform, rigid, and often arbitrary pollutant standards throughout the diverse geography and human landscape of the nation. In principle (though in recent years variably diffused in practice) these standards were designed to be met regardless of the economic consequences. The words "compliance," "violation," and "penalty" set the tone of regulations.
2. Aside from its research, coordination, databasing, and public information functions, EPA is essentially designated as a sheriff's summons server for the laws passed by Congress. It may delay some actions but has no power to create or authorize innovative solutions without Congressional action. This has prevented development of more effective methods of dealing with hazardous wastes and contaminated land sites than those procedures widely acknowledged to be extraordinarily inefficient and costly (Rosenbaum, 2007; Vig & Kraft, 2005).

3. Obscure animal and plant species have standing that takes precedence over human populations in the civil courts, inevitably creating public resentment rather than cooperation (Knopman & Fleschner, 1999).
4. The laws encourage local citizens and strongly motivated interest groups, often supported by grants from wealthy private foundations and even the federal government, to challenge any activity on environmental grounds using NEPA. The Act has been interpreted by the courts to include esthetics, and became (to the regret of its original sponsor) a major factor in evolution of entrenched NIMBYism that regularly defeats needed or desirable development throughout the United States.
5. Economic or other competing values, initially completely excluded from consideration, have gotten marginal consideration through benefit/cost assessments whose criteria and methods of calculation can be influenced by the politics of the moment, or the objectives of the agency involved.[144]

The rigorous rules and legal challenge provisions send a message that business and regulatory agencies are not trustworthy and will always be overrun by economic interests if given a chance. The stigma on economic and other private interests means that regulators or public officials who try to cooperate with business or infrastructural developments risk assumptions and accusations that they may be "captured" by economic interests.

When anger or resentment at impacts on industry has been expressed, these complaints have tended to be either ignored or treated as griping, overreaction, or failure to accept responsibilities. Counterreaction by industry in the form of self-serving advertising or lobbying campaigns, or questionable administrative maneuvers as in the case of the George W. Bush administration[145] are taken by leaders of the environmental community as proof of the irresponsibility and fundamental corruption of the business/conservative political axis (Kennedy, 2004) and taken to confirm the necessity of the draconian laws.[146]

Infrastructural development always intrudes on land or environment in some way. Consider bridges, airports, ports and harbors, transportation and power transmission corridors, and water treatment plants. Bold infrastructural proposals that envisage long-term future developments will have correspondingly great impacts on existing land uses. Could a Quabbin Reservoir (water supply for Boston) or the New York City aqueducts have ever gotten through the EIS gauntlets, overlapping regulatory jurisdictions, and NIMBY reactions that we have today?

The radicalizing trends in the past 30 years have created a spectrum of environmentalist movements that raise unspoiled environment to the highest level of human value and tend to reject human intrusion into or use of natural land. An extreme exponent of the radical environmentalism movement is Dave Foreman, founder of *Earth First!* He became famous for "ecotage," such as spiking trees so that they would destroy sawmills, disabling heavy machinery by use of silicon carbide-motor oil grit, or salting airstrips in wilderness areas so that "deer, elk and moose will come along and paw it up, leaving large holes."

Another end-member movement, "deep ecology," was initiated by Arne Næss in 1972. His book, *Ecology, Community and Lifestyle* (Næss, 1989), first published

in Oslo (as *Økologi, Samfunn, og Livsstil*), details the distinction between shallow
ecology, in which the environment is preserved mainly to serve human interests and
deep ecology, which is concerned with the whole biosphere. In the biosphere all
species, even the most insignificant, have equal rights. Næss, though contempla-
tive and apolitical, extended his ideas to sustainability issues in human society. He
advocated decentralized communities and an Earth that would ideally be limited to
100 million people.

One of the most articulate American wilderness environmentalists is Bill
McKibben, whose book, *The End of Nature* (McKibben, 1989), has been touted as
the first popular book to stress the danger of global warming. The book has had wide
influence and become something of a latter-day *Sand County Almanac* for environmentalists. McKibben argues for retaining untouched landscape even if few people
ever visit it. People, McKibben says, may:

> "....feel the need for pristine places, places substantially unaltered by man. Even if we do
> not visit them, they matter to us. We need to know that though we are surrounded by build-
> ings there are places where the world goes on as it always has."

McKibben's most recent book, *Deep Economy: The Wealth of Communities and
the Durable Future* (McKibben, 2007), is a handbook on how to stop global climate
change through political action.

The wilderness movements have had much greater traction in the United States
than in Europe, especially in Scandinavia. In the Scandinavian countries care in
fostering environment has long traditions, but the idea of walling off nature from
humans would be regarded as an anachronism. In Sweden the national tradition
of allowing people to cross private land to reach sea or lakeshore or forest areas
has been retained – with the provision that private rights are respected (i.e., gates
are closed after passing through, no littering, etc.). In Norway hiking or ski-
ing in mountains is pursued by a high proportion of the population. In Finland
and Sweden mining, appropriately conducted, is not regarded as antithetical
to environmental values. In fact, a statement of positions by Sweden's largest
environmental organization, *Naturskyddsföreningen*, includes recognition that in
many cases the revelation of some geological wonders is only possible through
exploitation, such as the creation of rock faces where sediment layers or fossils
may be exposed.

Born in 1892, the famed US forest conservationist and planner, Arthur H. Carhart,
took the approach that balanced forest conservation and wilderness with appropriate timber use. Carhart wanted creative public access to forests in order to educate
people and encourage respect for nature (Carhart, 1959).

4.3.6 Benefit/Cost Analysis

The first Reagan administration introduced requirements for regulatory impact
assessment and cost-benefit analyses.[147] Cropper and Oates cite estimates of 1–2.5%

for the cost of regulations for most polluting industries, and less than 3% for all industries except electric utilities (5.4%). However, requirements to use quantitative methods for computing the *benefits* of regulation resulted in numbers so unrealistically large as to imply that the United States was made fabulously rich by the CAA. According to an information release by EPA:

> "From 1970 to 1990, EPA estimates that the total benefits of CAA programs ranged from about $6 trillion to about $50 trillion, with an average benefit of about $22 trillion. These estimates represent the value of avoiding the dire air quality conditions and dramatic increases in illness and premature death which would have prevailed without the Act. By contrast, the actual costs of achieving the pollution reductions over the same 20 year period were $523 billion, a small fraction of the estimated monetary benefits (EPA, 1997)."

As the EPA analyses suggest, "benefits" calculated using multipliers that involve monetary value of avoided deaths or cancers can lead to fanciful and misleading results. They have come to receive less weight in serious evaluations than costs.[148]

Hazilla and Kopp (1990) pointed out that counting only the direct outlays of firms to meet regulatory requirements overlooks important impacts that are included in a "total equilibrium" analysis. The first of these is social costs of regulation. Other influences include the effect of regulation on microeconomic decisions by the public in production sectors, and secondary impact of effects on intermediate products.[149] Hazilla and Kopp found that whereas the baseline direct costs were estimated at $79 billion/year for the CAA, the truer costs could be as high as $320 billion.[150] An opposite influence is described in the Porter hypothesis, which recognizes that regulation can stimulate innovation and more efficient production.[151]

In the United States, major negative effects of regulation may result in attempts at remedial action, but the continuing everyday application of regulations has remained largely independent of economic consequences. Economic activities that become unprofitable simply disappear. Or new enterprises that might have been initiated under more favorable conditions do not start, or fail early without much notice. This is fundamentally different from the European system.[152]

4.3.7 "Sink or Swim" or "We're All in This Together"

In the EU, the "sink or swim" US approach is replaced by a "we're all in this together" approach. In the 1980s, to counter widely noted *Waldsterben* ("forest death" due to sulfur and other acidic air pollution), the German Government initiated a draconian campaign to reduce emissions. What it did *not* do, on the other hand, was simply write off heavy industry – a mainstay of the German economy. A cooperative campaign linking regulatory agencies, politicians, industries, and scientific and engineering expertise combined efforts to achieve the necessary advances while retaining a viable industry (Wätzold, 2004, and references cited therein).

In the later 1970s, Japan felt the traumatic impact of a low-cost Korean shipbuilding industry on dominant Japanese supertanker-building industry (Uriu, 1996; Vogel, 1985). To meet the catastrophe for the Japanese economy, representatives of

the government, industries, trade unions, and local authorities in affected shipbuilding areas of Yokohama, Kobe, Sakaide, and Kure came together in crisis efforts to
provide solutions (World Maritime News, 2001). With governmental support, new
industrial initiatives employed idled work staff of the shipbuilders at or near the local
sites. Innovative marine vessels included floating plywood factories that became posi-
tioned off the West Coast of the United States, buying US lumber, and selling quality
plywood to West Coast buyers. Japanese policy avoided subsidizing industries, but
was quick to identify economic problems and support new initiatives and solutions.

The almost complete absence of comparable concern for US economic development is a critical factor in the decline of US industry. Its insidiousness is due to the fact
that adverse actions and outcomes may occur in thousands of small steps, decisions,
delays, failures, and obstructions that never attract individual attention and which may
be difficult to analyze in traditional economic analytical models.

Table 4.4,[153] which relates competitive investment and environmental performance of the United States and other leading mining nations, reveals some astonishing results. Environmental performance and industrial activity are not inherently
a zero sum game, as has been the assumption in the United States (Gardiner &
Portney, 1999). Rather, two of the leading nations in environmental performance
and policy, Sweden and Finland, are also among world leaders in having national
conditions attractive to mining investment. The United States, in contrast, is rated
far down in environmental performance. It is strong in some features desired for
mining investment, such as stability, but ranks near the bottom, with Zimbabwe,
in permitting delays, and is at similarly low levels for "environmental issues
(policies)" and "land issues." Here, we can see how the operation of microeconomic principles on a day-to-day basis systematically, and over time, eroded what
was once a world-leading industry. The United States had far more remediation to
do in terms of areas with hazardous waste problems – which became the object of
CERCLA (Superfund) activities. But which would be likely to have attacked this

Table 4.4 Competitiveness and environmental performance of selected mining countries

Country	Invest. rank[1]	Envir. issues[2]	Land issues[3]	Permit delays[4]	EPI[5]
Finland	1	8	7	8	3
Australia	2	6	6	9	20
Sweden	3	8	7	7	2
Germany	4	8	6	7	22
Austria	5	8	6	7	6
Norway	6	8	7	5	18
Canada	12	6	7	5	8
U.S.	14	4	6	3	28

Source: Esty, Levy, et al. (2006, environmental performance index, EPI) and Transparency International (2005, other categories)
[1] Investment ranking, overall ranking for mineral investment competitiveness; 1 is best
[2] Environmental issues; efficiency of environmental management; high values best
[3] Leasing and property law; high values best
[4] High values denote shortest delay in obtaining operating permit
[5] Low values best

problem more effectively: a dynamic, cooperative, and financially robust industry or a demoralized ever-receding industry avoided by the brightest engineers, scientists, and business experts in favor of more respected industry sectors?

I am aware that the preceding points might be taken as placing blame on environmentalists and the system of regulation that they initiated and defend. That is much too simplistic. There is no doubt that the framers of the laws were ethical, thoughtful, and well-meaning persons whose actions could be justified by the circumstances of the times. I suggest that deeper-lying instabilities in the US system of law and policymaking – plus chance convergence of crises – helped bring about the current problems (see Chapter 9 for further discussion of this perspective).

4.4 Infrastructure

I end this chapter with a remarkable description, largely taken from the book by Robert Kagan, *Adversarial Legalism* (Kagan, 2003), of how the 1970s laws affected US infrastructure, now rated "D" by the American Society of Civil Engineers.[154] I am placing a short version of the Oakland Harbor case in the text of this chapter, rather than in a case study in Chapter 8, because its byzantine details illustrate problems that to various degrees have impeded infrastructural development throughout the United States since the 1970s.[155]

Kagan's second chapter, entitled, *The two faces of environmental legalism,* includes positive attributes of adversarial legalism. The independent, adversarial system "encourages Americans to regard themselves as rights-bearing citizens," and has conditioned judges to be open to considering grievances by even the lowest, most despised citizens. It has helped minorities achieve their rights in society, and in the early 1970s, it challenged Alabama prisons that were "unfit for human habitation." Adversarial legalism "energizes the American legal profession as a whole," hence "American lawyers are far more entrepreneurial and aggressive than their counterparts in other democracies (Osiel, 1990)." "In the competitive American economic system ... adversarial legalism often enables private litigants to bring dishonest businesses to justice (Kagan, 2003)."

But Kagan puts greater emphasis on the negative aspects of the "pathologies of adversarial legalism [which] are of immense social importance, for they are both unpredictable and enormously debilitating. They engender costly 'defensive medicine' .....undermine faith in the justice system, and invite political overreaction." Adversarial legalism of the type that Kagan describes from Oakland largely arose in the 1970s.

In 1958, the US shipbuilding firm owned by Daniel Ludwig broke the 100,000-ton barrier with the 104,000-ton tanker, *Universe Apollo.* But in the 1960s, leadership in supertanker construction was assumed by the Japanese, who by 1989 built record-setting 500,000-ton tankers that required harbor depths of 80 ft. The United States also pioneered in container ships; in 1956, Malcolm McLean, a trucking firm executive, launched the first cargo of truck trailers with a company that would be called the Sea-Land Corp. By the early 1970s, however, leadership in container ship development had moved abroad, first to Germany.

The port of Oakland in San Francisco Bay early recognized that trends to larger container ships were dramatically changing efficiency in ocean transport. In 1972, it applied for a permit from the US Army Corps of Engineers to deepen its harbor channel to 42 ft in order to accommodate the container ships. Unfortunately for Oakland, in that year Congress not only passed the Clean Water Amendments Act (which applied to dredge spoils), but also the Marine Protection, Research, and Sanctuaries Act, which placed constraints on ocean dumping, as well as the Marine Mammal Protection Act, the Coastal Zone Management Act, and in 1973, the ESA. There ensued delays and confusion over how many contaminants were in the dredged areas and where potentially contaminated spoils could be dumped under the new, more stringent conditions. By the early 1980s still more laws had been passed.

In 1986, the Corps issued an EIS, approving disposal of dredged harbor floor sediments at an established dumping site near Alcatraz Island, in San Francisco Bay, near Oakland. "Environmentalists, fishing interests, and state regulatory officials raised concerns about damage to water quality and fisheries."

The Corps conducted more tests and in September 1987 issued a supplemental EIS offering:

"... 'special care' and 'capping' methods for about 21,000 cubic yards of sediments."

But multiple challenges followed:

"...state and local regulatory agencies had the power to block in-Bay disposal under state law, and they preferred disposal in deeper waters of the Pacific Ocean. ... Port of Oakland officials in January 1988 proposed disposal at an ocean site designated 1M although that would double the cost of dredging."

Fishermen objected to the Port of Oakland's alternative proposal, "... claiming disposal at 1M would harm ocean fisheries."

"The Corps prepared another Supplementary EIS that disputed the fishermen's claims. Nevertheless, EPA refused to issue a permit ... and an environmental advocacy group that had opposed disposal at Alcatraz prepared to bring a lawsuit challenging the Corps' Supplementary EIS. They said the sediments should be dumped beyond the Continental Shelf, fifty miles out at sea."

"A compromise was negotiated between the Corps and EPA, agreed to by the environmental groups ... ..and a federation of fishing associations." But just before the start of dredging,

[The] "Half Moon Bay Fishermen's Marketing Association brought suit in federal court, alleging that the disposal decision violated a number of federal regulatory provisions and would disrupt fisheries. The District court judge and the court of appeals rejected the fishermen's claims, but after one day of digging, a state court judge held that the dredging permit had been issued without a requisite certification from the California Coastal Commission ... Shipping lines using the Port of Oakland were screaming for deeper water."

Desperate port officials announced an alternative plan that included using the first 500,000 cubic yards of dredgings to reinforce levees in the Sacramento River Delta, but

"After a year or so a California regional water quality agency approved the plan, but the Contra Costa Water District challenged it in state court. … Yet another year later, in July 1990, the court upheld the Delta plan, but the Port declined because regulatory conditions designed to safeguard Delta water quality had pushed estimated disposal costs to $21 per yard, ten times the cost of disposal at Alcatraz."

Another regulatory agency now entered the fray!

"The Water Quality Control Board for San Francisco Bay banned all new dredge spoils in the Bay, as did the National Marine Fisheries Service, and the latter objected to disposal off the Continental Shelf. EPA, having been burned by litigation, refused to. … approve any new ocean disposal site. Pending mapping of ocean bottom and currents, no decision could be made before 1994. An attorney for an environmental group rejected any ocean dumping at all and said spoils should be placed in land repositories (Transportation_Research_Board, 1994)."

Economic crises grew for Oakland.

"Costly dredging equipment stood idle. Big ships had to carry reduced loads … and wait for high tides. … Shipping companies scrapped plans to expand operations at the Port of Oakland. The previously successful port lost money, and the municipal government lost revenues needed to maintain social services. … "

"Finally, political pressures mounted, new studies … were funded, new impact statements were prepared and more regulatory hearings were held." Highly contaminated sediments would be deposited in a lined upland site, and officials allowed the port to dump the remainder near Alcatraz, just where the Port and the Corps had first proposed in 1986. Sampling and testing costs reached almost $4 million, more than double the actual barging and disposal operation. Another multimillion dollar analysis program finally enabled a decision for the remaining 6 million cubic yards in 1994. Funding for wetlands disposal required a $15 million appropriation from Congress and $5 million from the California legislature. In 1995, the port was finally dredged to 42-ft depth, 23 years after Oakland authorities first requested a permit.

As Kagan describes the outcome, legal provisions designed to protect the rights of interest groups and agencies "*seemed to fall into the hands of the Sorcerer's Apprentice.*"

"…. multiplying themselves beyond control. Regulatory officials, scientists and lawyers … debated and argued in endless series of forums. No proceeding ever produced a finding that the disposal plans were environmentally dangerous, but no single court or agency could designate an acceptable, economically sensible plan. … In the tangled web of adversarial legalism mollusks received more protection than human communities."

(In Chapter 5, we will see how the current regulatory system has had similar effects on renewable energy developments that are expected to reduce greenhouse gas emissions and provide millions of new green jobs. Case examples of the effects of regulatory policy on the nation, Milwaukee wastewater treatment system, and "Blue Revolution," will be presented in more detail in Chapter 8. Further examples include the effects of NEPA on home ownership costs in Alameda County and the Oakland Airport, California, and the effect of the Delta smelt controversy (ESA) on water management in San Francisco Bay.)

Chapter 5
Why Do Conflict and Polarization Matter?

5.1 Changing Energy Policies

The United States is the largest emitter of greenhouse gases per capita but has recently been passed by China in total emissions (EIA, 2007). The principal energy supply sources for the United States in recent years are shown in Table 5.1(values slightly different from EPA's Inventory of U.S. Greenhouse Gas Emissions and Sinks: 1990–2006). The energy in oil and coal is concentrated, transportable, and easily stored. It can be used flexibly to meet peak production needs – and production can be reduced when demand is low. Oil remains indispensable for transportation: automobiles, trucks, locomotives, ships, and aircraft. As a devil's advocate and good reference for the problems faced in moving from traditional fossil fuels puts it, "[Oil's] complex hydrocarbon chains are the basis of the petrochemical industry, which uses oil and natural gas as a component in over half a million products. Basically, if you wanted to invent an ideal energy source, you'd create oil. The biggest problem, of course, is that combustion of these fuels produces CO_2 (Friedemann, 2007)."

The next most important energy source, natural gas, methane (CH_4) is valuable for use in electricity generation because it delivers roughly the same amount of power as coal while yielding only half as much CO_2, because the hydrogen burns to form water:

$$CH_4 + 2O_2 \Rightarrow CO_2 + 2H_2O$$

Switching from coal-fired electrical utilities to gas-fired production immediately produces a 50% saving in CO_2 emissions, besides much lower sulfur, mercury, and other emission products.

The United States' growing dependency on oil imports (60%) means that import costs projected for 2008 will balloon to an estimated $450 billion – *over half of the anticipated foreign exchange deficit.* That produces a devastating spiral. Growing trade imbalances weaken the US dollar against the euro and other currencies. The weakened US dollar increases the cost of imports and increases the trade imbalance – and the cycle continues.

After Australia's ratification of the Kyoto Protocol on December 3, 2007, the United States remained the only major nation formally outside the international

F.T. Manheim, *The Conflict Over Environmental Regulation in the United States*, 111
DOI 10.1007/978-0-387-75877-0_5, © Springer Science+Business Media, LLC 2009

Table 5.1 US energy consumption by energy source, 2002–2006 (quadrillion Btu)

Energy source	2002	2003	2004	2005	2006
Carbon emissions (million metric tons)[1]	5,816	5,878	5,988	5,857	5,877
Petroleum	38.2	38.8	40.3	40.4	40.0
Natural gas	23.6	22.9	22.9	22.6	22.2
Coal	21.9	22.3	22.5	22.8	22.5
Nuclear electric power	8.2	8.0	8.2	8.2	8.2
Hydroelectric (conventional)	2.7	2.8	2.7	2.7	2.9
Renewable (other than conventional hydroelectric)	3.2	3.3	3.6	3.7	4.1
Total[2]	97.7	98.0	100.1	100.2	99.4

Source: Modified from Energy Information Administration and UNFCCC (U.N. Framework Convention on Climate Change)
[1] UNFCCC, million metric tons of CO_2
[2] Ethanol blended into gasoline included in petroleum and nonhydro renewables, but single counted in total

cooperation.[156] However, scientific and political developments, including positions of the presidential candidates Barack Obama and John McCain, presage changes in US policies toward accelerating emissions reduction beginning in 2009. Remaining scientific uncertainties (e.g., Sample, 2008) have not prevented EU (European Union) and other leading foreign nations from accepting global climate change as the number one problem in achieving "sustainability." This chapter assumes that the United States will take up a similar policy, adopting "hard" goals for curbing green house gas emissions in 2009.[157]

A prominent assumption for future policies is that capping carbon emissions (i.e., restricting use of fossil fuels) while investing in conservation and alternative energy sources will bring about a new "greener" economy without creating economic repercussions. This has been successful to a considerable degree in Europe, for example, Germany (Fig. 5.1). However, experience in Europe has shown that progress in replacing cheaper, concentrated, and convenient sources of energy like coal and oil by "green" alternatives is not simple and easy. It requires intensive research, technological and public policy adaptation and innovation, and cooperation of the whole society. When problems arise, as in the case of increases in cost of food (especially critical for developing nations) because of diversion of cropland from food to energy crops, the adjustments require even more intense effort.

The previous chapters suggest that while the United States in past decades appears to have maintained overall economic growth, under the surface its basic industrial strength, infrastructure, and environmental performance have deteriorated. Internal conflicts and national policies have created difficulty in achieving consistent or effective domestic policies, solving known problems, and preparing for future challenges. The United States has scrambled to cope with unexpected problems and failures like Hurricane Katrina *after* they have occurred – at great cost. Preoccupation with issues of the moment leaves difficult or unpleasant problems to grow into crises before they get attention. The next sections will review social and political issues that will need attention in order to transform US energy use.

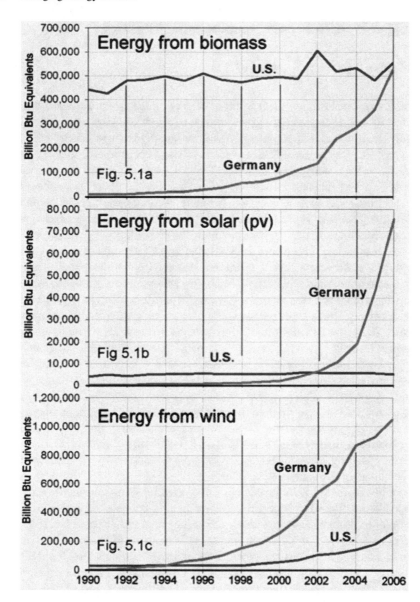

Fig. 5.1 European nations such as Germany are rapidly increasing their production of electricity from renewable resources; the U.S. is beginning to increase its energy from wind

Source: Data from OECD. Figure 5.1a, electricity from biomass; U.S. biomass is mainly wood; in Germany, other materials are included. Figure 5.1b, electricity from solar photovoltaic cells. Figure 5.1c, electricity from wind turbines

5.2 Good Politics Versus "Inspirational" Politics

5.2.1 Developments in the EU

A remarkably successful political union developed in Europe since World War II. It was built by stages around limited-scale trade and industrial agreements involving carefully coordinated and planned cooperation among diverse nations (discussed in Chapter 6). Special sensitivities were involved because the initial six nations included Germany, the aggressor in World War II, and nations that it had invaded.

In the Treaty of Amsterdam that took effect in 1993, the EU affirmed the concept of sustainable development that was formally recommended in the Brundtland Commission Report of 1987. The heads of all EU environmental agencies agreed in the Prague Meeting of 2005 (European Union, 2005) that good regulation could and should be compatible with competitiveness in industry and business. This affirmed Brundtland's observation that if economic and environmental policies were not integrated, neither could be effective.[158] The soundness of this thesis has been demonstrated by EU development. At the same time that EU nations have taken the lead in global climate policy and environmental performance, their traditional industries have survived against the formidable competition of Japan and "The Asian tigers," Singapore, Hong Kong, and Korea.[159] Germany, for example, did not abandon its former industrial specialties. German lenses are now widely used in Japanese digital cameras, Carl Zeiss optical equipment has developed thriving specialty uses in dentistry, ophthalmology, spinal- and neurosurgery. Leica still makes cameras and microscopes but has branched into surveying, geodetic, mining, scanning, and police forensic systems. The four major German automakers, Audi, BMW, Daimler, and Volkswagen made record sales in the United States in 2006, and Daimler is the world's leading truck maker (followed by Swedish Volvo). The list could be extended to other EU nations.[160]

The EU was never free of problems. But increasing economic stability and economic success achieved by successive enlargement of cooperation helped overcome many troubles along the way. Economies of leading nations benefited from expanded markets, while newer, less affluent nations gained in political and administrative sophistication and received developmental assistance. Some like Romania and Ireland have risen rapidly from former impoverished levels. New national policies proposed by political leaders are prepared and carefully developed with the help of professionally skilled specialists, and thoroughly discussed, analyzed, and reviewed by relevant and affected groups and the public before being submitted for political approval. A similar evolutionary policy process takes place at the larger EU level.

5.2.2 US Assets

The United States has many natural assets in developing energy policy: natural resources, an effective existing energy supply network and technical expertise, a large internal market, willingness to provide risk capital, entrepreneurial energy and creativity, an advanced higher education system, and numerous environmental NGOs and specialized research organizations. "Out of the box" innovation flourishes in the United States.[161] Examples include the Internet, Apple Computer with its online music sales system, and I-Tunes; the online bookseller, Amazon.com, that introduced revolutionary features like reader reviews and even comments on and ratings of reviewers. Other innovations include the credit card and eBay online auctions. US states are laboratories for testing new ideas. As the French cultural observer, Alexis de Tocqueville, noted in the nineteenth century (de Tocqueville, 1835), Americans readily form new associations, giving society great flexibility. Thus, a number of international social organizations got their start in the United States: the Salvation Army, Alcoholics Anonymous (which provided the model for many modern group therapies), Toastmaster's Club, various global help and outreach organizations, and political party organization. US entrepreneurship in quickly commercializing innovations (as in electrifying cities and rural areas) would be important, but constraints on this latent capacity should not be underestimated.

5.2.3 US Problems

Since the 1960s, "inspirational" or crisis-driven policies have often been initiated at the presidential or Congressional level. Politicized science and environmental policies have put the nation's professional science and other agencies, as well as states, in the position of having to work under policies hastily drawn up on relatively short notice, without careful planning and analysis or popular consensus.

One of the contributions to a recently released, comprehensive review of US energy policy, sponsored by the US National Academies, *Our Energy Future*, included a witty but telling review of major energy-related initiatives since 1970 by Senator Jeff Bingaman (D-NM). Bingaman introduced his historical observations of US policy patterns with the comment:

> "Focus is on advocating a particular technological solution, instead of solving an energy problem; excessively optimistic assumptions about technology costs and capabilities; limited consideration of interplay with other policy areas; and under appreciation of the scale of the energy enterprise (Bingaman, 2008)."

Technology Attention Deficit Disorder

- Vehicle Technology

  - Virtually Pollution-Free Car (Nixon, 1970)
  - Reinventing the Car (Carter, 1977–1980)
  - Partnership for a New Generation of Vehicles (Clinton, 1993–2000)
  - Freedom Car (Bush, 2003)

- Coal Utilization

  - Synthetic Fuels Corporation (1979–1985)
  - Clean Coal Technology Program (1987)

- Clean Coal Power Initiative (2001)
- FutureGen (2003)

- Nuclear Technology

  - Clinch River Breeder Reactor (1970–1983)
  - Advanced Liquid Metal Reactor Program (1989–1994)
  - Global Nuclear Energy Partnership (2006)

- Biofuels

  - Alcohol fuels (Energy Security Act 1980)
  - Oxygenated fuels (Clean Air Act Amendments 1990)
  - Biofuels (EPA Act, 2005; EISA, 2007)

The hallmark of each of the abandoned, stranded, or currently jeopardized programs was that they all began as enthusiastically advocated initiatives, often with large initial funding. They offered potential to fulfill a need, but there was little careful review of potential short-term and long-term economic or other impacts or interaction with other national factors. Some showed promise but were not carried through fluctuations in prices of energy, suffered from political polarization or other political swings of opinion, or were damaged by lack of transparency or failure to take real problems seriously.

5.2.3.1 US Lawmaking

A comparison between EU and US lawmaking offers startling and unflattering (to the United States) contrasts. Proposed laws in leading EU nations are taken seriously. They are carefully evaluated for short- and long-term impacts by relevant ministries before being submitted to Parliament for debate, amendment, and vote.

In the United States for energy policy alone, some 500 bills with relevant-sounding titles poured into the 110th Congress (2007–2008). Past experience indicates that less than a quarter of this number is "serious in intent." The subsequent fate of bills is highly dependent on the discretion of Committee chairs and numerous subjective factors. It is hard to believe that a leading nation of 300 million people could have such a haphazard system. It is not surprising that the results are so poor.

Since the 1970s, members of Congress have felt free to respond to any concern or development with detailed, highly intrusive laws in areas where Congress has no specific expertise.[162,163] A partial analogy might be individual members of a large city council creating guidelines for doctors in operating rooms, the police department, or school teachers.

5.2.3.2 Environmental Regulations

Daniel Fiorino points out in his book, *The New Environmental Regulation* (Fiorino, 2006) that the benefits of the "old" US environmental regime were substantial.

However, "that regime [the 1970s laws] was based on the assumption that commercial firms were 'amoral calculators' who would assert their own narrow interests over society's at every turn."

> "Given the incompatibility of private and public interests… the view was that only the blunt hand of legalistic and deterrence-based regulation could be effective in changing industry behavior…. The old approach to regulation was based on questionable assumptions about government as well. A primary one was that government had the cognitive capacity to determine not only what society's environmental goals should be but how, in some detail, they should be achieved."

Regulatory problems now obstruct goals of good environmental management and business community cooperation in transforming energy technologies (see Section 9.4 for examples).

5.2.3.3 Related Societal Problems

Academicization of social and natural scientific research. The United States is the undisputed global leader in production of conferences, books, research studies, and media reports on global climate change. This can be shown by 32,500 conferences retrieved by a query on Google Scholar, and the over 4,000 books on "global warming" retrieved by Amazon.com or Google Books. The bulk of these conferences and books are of American origin. There is little action to show for this research and communication.

> Too much talk. Too little action. The vast amount of academic research and publication contrasts with the United States' record in concrete action with respect to global warming.

A large part of the outpouring of publications reflects the legacy of the basic research paradigm introduced after World War II. Talented people were drawn to university or other research institutions where scholarly research flourished. The trouble was that growth in educational institutions was not linked to goals that involved national needs. Even where the subject of research is relevant, as in the case of global climate change, most researchers remain observers and analysts, not participants in meaningful progress. (The results of the peer-review publication system described in Chapter 2 are illustrated by a poll of earth science researchers regarding major federal agencies; see Table 9.1 for the Earth scientists' poll.)

Breakdown of social cohesion and responsibility. The internal rifts that arose during the 1960s and 1970s took a toll on the sense of national solidarity and social responsibility. Instead of cooperation group-oriented attitudes have grown on all sides.

Alienation in the business and financial community. A bias against business arose in the educational and intellectual community during the 1960s (and grew in the 1970s). One of the results was the rise of a new group of business leaders who were more narrowly or defensively focused on the affairs of the company than those who had, for example, participated in cooperative efforts with government and academic scientists during World War II. After 1980, economic theories championed self-seeking behaviors in the business and financial community. Along with other factors, they may have helped foster the later wave of financial abuses unprecedented in the twentieth century. In little more than a week during April 2005, *The Washington Post* reported an astounding list of major corporate ethics scandals (see Section 8.5, Corporate Scandals).

Loss of respect for ethical consistency. In both federal and state policies short-term considerations trump ethical consistency. For example, the federal government requires tax payments for winnings in gambling that is illegal in the states. It holds information confidential. In turn, states withhold information from the Internal Revenue Service about income from illegal aliens, even guaranteeing them that the information will remain confidential. In both cases money talks; ethics are put aside.

Adversarial developments in environmental organizations. Deborah Schreurs (2002) has pointed out that the United States has by far the largest, best funded, and most politically powerful environmental NGOs of any nation. US progress in global climate change issues is well behind that of Japan, which has the smallest and weakest network of environmental organizations. Ironically, some of the nation's richest foundations, built from the fortunes made in big industrial companies, like the Ford Foundation, the Pew Trust (based on money from Sun Oil Co. founders), Carnegie Corporation of New York (steel for railroads), Rockefeller funds and foundations (Standard Oil), Doris Duke Charitable Foundations (tobacco and electric energy) – now support campaigns of anticorporate legal warfare mounted by activist environmental organizations.

In the past 20 years, a number of environmental organizations with distinguished earlier traditions have become radicalized. Focusing on campaigns for political objectives, exaggerated, or slanted claims about industry projects for energy development have become commonplace on web sites and in massive mail and Internet campaigns (see Section 8.6 environmental campaigns). Organizations have come to pride themselves on the number of lawsuits filed against industry or government (see Table 1.2, litigation index). Minimal concern is shown about economic or other societal issues involved, and leaders have become desensitized to the increase in domestic conflict caused by demonizing and stigmatizing organizations on which society is dependent for economic development, fuel, and other essential services. Drilling for oil has become a metaphor for profit-seeking activities that are regarded as harmful to the local and global environment. (Environmental NGOs with more positive agendas are mentioned in Chapter 7.)

Litigiousness. The legal profession, nominally dedicated to justice and rule of law, has used health hazards from pollutant hazards as a milch cow for mass production class action law firms. The results have included huge awards at the cost of destroying or crippling whole industries such as the asbestos industry.[164] The selective suit approach is also capricious in its benefits to victims. It diverts business and industrial talent and investment away from areas that are critical and needed – for example, dealing with hazardous wastes and difficult or dangerous processes. Owing to legal liabilities and insurance risks associated with new nuclear energy plants, new nuclear startups – envisaged as a part of the future energy portfolio by the two current presidential candidates – are only feasible through federal guarantees and limitation on liabilities (the Price–Anderson Nuclear Industries Indemnity Act).

NIMBY. The "Not In My Backyard" syndrome emerged in the 1970s through a combination of factors. One was the elevation of the status of the environment by 1970s environmental laws, excluding formal consideration of economic or other competing values in management or enforcement. The second was the empowerment of citizens to challenge developments or activities on environmental grounds through NEPA (National Environmental Policy Act) and get legal fees reimbursed for successful suits. Court interpretations added esthetics to the environmental grounds for suit. The actual number of NEPA suits has been relatively small, but the risk posed at the environmental impact statement stage is magnified because uncertainties associated with the EIS process first enter the picture only after investments of time and money have been made. In many cases, no actual litigation is needed for the potential of such action to create delays or ultimate abandonment of initiatives where a minority of residents show determined opposition. (See Section 5.3.4 for more information on ocean tide and wave energy.)

The public information problem. It is difficult to gain solid, reliable public information on controversial areas such as energy policies because of the proliferation of shallow reportage, political polarization, one-sided analyses, and massive but over intellectualized scholarly publication.[165] Although data from federal agencies is usually reliable, interpretations, policy discussions, or progress reports by some governmental organizations can be influenced by political and turf considerations. US federal web sites tend to offer more elementary explanation, statement of goals, projections, or high motivations, and pictures showing hard-working government employees on the job, than are found in web sites of corresponding European governmental organizations. Major industrial firms who have to deal with contentious policy issues often keep meaningful communications on important subject internal or reserved for networking and lobbying. Absence of meaningful, candid communication from business fuels suspicions.

There is a deluge of media reports about the glowing promise of green technology in the United States. But on examination, most reports show that the United States is at an early stage of development in comparison with Europe (see Green Jobs, US and Sweden, Section 8.9). A profusion of articles may stem from a single report – itself only a projection of possible future developments, such as the Department of Energy's projection of 20% wind energy use by 2030 (Department of Energy, 2008a). Environmental organizations project bright scenarios for future green development and millions of jobs, once we get rid of bad old technologies (Energy Efficiency and Renewable Energy, 2008b). But many visions lack reality checks. They focus on theoretical options, limited data, or they may be ideologically oriented toward strictly environmental protection and health, leaving it to "someone else" to deal with the economic development.[166]

The point of raising gritty and unpleasant issues of domestic conflict, barriers, and lack of social solidarity is not to take partisan shots at groups of people, political philosophies, or institutions. Rather, it is to emphasize their potential effects. If these effects are not recognized, they may wreck the chance of success in future programs.

Rebuilding national trust may be seen by cynics as a high-minded goal not likely to be achieved in America. However, the alternative has received extensive testing and few groups seem satisfied with its results.

5.2.4 Examples of Perspectives of Social Scientists

- *Sharon Beder.* Liberal Australian political scientist and writer Sharon Beder is highly critical of US private industry. She cites a Harris poll to the effect that between 1967 and 1977 the percentage of people who had "great confidence" in major companies fell from 55 to 16% (Beder, 2006a). Beder claims that throughout the twentieth century business associations and coalitions coordinated mass propaganda campaigns for an extreme free market ideology. Beder asserts that this free market propagandizing frees the market of responsibility for consequences (Beder, 2006b).
- *Mickey Craig.* According to conservative economist Mickey Craig (2008), President Reagan took office in 1981, at a time when "Americans were unnerved by the fact that the country had suffered double-digit inflation in 1974 and then for three years running in 1979, 1980, and 1981. They were bewildered by sky-rocketing interest rates, as the prime rate hit 19.77% in April 1980 and peaked at 21.5% in December 1980. Unemployment was running uncharacteristically high as well .... The Gross National Product had a negative growth rate of 0.2% in 1980. These trends all changed dramatically during the Reagan years." In short, Craig claims that the Reagan administration restored legitimacy to commercial enterprise and the profit motive, and vitality to the economy.
- *Paul Krugman.* According to liberal economist Paul Krugman (1994), "supply-side" economic theories that gained ground in the 1980s did not have an anchor in academic economic analysis, but were mainly spread by economic journalists with special encouragement by Robert Bartley at the *The Wall Street Journal.*[167]
- *Francis Fukuyama.* Political scientist Francis Fukuyama pointed out the importance of "social capital," i.e., individuals' willingness to value societal cooperation and solidarity beyond individual benefit. (see also Harisalo & Stenvall, 2004).

"Japan, Germany, and the United States became the world's leading industrial powers in large part because they had healthy endowments of social capital and spontaneous sociability (Fukuyama, 1995, p. 150)."

5.3 Exploring Methods to Reduce CO_2 Emissions and Their Effects

A widely used scheme to visualize the various approaches to reducing carbon emissions is the "wedge" system introduced by Pacala and Socolow (2004) and adopted by the IPCC in its 2007 Report (IPCC, 2007) (Fig. 5.2). The wedge identifies a

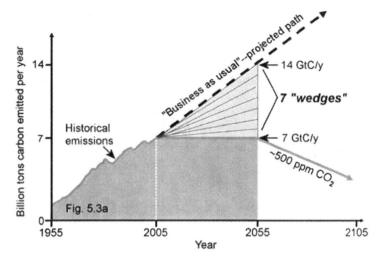

Fig. 5.2 Stabilization wedge concept, a policy and planning tool for mediating global warming through reduction of CO_2 emissions with existing technologies (Pacala & Socolow, 2004)
Source: Pacala and Socolow (2004) and the Cambridge IMPEE project (University of Cambridge, 2006)

"business as usual" line of increasing carbon emissions, and a series of measures to bring down CO_2 emissions to acceptable levels (e.g., 500 parts per million in air). The American Council for and Energy-Efficient Economy (ACEEE) recommends 15% electricity savings per capita by 2015 relative to 2007 (Eldridge, Elliott, et al., 2008). Energy is the indispensable multiplier that enhances human productivity and national economy, as one can see by a plot of GDP versus energy consumption (Fig. 5.3). Simple reduction of energy use without compensating increases

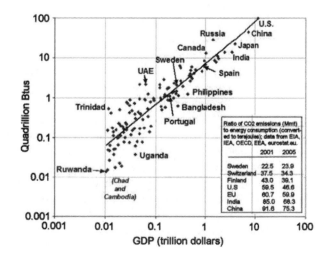

Fig. 5.3 Annual energy consumption (Btu equivalents) plotted against GDP for 131 nations
Sources: Data from EIA, IEA, OECD, EEA, and eurostat.eu. Data for 2003–2005, latest available

in efficiency produces economic decline, which tends to disproportionately affect the poorest people in society (and the world). Therefore, efficient replacement of carbon-containing energy sources is the goal. But replacing the cheapest and most convenient energy sources is not simple. It requires intensive technological and policy innovation and broad societal cooperation. Compared with leading nations in Europe, the United States has made meager progress in promoting alternative energy sources (Fig. 5.1). Where it has moved aggressively, as in the production of ethanol from food corn – but without adequate analysis of long-term effects – there have been serious negative consequences (see Section 5.3.5).

5.3.1 Energy Conservation and Efficiency – Costs and Complexities

The United States has made major progress in achieving energy and emission savings in federal and state agency operation, and new building construction and appliances.[168] A considerable effort by environmental organizations is devoted to promoting conservation and efficiency through cooperation between organizations like World Resources Institute with business and government is probably greatest in this area (World Resources Institute, 2007). Large costs will be involved in retrofitting existing low-income residences, which may result in variable (possibly zero) true energy savings, given the energy outlays involved in the remedial work, the expected life of the buildings, and other factors. A rough idea of direct costs is obtained by dividing federal DOE funds allocated for weatherization of low-income residences by the 300,000 residences reported by DOE for the National Energy Plan (Department of Energy, 2005). This yields a figure of about $3,000 per dwelling. Scaling a figure like this to the 20 million homes mentioned by Senator Hillary Clinton yields a figure of $60 billion in current dollars.

Without change in existing regulatory conditions, environmental regulations may significantly impact the cost of building renovation. An Internet search on the *Office of Information and Regulatory Affairs* (OIRA), mandated to maintain a database of regulations by the Paperwork Reduction Act of 1980, found that the largest number of "economically significant" regulations (having greater than $100 million impact) in the 50 US Codes of Law belonged to the environmental sector (613), followed by agriculture (444), health (404), and transportation (186). The environment sector had just under 30% of all regulations (Logomasini, 2007).

"Real" effects, including microeconomic and indirect, are certainly larger than nominal regulatory costs computed from direct outlays, as discussed in Section 4.3.5. These costs are at the heart of past acrimony and disputes, with wide disparities between claims of supporters and critics of regulation. Just as the computations of the "benefits of lives saved and cancers avoided" by regulations can reach unrealistically large numbers; the cost of regulation to the national economy can also be exaggerated by various assumptions. Of indisputable concern is the fact that environmental regulatory costs have much greater impact on small and medium-sized firms than on large firms: "Compliance with environmental regulations costs small firms up to 364% more than large firms" (Crain, 2005). Major difference in costs is understandable, given the ability of very large firms to have dedicated and experienced staff to handle regulations, whereas for smaller firms that make up 99% of all businesses much of time spent conducting monitoring, special studies, meeting reporting requirements, etc. will have to come out of "business time." This factor may account for increasing tendency in time for a few very large companies to dominate fields that involve environmental regulation.

Potential problems related to asbestos in older buildings raise large issues. The entire US asbestos industry (some 78 companies) has gone into bankruptcy because of litigation over mesothelioma and related illness, with some 720,000 cases filed against 8,400 defendants as of 2002 (U.S. Chamber Institute for Legal Reform, 2007).[169] At least half of the $70 billion spent was related to individuals having nonmalignant conditions. Estimates of future total costs for asbestos litigation range from $145 billion to $210 billion (Carroll, Hensler, et al., 2005).

Like asbestosis, silicosis is currently a surprisingly rare condition in the United States, based on the records of the National Institute of Occupational Safety and Health. Of the 342 cases of all injuries and occupational diseases reported to the Mine Safety and Health Administration in 2005,[170] only two were silicosis, and one was asbestosis (National Institute for Occupational Safety and Health, 2007).[171] Because costs for potential litigation are reflected in insurance and other expenses, and in the bidding for work supported by government, it would appear that establishing controls and oversight over potential litigation should be a central element in any rational plan. Success in rationalizing and simplifying contracting and oversight could open up renovation to innovative firms that might otherwise refuse to bid. Because trial lawyers have been active in partisan politics mainly supporting Democratic candidates, successful reforms would probably have to be led by Democrats.

Cogeneration, or CHP (cogeneration of heat and power) is potentially available from any electrical power generating source or other industrial site that operates at high temperature, i.e., coal, gas, and nuclear power plants, blast furnaces, and other smelters. In the United States, cogeneration was, in fact, the dominant source of electrical power at the turn of the century, but became abandoned with cheaper and more efficient delivery of power from centralized systems. In recent years, European nations have made far greater use of cogeneration (up to 50%) (Elliott & Spurr, 2008) compared with 8% for the United States (Pullenger, 2008). Elliott and Spurr point out the following inhibitions to greater employment of CHP:

- "A site-by-site environmental permitting system that is complex, costly, time consuming, and uncertain."
- "Current regulations do not recognize the overall energy efficiency of CHP or credit the emissions avoided from displaced grid electricity generation."
- "Many utilities currently charge discriminatory backup rates and require prohibitive interconnection arrangements. Increasingly, utilities are charging

(or are proposing to charge) prohibitive 'exit fees' as part of utility restructuring to customers who build CHP facilities."

- "Depreciation schedules for CHP investments vary depending on system ownership and may not reflect the true economic lives of the equipment."
- "The market is unaware of technology developments that have expanded the potential for CHP."
- "In addition, development of new district energy systems as part of a CHP implementation faces some additional barriers."

The authors report a DOE plan to double use of cogeneration by 2010. To accomplish this they recommend the following:

- "Reform of environmental permitting regulations and the permitting process to provide credit for the inherent efficiency of CHP systems."
- "Reform electric utility regulations to provide fair and open access to the grid for procurement of standby power and excess generation sales."
- "Modernize the depreciation schedules for CHP equipment to reflect current markets and technologies."
- "Provide financing opportunities and incentives, such as tax credit, to spur interest in CHP systems."
- "Develop educational and technical assistance programs to increase awareness of CHP opportunities and technologies."
- "Initiate research and development activities to expand the range of CHP technologies, especially for small-scale systems."
- "Installation of CHP systems in government facilities to demonstrate the benefits and provide market leadership."

5.3.2 Hydropower

In 2003, there were about 80,000 dams in the United States producing about 10% of the electricity (Energy Efficiency Renewable Energy, 2003). During the decades of major large dam construction, which peaked in the 1960s (O'Connor, Major, & Grant, 2008), the environmental impact of the structures was not considered as important as the cheap and reliable power they produced. Now more dams are being decommissioned and removed than are being built in the United States. But hydropower is having a renaissance in many forms, spurred on by the need for curbing greenhouse gases, reducing dependence on imported oil, and lowering the cost of energy.

Many of the "large" dams (over 7.6 m in height) in the United States are up for relicensing by the Federal Electric Regulatory Commission (FERC). During that process, various technological and structural upgrades may be required, and the expense may not be warranted or feasible from the point of view of the public or private utility. Many new types of in-river turbines have been designed and are currently being tested. Some could boost existing hydropower dam electricity output so that relicensing would be cost effective; with some additional infrastructure, similar turbines could be installed in existing dams that do not currently provide power. Other designs, such as barge-mounted turbines, could be moved to locales where electricity is currently provided by fossil-fuel power plants – without affecting the river flow. Association with existing power-generating facilities would minimize the infrastructure costs.

A continuing issue affecting dams on rivers flowing to the Pacific Ocean in the northwestern states is the longstanding battle between supporters of salmon and large power-producing dams. A letter to the National Marine Fisheries Service (NMFS) from *Save Our Wild Salmon, Taxpayers for Common Sense, Trout Unlimited, American Rivers, Idaho Rivers United, Earthjustice, National Wildlife Federation, Sierra Club*, and a nationwide coalition of conservation, business, and fishing groups, and supported by 92 Congressmen, demands "giving priority to most cost-effective, scientifically credible recovery solutions" (Save-our-wild-salmon, 2008). NMFS's earlier science plan was rejected by the US Court of Appeals, 9th Circuit. Some groups want the dams entirely removed – which would lose a total of 3.03 GW nameplate-installed power capacity, roughly 5% of the power needs of the Pacific Northwest. The Corps of Engineers has scheduled $400 million in upgrades but this may not save Chinook salmon, threatened with extinction.

The issue here is a classic one involving major natural resource management options and the federal government. What is the real story and what would the best options be? We have the strongly motivated save-the-salmon groups who want to save salmon. They are only interested in salmon. We have the Corps of Engineers, who issues only official opinions. We do not know whether there are bright engineers there – or elsewhere – who might have better solutions than the official proposals. Then we have the National Marine Fisheries Service (NMFS) that is locked into an impossible dilemma, with scientific fisheries opinion constrained by the fact that NMFS answers to the NOAA (National Oceanic and Atmospheric Administration) Chief Scientist, who in turn reports to the Secretary of Commerce – one of the most politicized executive departments.

If it were in Sweden NMFS scientists might give a straightforward opinion, followed by an opinion by the NOAA Chief Scientist. Because more than fish was at stake the "Minister of Commerce" would then consult widely and render a decision, probably giving the real, not political reasons. The Swedes like straight talk. It seems like a stretch to believe DOC will give up 3 GW of cheap, clean power. Question: How much power could in-line, nonobstructive water turbines produce?

5.3.3 Wind Power

Wind energy is currently the most widely installed and highly regarded renewable energy in active development in the United States, with 18,302 MW installed capacity in the first quarter of 2008 (American Wind Energy Association, 2008). A recent report by the Department of Energy projected that 20% of US electrical energy requirements in 2030 could be met by wind energy (Department of Energy,

2008a). The report provided extensive documentation of potential wind areas and assumptions of the projections, noting that a requirement would be the construction of a major network of 765 kV transmission lines, estimated to cost $60 billion.

Brief generic caveats are offered for other requirements to achieve large-scale development of wind power. "Strategies are needed to identify sites that are highly favored for wind energy, but also to avoid potential ecological risks and minimize community conflict" (American Wind Energy Association Siting Committee, 2008). These are short statements for very large problems, as we will soon see from descriptions that follow.

Multiple problems with wind power are often omitted or glossed over. In America as well as in Europe, wind power development has only been possible through substantial subsidies (Table 5.2). Denmark, which is exceptionally favorable for wind energy by virtue of its flatness and long shoreline, now gains 20% of its electrical energy from wind energy, but this has been facilitated by a 50% cost subsidy, partly offset by Denmark's leadership in wind turbine manufacture. High subsidies have resulted in electricity costs three times those typical in the United States (30 cents/kWh). Germany, another leader in wind energy development, has among the highest electricity costs in Europe. Power transmission adds to costs, and supply patterns mean that wind can only be used to supplement a dominant flexible power source that can meet peak requirements.

The most favored areas for wind power shown in the DOE study for wind energy potential lie off the Atlantic and Pacific coasts and in the Great Lakes, with wind power densities greater than 600 W/m^2 at 50 m height (Fig. 5.4). Favorable conditions here are provided by long unbroken fetch associated with ocean and lake surfaces. But environmental regulatory barriers and NIMBY problems for these areas have proven formidable.

A major signal of regulatory/NIMBY problems that would become commonplace in the United States for wind turbine farms was provided by the Cape Wind project

Table 5.2 Subsidy for renewable energies and other fuels

Fuel	Billion kWh	Subsidy million $[1]	$ Subsidy/mWh
Refined coal	72	2,156	29.81
Solar	1	14	24.34
Wind	31	724	23.37
Nuclear	794	1,267	1.59
Landfill gas	6	8	1.37
Geothermal	15	14	0.92
Biomass, biofuels	40	36	0.89
Hydro electric	258	174	0.67
Coal	1,946	854	0.44
Natural gas, petroleum	919	227	0.25
Municipal solid waste	–	1	0.13
Unallocated renewables	NM	37	NM
Total	2,138	4,208	–

Source: EIA

NM = not meaningful

[1] Fiscal Year 2007; 2007 dollars

Fig. 5.4 Wind resource map
Source: American Wind Energy Association, Wind Powering America project and Department of Energy (2008b). Numbered states (1 highest) have more than 1,000 MW of installed wind energy capacity; 17 other states have less capacity and 27 states have no installed wind energy capacity. *Asterisks* indicate the areas with the highest wind energy potential; none of these areas have installed capacity yet (2008). Shaded land areas have enhanced wind energy potential

off Nantucket shoals, MA. Some 130 turbines were proposed to supply three quarters of the electrical energy needed by Cape Cod and the Islands. The $900 million development was well backed by capital and federal construction grants. It was supported by an array of environmental and other organizations including Greenpeace and the Conservation Law Foundation, environmentalists Bill McKibben, George Woodwell, and Lester Brown, and local citizens. But it was strongly opposed by powerful opponents, among whom were Mitt Romney, then governor of Massachusetts, and his Democratic Attorney General, Ted Kennedy, and Robert F. Kennedy, Jr.,[172] an environmental activist and senior attorney with the NRDC (mentioned in Chapter 1), and a vocal minority of other Cape Cod residents.

A stupefyingly detailed, 3,000-page Draft EIS (DEIS) by the Corps of Engineers in 2005 (Cape Wind Associates, 2005) did not secure favorable readings by EPA, although a former EPA head for the Massachusetts District office favored the project. FWS also stated the DEIS was inadequate, requiring more studies regarding bird populations and migration routes. After 7 years of intense controversy[173] and 40,000 comments to MMS, the fate of the development remains uncertain (Cassidy, 2008). The opposing *Alliance to Protect Nantucket Sound* funded a massive critique of the project.

Large wind project proposals in Buzzard's Bay, MA have also been rendered problematic because of public objections and Massachusetts law that developments in ocean sanctuaries (including Buzzards Bay) are only possible if they are deemed "a public necessity." These sanctuaries were quickly enacted by the Massachusetts legislature after an oil barge spill in Buzzard's Bay in 1969, following the famous spill off Santa Barbara. After local opposition defeated several wind projects in Maine, a 42 MW wind farm on Mars Hill is operating with others planned or under construction, with the support of the governor.

Earlier offshore wind projects along the Atlantic seaboard have been abandoned, notably a proposed wind farm off Long Island in advanced planning stages. This leaves the most promising wind generation area in the United States, based on wind strength, water depths, and population densities, totally without active installations. The reasons clearly involve a gauntlet of multiple permits, NEPA, and NIMBY. NEPA never refuses a proposal – but it can be used to declare any proposal "inadequate," and inadequacy can continue forever – given resourceful lawyers, a vocal public group, and federal court judges.

Governor Jon Corzine of New Jersey has strongly advocated wind development off New Jersey, and bids for developments supported by state funds have been solicited. New Jersey PIRG cautiously approved a test installation with major evaluations and safeguards (New Jersey Public Interest Research Group, 2005). The same scenario has been seen in the watershed proposal for tidal turbines in the East River and in San Francisco Bay. If high-level political support is available, operations can gain at least experimental status. Given the greater costs of offshore installations than those on land (which recently caused Shell Oil Co. to sell its share in a massive, offshore wind farm development near London, England), and which has contributed to dwindling sales of offshore turbine installations in general, it is especially critical that the regulatory and permitting regimes for offshore developments be as friction-free as possible.

On land along the Atlantic seaboard progress has been meager. The vignettes cited below suggest that wind turbine siting is rarely trouble free. An organization named *Citizens for Responsible Wind Power*, that appears to oppose wind developments, reported a request by Representatives Alan Mollohan (D-WV) and Nick Rahall (D-WV) to the General Accounting Office (GAO) to investigate wind projects and their potential impacts, and listed newspaper reports along the East Coast and elsewhere in the United States (ResponsibleWindpower.org, 2006).

- *Virginia.* "Monterey Highland supervisors have said for months they expect to be sued following their upcoming July 14 vote on a utility permit either by the wind energy developer or by those who oppose the project." Fairfax County has planned to acquire 24 MW of wind power from a farm in Pennsylvania with a capacity of 68 MW for 2009 in order to increase its renewable energy portfolio (personal communication, Utility Commission, Fairfax, May, 2008).
- *Pennsylvania.* "Freedom Wind ran into controversy earlier this year after administrators proposed a wind farm at King's Mountain in southwestern Somerset County. Residents there rebelled and eventually sued."

- *Maryland.* "Wind power isn't looking popular in Maryland right now. Meanwhile, nuclear power has picked up strong local support. That might seem backward in the minds of some environmentalists, who portray wind turbines as a symbol of good and nuclear reactors as an emblem of evil. Some have called this one of the most liberal states in America. So why is the expected symbolism falling apart here?" (Pelton, 2008)
- *West Virginia.* "Environmentalists so lamented the pollution that comes from coal-fired power plants that Congress finally responded with tax incentives for alternative electricity generation, including the ultimate in clean – wind power .... But we now have a situation where speculators are staking claim to some of our most scenic areas and erecting these monstrosities that produce little energy and are made possible only by a tax credit" (The Charleston Gazette, 2004).

Following the coastal areas, the next most favorable areas for wind development are large flat areas in a north–south belt in the central United States, extending westward to eastern Montana (300–500 W/m^2) (Fig. 5.4). High mountain ridges are windy but are rarely acceptable wind power sites in either Europe or America because of esthetic issues.

Significant parts of the favorable areas are sparsely populated and distant from power lines. Six states have installations greater than 1,000 MW: Iowa, Minnesota, Colorado, California, Washington, and Texas. Texas has passed California (2,400 MW) with 4,300 MW (AWEA, 2008). Inspection of the developments offers indications that the downsides and costs of wind power development in the United States have been seriously underrated and underpublicized. The large California wind farm developments were built in elevated areas in the early 1980s, and little development has taken place since. Multiyear lawsuits about bird deaths at the Altamont wind farm were only settled in 2007, with agreement by the developers to 50% reduction in raptor deaths, "removal of the deadliest turbines and closedown during winter (Metinko, 2007)."

Texas made headlines when eight of 11 proposed coal-fired power plants were cancelled in a deal partly brokered by Fred Krupp of Environmental Defense, and William K. Reilly, former EPA Administrator in the G.H.W. Bush administration. Compensation for the power that would have been supplied by the coal-fired plants was to be provided by wind turbines, efficiency, and other means. However, some experts suggest that the actual proportion of power capacity that will be available from the new wind turbines may be as little as between 12.5 and 25% of the planned coal-fired plants (Schleede, 2005, and written communication, 2008).

Schleede also identified a different kind of green power in Texas oil investment wizard: T. Boone Pickens. Pickens announced on May 15, 2008 his decision to invest $10 billion in 2700 GE wind turbines for a huge 4,000 MW wind farm development in the Texas panhandle. Congress has just renewed for 1 year the $0.015/kW Production Tax Credit for wind energy. If renewed for an additional 9 years, this tax credit would reduce tax liability by an estimated $2 billion. Moreover, Pickens' investment would qualify for special accelerated depreciation for federal income tax purposes, which, depending on the amount of loans funding the investment, could allow him to recoup most of his investment within as little as 2 years, not including supplementary Texas benefits. Pickens was reported to anticipate making 25% return on his investment.

US and German growth in wind energy has been supported by government subsidy. Unlike the long-term commitment of German subsidies, which have fueled continuous growth, US subsidies have been on and off. Without subsidies installations have been less than 500 MW per year, whereas during subsidy periods installations ranged from 600 to nearly 2,500 MW through 2006 (American Wind Energy Association, 2008), not including state subsidies.

In summary, US wind development has promise. But European experience indicates that it will not be cheap, nor replace baseload capacity. Moreover, it may not see systematic development until the United States is able to reduce regulatory barriers and substitute realism, holistic planning, and benchmarking for serendipitous decision making.

5.3.4 Ocean Energy – A Modest but Important Symbolic Regulatory Breakthrough

In the 1970s, the NOAA and other federal agencies invested substantial effort and money in research on ocean thermal energy conversion projects. Congress passed the Ocean Thermal Conversion Act of 1980 (OTEC, 1980) to authorize and regulate promising applications for this source of electrical power. However, decrease in oil costs after the Iran crisis of 1978–1980, as well as sand in the federal gears deterred commercial implementation (Cohen, 1994). No applications for permits had been submitted before 2005, when two projects in Hawaii entered the planning stage (Thompson, 2006).

The United States' renewable ocean energy potential from physical movement of water (kinetic energy) is great, commensurate with its long coastline length of 12,383 miles. A calculation by Roger Bedard of the Electric Power Research Institute (EPRI) indicated that the amount of wave energy potentially available off the coasts of the United States is nine to ten times the energy currently generated by the country's hydroelectric dams (Bedard, 2005). The practical electrical power yield of ocean tides and waves has been calculated to be roughly comparable to or greater than that for wind power in coastal areas (Fig. 5.5).

Successful demonstration of electrical power delivery to a coastal grid by a UK wave energy conversion system in 2004 sharply increased interest in ocean wave and tide energy projects both in the United States and Europe. A research study I undertook in late 2006 showed that by late 2006 FERC had 11 approved plans and more than 40 proposed plans in pending status. Active or proposed experimental installations were distributed in multiple sites on the Atlantic and northwest Pacific coasts and Alaska. A common type was an underwater turbine with much slower-moving blades than a wind turbine.

State and regional organizations encouraged and provided funding to promising experimental installations. However, it soon became clear that beside a host of parallel

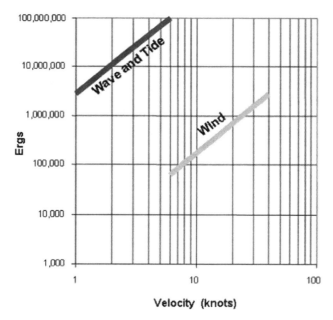

Fig. 5.5 Energy of motion for wind and water
Source: F.T. Manheim. Energy is given by $E = \frac{1}{2}mv^2$, where E = energy in ergs, m is mass in grams per cm^2, and v is velocity in cm/sec. The mass (m) for air and water is taken as 0.000643 and 1.032 g/cm^3, respectively

permitting hurdles and unclear regulatory jurisdictions, a massive regulatory barrier loomed to bar virtually any commercial implementation of these promising energy sources. At the point when installations became "commercial," defined as supplying power to regional grids – even though they might still be in an experimental stage, they became subject to a hydropower license. Such licenses are typically issued to dams and other major installations, and among other things require EIS statements. FERC guidelines for preparing EIS statements *alone* ran to 80 pages. The Fish and Wildlife Service posed requirements for conducting fish population inventories and pathways, and test results relating to potential fish mortalities because of the underwater equipment. Neither FERC, which has jurisdiction over ocean wave and tidal projects within state waters, nor the Department of Energy had the funds or mission to aid projects in overcoming policy and environmental problems. The same was true for other agencies with regulatory responsibilities, such as the FWS and state departments of environmental protection.

5.3.4.1 Light at the End of the Tunnel: FERC Issues First License for Hydrokinetic Energy Project

After my research study, FERC announced a new licensing system and issued its first license for a hydrokinetic energy project (i.e., a project that uses moving water to produce energy) on December 17, 2007 (Federal Energy Regulation Commission, 2007b). The license was granted to Finavera Renewables Ocean Energy Ltd. for a pilot project in Makah Bay off Washington State. The project had experienced multiple hurdles since upbeat press releases and media reports around the country greeted its 2001 permit request for a project to be anchored 3.2 nautical miles offshore in water depth up to 250 ft. It was to have an electrical capacity of 1,500 MWh annually. The project had received strong support from Representative Jay Inslee of Washington, an ardent backer of alternative energy sources.

"Hydrokinetic projects have tremendous potential," FERC Chairman Joseph T. Kelleher said in announcing the license. "The issuance of today's license is a major step toward realizing that potential, by authorizing a pilot project to demonstrate this promising new technology." (Federal Energy Regulation Commission, 2007b)

Commissioner Philip Moeller agreed. "Today is historic as we enter a new energy frontier. For the first time, we allow the harnessing of electricity from wave energy-power that results from the gravitational pull of the moon."

"Consumers are demanding more renewable energy options, especially those sources that are domestic, renewable, and carbon-free," Moeller added. "I am pleased to approve today's order because it meets these criteria and demonstrates this Commission's proactive approach to enable the development of this and other sources of hydropower."

The hydrokinetic license gave Finavera Renewables Ocean Energy Ltd. a 5-year license on the condition that Finavera obtain all necessary federal permits before they begin construction. In the meantime, the company may move forward with those portions of the license that do not require any type of construction. In a November policy statement FERC said:

"Issuing conditioned licenses for hydrokinetic technologies will have no environmental impact, will not diminish the authority of the states or other federal agencies, and will improve the ability of project developers to secure financing of demonstration projects (Federal Energy Regulation Commission, 2007a)."

To protect against any potential adverse impacts, the license contains a provision allowing FERC to shut down or remove the project should it find that operation unacceptably affects the surrounding environment. Various additional mitigation measures were also stipulated, including conducting a noise assessment and monitoring marine mammals to evaluate any noise effects and interactions with the buoys. The approved project was to be located in Makah Bay 1.9 nautical miles offshore of Watch Point in Clallam County, WA.

Hydropower[174] license requirements would have probably doomed all current ocean energy production of commercial scale. However, FERC has flexibility under the 1930 Federal Power Act to create a more appropriate license type and take the initiative to assist in the environmental impact statement process. Although ocean power operators still have to get all the other licenses, FERC can help in the coordination process. This role would be much less feasible for EPA and Fish and Wildlife because they are constrained by thousands of pages of laws and regulations.

The *hydrokinetic* license step is thus a small but significant development that points the way to reforms need on much larger scales.

5.3.5 Biofuels and Biomass

Biofuels have important potential as low-emission fuels. Such potential may be enhanced by new agricultural techniques like biochar (Lehmann, Gaunt, & Rondon, 2006) which provides a long-term carbon sink while improving soil fertility. But unrealistically optimistic projections in the United States and impulsive, short-range policies like subsidization of food-corn based alcohol production jeopardize meaningful development.

Claims like these have been made and continued to be heard:

"Biofuels can slash global warming pollution. By 2050, biofuels – especially cellulosic biofuels – could reduce our greenhouse gas emissions by 1.7 billion tons per year. That's equal to more than 80 percent of current transportation-related emissions."

"Biofuels can be cost competitive with gasoline and diesel. Economists estimate that by 2015, we could produce biofuels for sale at prices equal to, or lower than, average gas and diesel prices."

"Biofuels will provide a major new source of revenue for farmers. At $40 per dry ton, farmers growing 200 million tons of biomass in 2025 would make a profit of $5.1 billion per year. And that's just the beginning. Experts believe that farmers could produce six times that amount by 2050 (Natural Resources Defense Council, 2008)."

A detailed projection by US agricultural engineers providing quantitative estimates of rise in food costs and limitations of biofuel development throws cold water on the extravagant claims for biofuel potential, even in a nation with usually large cropping areas (Birur, Hertel, & Tyner, 2007). EU officials now recommend that European nations reduce or not increase agricultural areas now devoted to diesel biofuel production (Cendrowicz, 2008).

Switchgrass, a promising source of cellulosic biofuel, has been described as a native, perennial prairie grass that does not require a lot of pesticides and fertilizers. It uses water efficiently, has low nitrogen runoff, very low erosion, increases soil carbon, and provides good wildlife habitat.

Aside from the competition for land use between food and biofuel production, enthusiasts underestimate problems of longer-term use of crop-related biomass. If switchgrass, corn stover, or wheat stubble are removed from crop area to produce biofuels, nutrients and organic matter will be systematically depleted. Crop residues are needed to sustain soil organic matter and retain water. Switchgrass may improve land if allowed to grow undisturbed, but fertilizer (nitrogen), now produced through fossil fuel use, would, among other things, be required to sustain production.

Federal programs led to rapid growth in US conversion of corn (food grain) to ethanol. The United States recently passed Brazil as global leader in production of

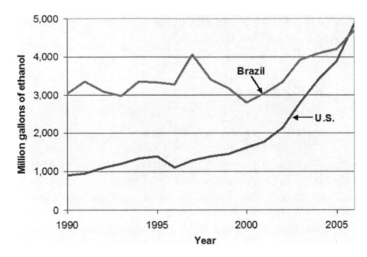

Fig. 5.6 Ethanol production in the U.S. and Brazil
Source: Renewable Fuels Association and the São Paulo Sugar Cane Agro-industry Union (UNICA)

alcohol (Fig. 5.6). However, combined with other world trends the 25% decrease in US food corn production diverted to alcohol has had serious consequences.

"Turning one-fourth of our corn into fuel is affecting global food prices. U.S. food prices are rising at twice the rate of inflation, hitting the pocketbooks of lower-income Americans and people living on fixed incomes. Globally, the United Nations and other relief organizations are facing gaping shortfalls as the cost of food outpaces their ability to provide aid for the 800 million people who lack food security. Deadly food riots have broken out in dozens of nations in the past few months, most recently in Haiti and Egypt. World Bank President Robert Zoellick warns of a global food emergency (Brown & Lewis, 2008)."

Some 40% of currently available railway freight cars are used to transport coal – approaching full use. Major increases in transport of biofuels from dispersed areas of the nation would probably require expansion of transportation capabilities. The major technical innovation and infrastructural changes have not begun to be modeled. Although Midwest farmers are already exporting hay pellets to Scandinavia, the extensive experience of Finland and other nations in biomass use seems not to be utilized in DOE and Department of Agriculture analyses.

One may also note that some 107 million acres of federal lands have been sequestered for Wilderness since passage of the *Wilderness Act* in 1964 (wilderness.net, 2008). An interesting question is whether selective utilization of dead or weakened trees or forest litter could add significant biomass to the renewable energy supply, without significantly impacting forest ecology or esthetics. Supporting such potential is the fact that leaves and needles are nutrient rich and have the most soil-enriching potential in forests, whereas wood is nutrient poor. On the other hand, if biomass potential were shown to be significant the current highly restrictive provisions of the *Wilderness Act* might require amendment to allow such use. Under current conditions any such action would probably encounter fierce resistance by environmental NGOs.

5.3.6 Other Renewable Energies and Carbon Capture and Storage

Geothermal and especially solar energy were among the areas that gained attention during the Carter administration, and were also benefited by research breakthroughs gained by the short-lived applied science directorate of the National Science Foundation (1969–1978) through the RANN program (Research Applied to the Nation's Needs). (Further discussion of these potential energy sources and compensating technologies, especially carbon capture and storage (CCS) is provided in Section 8.10.)

5.4 Discussion

A critical issue regarding oil imports is highlighted by oil prices that have surged to more than $140/bbl at the writing of this book. This means that oil import costs of some $350 billion for the United States in 2007 may rise to $700 billion in 2008 – twice the cost of the Iraq war. Added to already great projected needs for infrastructural development, health care, pension fund, and other domestic requirements, such open-ended hemorrhaging of money for indispensable fuel and petrochemical raw materials for the United States puts a spotlight on the character of US conflict that makes any vote on "drilling" a matter of commitment to conscience and principle, rather than a rational choice.

5.4.1 Offshore Oil and Gas

The current controversies over "offshore drilling" have less to do with real hazards than with the products of historical conflict – and possibly with bad psychological images generated by the conflict. The last realistic discussions about the offshore drilling questions known to me took place in the 1970s. Statistical studies, most recently a National Academy of Sciences study (NAS, 2003) have long established greater oil spill hazards from tankers and surface supply of oil than offshore production facilities. The automatic shutoff of Gulf of Mexico producing wells during the extreme conditions of Hurricane Katrina in 2005, referred to earlier, is a testimonial to the extraordinary advances achieved by the offshore oil industry. Unlike coastal tankers, undersea pipelines would be essentially out of reach to potential terrorists. (Norway's offshore oil production is discussed at length in Chapter 6, Section 6.3.3.)

Informal discussions by me with leaders of a major environmental coalition in the late 1980s revealed another dimension beneath the then debates about offshore oil leasing and production in the Atlantic offshore area. One leader told me privately that he had no problem with offshore drilling per se. Blocking offshore petroleum production was the only major bargaining chip seen as available to achieve a national energy policy – that the G.H.W. Bush administration was otherwise reluctant to take up. Allowing offshore exploration and production – "business as usual" – would simply encourage continuation of existing policies to achieve reduced dependence on imported energy by production rather than conservation and coordinated planning. However, formal legal challenge based on NEPA was not practical on energy policy grounds, so this made it understandable that tactical or legal arguments used in public debate or potential litigation would be environmental issues.

Could "mediator-leaders" have created constructive agreements? Because "conservation" issues would have had to include the big energy users, automakers and power companies, the job of reaching consensus strategies might have been very difficult. An opportunity was lost during debates in 2006 over partial relaxation of the current moratorium on offshore oil and gas exploration and leasing, extended to 2012 by President Clinton. Although the current hostilities are greater than they were in 1989, much higher stakes for the United States – and acute crisis for oil prices and supply – has brought the offshore oil issue up again in proposals by Republican presidential candidate John McCain in June 2008.

The most important contribution, regardless of whether leasing was agreed on or commercial production was achieved, would be to replace the counterproductive polarized approaches that have characterized past debates with serious dialogues on planning and policy, mediated by respected individuals. The issues are now too important to allow emotionally driven media and political campaigns – or prejudice attached to individuals or groups associated with advocacy – to prevent resolution of this costly outgrowth from misguided zeal of the Watt administration 35 years ago.

5.5 Summary

Examination of background and recent history of a number of different renewable energy types has revealed persistent patterns of poorly planned or coordinated federal government initiatives, including impulsive or serendipitous Congressional action. These patterns have often left behind stranded investments and disillusioned participants. Promising initiatives in wind, solar, geothermal energy, and ocean wave and tide energy have foundered on regulatory barriers and NIMBY that developed under, and is fostered by the distorted application of NEPA. This act is now often not being used as a policy model to foster well-considered development, but merely to block or delay undesired projects. The United States, once a model for rapid adaptation of new technology has become a model for endless delays and failed plans. The sole US plan for a holistic demonstration project for carbon burial seems to have foundered – while operations are working abroad, and new initiatives are even under way in China.

Realistic assessments and dialoguing are often omitted in projections for energy-related programs and plans in the United States, unlike planning processes in the EU. The difference between optimistic projections and reality is noteworthy for wind energy. If problems and disruptions related to erratic political processes – especially in Congress – are not considered, they may affect future energy programs just as they have past programs.

Finally, potentially ruinously costly increases in energy imports force attention to the problem of how major public policy plans have been and are prepared in the United States. A witches' brew of some 500 energy bills proposed in the 110th Congress in the House and Senate is now being stirred up. This "inspirational" approach to public policymaking bears little resemblance to the thoughtful way critical policies have been developed in the EU. A change of the way major national planning is undertaken may do more than anything else to bring facts and reality into play, reduce hostilities, open up cooperation, new resources, technologies, creative energies, and productivity toward energy policy transitions.

Chapter 6
Foreign Experience

6.1 The European Union and Other Nations Take the Lead

"The EU has pioneered a new form of post-national government, in which nation-states pool some of their sovereignty for the common good. Many of its admirers see this as a useful potential model for Southeast Asia, the Indian subcontinent, China-Taiwan, Latin America, parts of Africa and so on. The EU takes some issues, like human rights, global warming and the fostering of an international system of justice, with admirable seriousness ...... Considering the kind of Europe it replaced, the EU has been an almost miraculous success (Walker, 2007)."

In the bitter ashes of World War II, seeds of hope were sown, survived, and grew among former enemies. With fascism crushed, visionaries in Western Europe had the opportunity to chart a new course that would avoid the blood feuds of the past by laying economic and social groundwork for what evolved into the European Union (EU).[175] The skills acquired in overcoming many problems on the road to cooperation would prove effective in taking on environmental problems much larger than could have been anticipated by the early European pioneers.

European federalists began organizing conferences in cities throughout Western Europe, starting at The Hague in 1948 (Table 6.1). A few sophisticated and experienced visionaries constructed an ever larger network of cooperation that has become one of the most stable and powerful economic and political unions in the world. The founders insisted that all voices be heard. With careful steps and planning, the EU has gained wide respect among its diverse membership. Along the way it has addressed human rights, international justice, global climate change, and competitiveness.

Inspired by Frenchmen Jean Monnet and Robert Schuman, the Treaty of Paris in 1951 established the European Coal and Steel Community (ECSC), based on the idea that economic partners would have incentives to cooperate with rather than fight each other. Signatories France, Germany, Italy, Belgium, The Netherlands, and Luxembourg (Table 6.2) began to thrive, no doubt helped both economically and philosophically by the generosity of the US Marshall Plan. The ECSC managed the steel and coal industries through a High Authority and a Common Assembly, bodies that were the precursor to the European Parliament. A Council of Ministers represented the individual interests of the member states.

F.T. Manheim, *The Conflict Over Environmental Regulation in the United States,*
DOI 10.1007/978-0-387-75877-0_6, © Springer Science+Business Media, LLC 2009

Table 6.1 Major events in formation of the European Union

Year	Event	Description
1948	The Hague Conference	European federalists begin to meet
1949	NATO	North Atlantic Treaty Organization
1951	Treaty of Paris	European Coal and Steel Community (ECSC)
1957	Treaty of Rome	European Economic Community (EEC) and European Atomic Energy Commission (Euratom)
1960	EFTA	European Free Trade Association; not part of EEC
1967	Structural merger	EEC, ECSC, Euratom merged
1971	Bretton Woods abrogated	US Dollar floats; EEC currencies destabilized
1979	European currency	Initial moves toward euro (realized in 1999)
1986	Single European Act	SEA; common market agreements completed
1992	Maastricht Treaty	European Union (EU) established (upon ratification)
1995	Schengen agreement	Led to the Charter of Fundamental Rights
1999	European Central Bank	ECB and European monetary system begin
2002	Euro	First euros circulated; UK, Sweden, Denmark[1] abstain
2004	EU constitution	Ratification failed; EU continued under treaties

Source: Piper (2005) and Wood and Yesilada (2004)
[1]Denmark will hold a referendum in late 2008 and is expected to adopt the euro in lieu of the krone

Security was surely one of the unstated goals of the ECSC. But as long as the US-led NATO, established by treaty in 1949, was willing to keep the peace, European nations showed little interest in raising armies. The cost of the Korean War (1950–1953) eventually prompted the United States to require Europe to support more of its own defense. A move by European nations to create a European Defense Community eventually failed when the French parliament rejected it in 1954.

Table 6.2 Member countries of the European Union[1]

Country	Year joined
Belgium	1951
France	1951
Germany	1951
Italy	1951
Luxembourg	1951
The Netherlands	1951
Denmark	1973
Ireland	1973
United Kingdom	1973
Greece	1981
Portugal	1986
Spain	1986
Austria	1995
Finland	1995
Sweden	1995

Source: Wood and Yesilada (2004)
[1]The EU currently (2008) has 27 members; since 1995, the following countries have joined: Bulgaria, Cyprus, Czech Republic, Estonia, Hungary, Latvia, Lithuania, Malta, Poland, Romania, Slovakia, and Slovenia

Remarkable concessions during early years of the ECSC led to the success of a novel idea: cooperation instead of competition among nations. A mere 6 years after the ECSC was formed, the Treaty of Rome (1957) initiated the European Economic Community (EEC) – the heart of what became a true common market, free of duties, with capital and labor flowing freely across national boundaries, in 1986 (Table 6.2). Subsidies remained in some agricultural regions – the price of ratification. The Treaty of Rome also founded Euratom – the European Atomic Energy Community – for the development of nuclear energy.

The controlling bodies of the EEC, a commission, a council of ministers, and an advisory parliamentary assembly, like the hierarchy of the ESCS, foreshadowed the structure of the EU. In addition to the administrative and legislative branches, a judicial branch, the European Court of Justice, was created to settle disputes involving the Treaty of Rome and the decisions that were made by the EEC.

Nations that remained outside the EEC formed the European Free Trade Association (EFTA) in 1960. EFTA had little of the structure and none of the glamour of the EEC, and some of the EFTA members applied for membership in EEC in 1961. It took until 1973 for some applicants to overcome challenges posed by EEC members and failed referendums at home; by then, the treaty combining structures of European communities and the treaty establishing the European Community customs union had been ratified.

French President Charles de Gaulle, presidents of the Bundesbank, and British Prime Minister Margaret Thatcher were among the strong nationalists that put the brakes on EU progress. They objected repeatedly to the strengthening federal government of the EEC and then the EU. De Gaulle wanted a permanent leadership role for France. The conservative, semiautonomous Bundesbank resisted relinquishing Germany's position as the European central banker to the new EU central bank. And Thatcher could give up neither Britain's borders nor its currency. Balancing the nationalists were pan-Europeanists – the preeminent peacemaker, German Chancellor Konrad Adenauer; author of the EU constitution, French President Giscard-d'Estaing; and early visionaries Jean Monnet and Robert Schuman. But all three nations joined and remain staunch members of the EU, working around their differences with compromises interspersed with periods of proposals, discussion, and debate.

The path to a single European currency began in 1979 and was realized in 1999 when 11 nations replaced their national currencies with euro notes and coins. Along the way other nations joined the club, but holdouts remain, notable the United Kingdom and Sweden. The strength of the EU is attested to by its ability to allow different practices without discussion escalating into destructive rhetoric. A common monetary system had the effect of allowing all nations to see each other's progress on a level playing field – and the differences in gross domestic product became a problem in the late 1980s. It was solved by the richer nations increasing their internal EC aid to the poorer nations. Equally obvious was the fact that some nations were better run than others; that difference was addressed through pressure from the central bodies – and from neighbor states – to improve performance.[176]

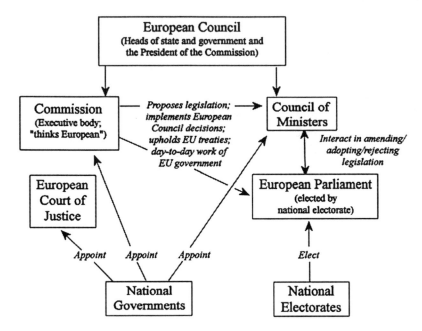

Fig. 6.1 Basic European Union government structure
Source: Modified from Wood & Yesilada (2004)

An important key to progress was introduction of regular twice-yearly summit meetings held in different member states, and rotation of chairs of key governing bodies. At each summit, problems were discussed and new steps were planned. The mixture of supranational and intergovernmental bodies (Fig. 6.1) assured full knowledge and agreement before each new step was taken.

In 1991, the Cold War ended. Germany was reunified and Eastern European members of the Warsaw Bloc – nations forced into the Soviet Communist sphere after World War II – asserted their independence. These events opened the way to expansion of the EC. With ratification of the Maastricht Treaty (1992), the EC would become the EU. The treaty ran into opposition in some nations when it was submitted to the citizens in referendums. After 2 years ratification was achieved with reservations. Nonetheless, treaty provisions for common monetary and social policies strengthened the EU, and benefits attracted new applicants.

An essential element of the success of the EU has been the benefit derived from it by all member nations. Richer nations profit by the interchange of ideas, technologies, and expansion of markets for their products. Poorer nations gain discipline and more effective administrative and fiscal policies needed to meet membership requirements, and financial and technical assistance from the EU. The transparency introduced to governmental operations through EU member-ship has been a major factor in rapid growth in nations like Ireland, Estonia, and Romania.

Laws and regulations decided by the European Parliament involve matters of principle and policy. The details are left to individual nations to work out and implement. The EU does not act as a policeman, but the broad consensus behind its policies and laws allows it to exert substantial pressure on nations that lag.

The Amsterdam Treaty of 1997 anticipated the influx of new member states from Eastern Europe that did not have long experience with democracy. In the treaty, the EU upgraded its laws and social policies and gave the Schengen agreement of 1995 the force of law. The Schengen agreement eliminated border controls within the EC and was the starting point for the Charter of Fundamental Rights.

When the euro made its formal debut in 2002, the logical next step was writing a constitution for the EU. The chief author of the presentation draft of the EU constitution was former French President Valery Giscard-d'Estaing. His conscientious and cooperative effort stood in stark contrast to President de Gaulle's repeated efforts to give France permanent EC leadership, out of keeping with the spirit of partnership and equality sought by his countrymen, Schuman and Monnet. As has often happened with EU documents, the constitution needed more work before it could be accepted by all parties. National referendums again failed to achieve full ratification in 2004. The EU, however, is resilient. It appears to be able to continue to expand and to serve its member states effectively without a permanent constitution, operating on existing treaties.

The EU government grew from the ECSC and the EC, adding bodies as needed to carry out designated responsibilities. Checks and balances ensured accountability. Representation of both the people and the governments of the member nations helped gain acceptance for decisions by the EU's governing bodies. Publications by EU research and fact-finding bodies tend to be solid and straightforward. Their authors have the kind of ethnic mixture of names that were formerly found mainly in US publications. Though the EU is a looser confederation than is the United States, it has evolved policy developing and political mechanisms that have demonstrated effectiveness, while maintaining flexibility and respect for the concerns of all affected groups. EU experience shows that cooperation and diverse viewpoints are not mutually exclusive. It offers insights that can help the United States resolve its political gridlock – if its experience becomes known and understood.

6.1.1 Lawmaking in EU Nations and in the EU

6.1.1.1 National Lawmaking

Most EU nations have a parliamentary system, meaning that the governing party or coalition initiates laws. With variations, the development of law can be described in the following schematic steps.

1. *Initiative.* Most legislative proposals are initiated by the government, i.e., the ruling party or coalition. More rarely bills may be proposed by members of parliament, private citizens, or other bodies.

2. *Inquiry.* For major laws the proposal is given to the appropriate ministry (lesser or minor laws may be given to a commission or a one-person committee). The ministry task force assigned to the inquiry submits the proposal to thorough evaluation, comment, and criticism within the ministry.
3. *Referral.* The revised inquiry is submitted "across ministries," to interest groups and to all other relevant stakeholders including local political units, in order to gauge the response to the bill. Consultant reports or contacts with other EU nations may be included if appropriate. If multiple negative responses come out of the process the Government takes back the proposal and revises its approach. Germany was the initiator for the *Elektrische Gesetz*, i.e., Electrical (WEEE) Law[177], which was submitted to the EC through the EU Council.
4. *Government bill.* The ministry drafts a bill to be submitted to the Parliament. If the subject involves potential interactions with or potential conflict with existing law it is sent to a legislative advisory committee for refinement. EU law is also considered. Norway is not an EU member but generally aligns its legislation with EU law.
5. *Parliamentary Committee evaluation.* The Draft Bill is given to an appropriate parliamentary committee for refinement. When deliberations are completed the committee submits a report. In Sweden any Member of Parliament may propose amendments, and the government must accept Parliamentary decisions
6. *The Bill* is given to Parliament for debate, amendment, and voting. If passed the law is published and encoded in the book of law codes.

6.1.1.2 Supranational (EU) Lawmaking

New laws originate in the EC, the executive body for the EU. Legislation may be initiated within the EC or suggested by EU Council members. Once proposed, they are subjected to an Impact Assessment (IA), i.e., thoroughly vetted for their potential impact from all perspectives and by dialoging with all potential stakeholders, including consumers, labor unions, industry, NGOs, political bodies, etc. A draft law is then prepared for submission to committees of the European Parliament that examine potential conflicts and otherwise prepare the laws for a vote. For example, Germany proposed its Elektrische Gesetz or law governing recycling of electrical appliances to the EC. After due process The Waste Electrical and Electronic Equipment (WEEE) legislation was passed by the European Parliament and the European Councils on January 27, 2003 (euractiv.com, 2003). Among other provisions, WEEE required replacement of potentially toxic metals like cadmium, mercury, and lead in electrical and electronic equipment to the extent practicable.

Four years of intensive analysis, discussion, and negotiation preceded the introduction of the *Registration, Evaluation and Authorisation of Chemicals* (REACH) law, approved by the European Parliament and European Council in December 2006. Under this law a chemical management agency is set up in Helsinki, Finland, to which producers or importers of toxic chemicals to the EU must submit applications (Center for Sustainable Production, 2008). The REACH law places responsibility for assessing toxicities and describing safeguards for use of toxic chemicals on producers and importers, not on the agency. Applications for chemicals involving volumes exceeding 1,000 metric tons must be submitted within 3 years; volumes between 100 and 1,000 tons must be submitted within 6 years, and volumes between 1 and 100 tons must be submitted within 10 years, i.e., by 2018. The importance of this law is indicated by the fact that chemicals are the EU's single largest manufactured product, and account for 31% of world production. Because it affects sellers to the EU, REACH has become a de facto world law, to which Japan has already adapted, and regarding which US laws are being proposed. REACH will have profound effects on the use of chemicals in society.[178]

Unlike the US laws of the 1970s and 1980s, REACH locks in no uniform standards or procedures, but leaves open both to the judgment of professional regulators, who are empowered to evaluate and accept or reject applications for use of chemicals (subject to broad guidelines in the REACH Law), allow provisional use subject to future submission of supplementary data or research, and consider requests for special waivers.

Unlike US law proposals, the great majority of which never get through committee or fail to be enacted, meticulous advance preparation for EU legislative proposals means that a large proportion gain approval – though sometimes only after extended reexamination and modification. New laws (like *WEEE*) have often undergone testing in one or more EU nations, so drastic revisions or changes in policy are less likely to be needed. Criticisms of excessive environmental regulations in the EU arise regularly and may have been among factors leading to the 2005 Prague summit of environmental agency heads. According to EC Environmental Counselor in Washington, Malachy Hargadon (presentation at EPA Environmental Partnership Summit, Washington D.C. May 20, 2008), the current weight of emphasis in the EC is now on the urgency of moving forward in global climate change action.

6.2   Environmental Policies

"Sustainable development seeks to meet the needs of the present without compromising the ability of future generations to meet their own needs." Gro Harlem Brundtland, Oslo, Norway, March 20, 1987 (Brundtland, 1987).[179]

Apollo 17's crew took the first full color "Blue Marble" photograph of Earth on December 7, 1972, as they left the Earth's orbit for the moon (Fig. 6.2). In the early 1970s, we were enthralled by the new views of our beautiful planet,[180] but we were also increasingly aware of how fragile it was. A decade earlier, Rachel Carson shocked Americans with her book, *Silent Spring*, in which she let us know

Fig. 6.2 Eastern hemisphere
of Earth, viewed from space,
2002
Source: National Atmospheric
and Space Administra-
tion Goddard Space Flight
Center image by Reto Stöckli,
enhancements by Robert
Simmon

how we were damaging Earth's biosphere through overuse of pesticides (Carson, 1962). Fifteen years after Apollo 17, the UN's World Commission on Environment and Development, chaired by Norwegian Prime Minister Brundtland, sounded a new alarm: the Earth could not sustain the lifestyles of Western nations without jeopardizing the future of poorer nations, and future sustainability for all (Estes, 1993). The UN report, aptly titled Our Common Future, had sustainable development as its main theme. Subsequent EEC summit meetings adopted the Commission's recommendation as a common policy and goal for the future. Though they were relative latecomers to the EU, Scandinavian nations have typically been leaders in commitment as well as performance in sustainable development (see, for example, the Environmental Performance Index ranking of nations by the Yale–Columbia consortium (Esty, Levy, et al., 2008) in Table 4.4).[181]

6.2.1 Germany and Austria

Environmentalism based on deep identification with the natural environment has especially long roots in Scandinavia and in Austria. In Germany (and Japan), where industry played an especially important role, environmentalism gained new impetus in the post-World War II period around the same time as in the United States but along different pathways.

Germany's science-based industry propelled it into international leadership in the nineteenth century (see German Science, Section 8.1). In the post-World War II period it created much of the nation's wealth, but also generated significant air and water pollution, especially in the highly industrialized Ruhr Valley. During the 1970s, citizens organized local proenvironment groups to push the Länder (states) and federal governments to control pollution. The transition in 1973 from the environmentally conscious administration of Willy Brandt (Socialist, or SPD) to the probusiness administration of Helmut Schmidt (Christian Democratic Union, or CDU) triggered a sometimes violent radical green movement.

Antinuclear power groups joined the environmentalists after the Three Mile Island nuclear power plant accident in the United States in March 1979. By the 1980 elections, the various factions had coalesced into the Green Party, and elected members of Parliament. Once the "Greens" entered formal political coalitions they largely ceased being radical gadflies.

Waldsterben or death-of-the-forests, caused by acidic rain due to emission of sulfur dioxide from combustion of coal, became a great concern in Germany in the 1970s. The federal and state governments stepped in and, after evaluation and discussion, imposed draconian restrictions on sulfur emissions. This intervention followed what Miranda Schreurs called the "green social welfare state approach to pollution control" (Schreurs, 2002). Where Germany differed radically from the United States was in its commitment to the industrial underpinnings of its economy. Many industries were effectively monopolies that could both be more easily controlled by federal and state environmental regulations than was the case in the United States, and could pass the increased costs due to pollution control to society. By the 1990s, environmental concerns were part of a consensus agenda for government and no longer generated major opposition. Germany has become a leader in green energy – particularly in wind, solar, and biofuels (see Fig. 5.1).

The recognition of industry and economic considerations as inseparable from environmental policy may be the most profound difference between European and American approaches to environmental regulation. It is starkly illustrated in an interview I conducted with leaders from *Deutsche Greenpeace* in Hamburg in September 2007. The director of energy initiatives for the German Greenpeace organization, a vigorous activist in social and political initiatives to curb greenhouse gas emissions and global warming, told us of a recent visit to New York. After a conference with his American Greenpeace colleagues, the German leader asked to meet with an executive of an American oil company. His US counterpart replied, "We don't talk with those people!"

Discussions with EU environmental officials and experts from five nations in 2007 brought out a consensus assessment among leading EU nations. Germany has the most formal regulatory system, whereas the United Kingdom favors more voluntary approaches. However, the general trend is toward consolidation and simplification of national laws. Within the last 5 years, the Austrian Parliament consolidated previous laws into a single statute. Austrian authorities take pride in the fact that the new unified law system calls for the lead agency to coordinate all other required permitting into a single step. Though requiring much planning and effort in preparation, the new system has the benefit of fostering communication among governmental units, and simplifying permit applications for smaller applicants that would have been turned back by a complex applications process. Fritz Kroiss, an official of the Austrian

Umweltbundesamt GmbH (Office of Environment, within the Environmental Ministry, personal communication 2007) thought that Germany had taken interest in the Austrian approach and was considering consolidation of its own law framework.

6.2.2 Japan

Focus on environmental pollution issues in Japan was intensified by two major post-World War II developments. From 1933 to 1959 the Chisso Company had dumped mercury-containing wastes into Minamata Bay. The spread of methyl mercury poisoning through consumption of mercury-containing fish and shellfish caused death or severe illness in more than 2,000 persons, with wider effects on animals as well as humans. This disaster led to widely publicized compensatory actions and increase in governmental mechanisms to control pollutants after 1969 (Ui, 1992). The second major development was caused by rising industrial development and automobile traffic. By the late 1960s air was so polluted in Tokyo that traffic police used oxygen masks. Beginning with the 1967 Air Pollution Law, a series of escalating measures was introduced:

- New air pollution reduction technologies in 1970s: Japan introduced these in the 1970s, ahead of Germany (1980s) and the United States (1990s);
- Small, fuel-efficient internal-combustion cars;
- NOx-reduction regulations mandating catalytic converters;
- Hybrid electric cars;
- Electricity production from solar energy.

Japan showed significant care and foresight in its environmental protection process, which works more smoothly than in Germany or the US citizen and NGO litigation and politicking, a pillar of US environmental protection, declined to low levels in Japan after the 1960s.[182] Japanese government at all levels works hard to avoid unnecessary conflict. Consensus decision making is a key in this style of governance. This often means time-consuming processes for arriving at a consensus, where the government has gone to considerable effort to resolve problems. Local concerns have led to protests and demonstrations by local populations or affected groups. For example, in response to Japanese citizens' opposition to siting of nuclear power plants near centers of population, the government has come to build them in sparsely populated areas where citizen opposition is minimal.

6.2.3 Canada

Canada, like the United States, has vast natural resources. As "Crown land," Canadian mining and land use was subjected to greater oversight than in the laissez-faire exploitation period in the US Western states. However, conservation and environmental policies evolved gradually and severe impacts have occurred. For example, the huge Sudbury, Ontario copper and nickel-mining district, first discovered in 1883, experienced a major boom once the United States began to stockpile nickel to diversify its supply of noncommunist nickel after World War II. Sulfur dioxide emissions from smelting of ores had a severe effect on the area:

> "Falconbridge was the chief beneficiary of this policy and grew significantly [INCO was another major company involved in nickel mining]. Between 1946 and 1961, the region's population grew from 70,000 to 137,000. ... Sulfur dioxide emissions from the smelting process so damaged the local landscape that NASA astronauts rehearsed their lunar landings in the area. ... .These concerns led to the construction of the Superstack [a very high smokestack] in 1972, which dispersed the smelter's emissions into the jet stream. Over the past 20 years, the city has replanted millions of trees, and the region has regained much of its original greenness (FoundLocally, 2008)."

The Canadian National Environmental Protection Law (CNEPL) of 1999 is now the principal national environmental Law. It is supplemented by laws for the Arctic area and fisheries. The CNEPL is overseen by the federal agency, *Environment Canada*, but detailed environmental management is delegated to the provinces. The Canadian system includes special environmental courts and has low levels of litigation (Manheim, 2006). The gradualist and province-based approach to environmental management allows for considerable variation in environmental rigor in mining and logging operations, and is interpreted by some American environmentalists as lax. However, standards have systematically increased while Canada has avoided the polarization and costly and inefficient regulatory regime that has accelerated decline in the US minerals industry and increased imports of products formerly produced domestically. Canada's mineral industry has experienced a major boom since 2003, fueled by Chinese demands for raw materials (McCartan, Menzie, et al., 2006). The economic viability of technically advanced minerals industries is an asset in terms of rehabilitating earlier mining areas exploited with less environmentally effective methods.

6.3   Scandinavian Nations: Emergence of Post-environmental Societies

The term "post-industrial nations" was coined by Bell (1973) to refer to societies in which industrial processes had become fully integrated into national life and flexibly or automatically adapted to needs in society. I propose that the Scandinavian nations have integrated environmental principles into the national life so thoroughly that need for formal regulations is diminished.

As leaders in both environmental protection and industrial and technical innovation, Sweden and Finland suggest where the EU as a whole is headed. Both nations are successfully adjusting to the global environmental challenge and are maintaining strong economies and export industries as they respond.

- "Sweden has made a radical change from oil to nonfossil based energy sources including biofuels, which has led to a reduction of its greenhouse gas emissions by more than 40 per cent since the mid-1970s. Between 1990 and 2006, emissions *fell* by almost 9 per cent. At the same time, GDP *increased* by 44 per cent. But much remains to be done (Swedish Ministry of the Environment, 2008)." (*Emphasis added.*)
- "The national action plan for environmental business in Finland was launched in February 2007. The aim of the plan is to make clean technology a cornerstone of Finnish industry and Finland the leading country in environmental business by 2012. Environmental challenges must be addressed through market-based mechanisms so that companies can be closely involved in the development work. The new government should take up the opportunity by creating a programme in which energy, environmental and innovation policies are further integrated and support the creation of new industry and jobs (SITRA, 2007)."
- "The Academy of Finland's Sustainable Energy (SusEn) research program for the years 2008–2011 aims at generating new and innovative scientific knowledge on technologies for energy production, effective energy systems and energy efficiency. In addition, the program directs research to develop sustainable energy solutions as well as know-how in identifying future energy systems alternatives (Academy of Finland, 2008)."
- "For the first time in Finland, the activities of municipalities and the businesses and people within them are examined in the same study from the perspective of the reduction of greenhouse gas emissions. The objective is to reduce emissions more than what is required by EU goals and more quickly than has been agreed (Finnish Environment Institute, 2008)."

Norway is the clear European leader in per capita use of renewable energy (Table 6.3); its vast hydroelectric power network supplies almost all of its domestic needs.[183] Table 6.3 ranks ten nations by per capita share of total renewable energy.

Table 6.3 Renewable energy production[1] in 10 nations in 2005

KWh/person × 10^3	Country	Nation ranks 1st	Nation ranks 2nd	Nation ranks 3rd
30.5	Norway	H		
9.2	Sweden		MW, PSB, H, LBF	
5.3	Austria	IW		H, LBF
4.6	Finland	PSB		
2.0	Denmark	MW, W		BG, W
1.4	U.S.	GT	IW, ST	
1.2	Spain		W	
1.1	France	O		
0.9	Japan[2]	SPV; BG	GT	
0.9	Germany	SPV, LBF	BG, LBF	W

Source: IEA (energy data) and CIA Factbook (population)
[1] H, hydro; MW, municipal waste; PSB, primary solid biomass; IW, industrial waste; W, wind; BG, biogas; GT, geothermal; ST, solar thermal; O, ocean wave and tide; SPV, solar photovoltaic; LBF, liquid biofuel
[2] Other data sources list Japan as a world leader in solar photovoltaic electricity production

The first, second, and third leading producers of 11 renewables are shown where they clearly stand out.

Sweden has a distributed environmental management policy. The federal law governing the environment describes general policies and classifies specific activities according to their potential for environmental hazards. The law then assigns responsibility for administering the policies to state, regional, and local entities. The federal Swedish environmental code is 177 pages; the environmental protection act in Finland is 54 pages including amendments through February 2000. Detailed guidelines are developed by the state, regional, and local authorities tasked with implementing the federal laws.[184]

Non-European environmentalists and law makers are usually incredulous at how short Scandinavian environmental laws are. Let us compare forestry laws, for example. About 84 US laws have titles or brief summaries that contain the word "forest." Fifty-six of these laws were passed since 1970.[185] The statutes contain thousands of pages. In contrast, the *full* Swedish Forest Law has only *nine* pages with 36 sections – and this for a nation for whom forests are very important for timber production, biodiversity, recreation, and cultural resources. The opening section can be considered a quintessential example of the Scandinavian penchant for no-nonsense brevity. Sections selected from the Swedish Forest law (Swedish Forest Agency, 1994) that follow *have not been shortened.*

- "*Section 1.* The forest is a National resource. It shall be managed in such a way as to provide a valuable yield and at the same time preserve biodiversity. Forest management shall also take into account other public interests."
- "*Section 4.* This Act or regulations issued under the provisions of this Act shall not be applied to the extent that they conflict with provisions in the second paragraph of section 11 of the Environmental Code or with regulations issued under the provisions of the same Act or of other laws."
- "*Section 10.* Felling on forest land shall be performed in order to promote the establishment of a new stand, or to benefit the development of the existing stand. In order to protect young forest, the Government, or public authority designated by the Government, may issue regulations to protect forest stands under a certain age from felling and may regulate how fellings shall be done, in order to assure fulfillment of the requirements of the first paragraph. To facilitate experimental activities, the Regional Forestry Board may grant exemptions from the first part of this section."
- "*Section 16.* Felling on forest land with adverse regeneration conditions or on protected forest land, may not take place without permission from the Swedish Forest Agency. When applying for felling permission, the forest owner shall describe planned measures to satisfy nature conservation and cultural heritage preservation interests. In connection with the granting of felling permission, the Swedish Forest Agency may decide on measures to limit or minimise disturbance, and to ensure the establishment of new stands. Permission is not required for cleaning or thinning which benefits development of an existing stand."

- *"Section 35.* The supervisory authority may, where necessary, prescribe or prohibit certain action, to ensure compliance with this Act or regulations made
pursuant to this Act. Certain action may be enforced or prohibited when it has
become clear that the advice and directives from the supervisory authority have
not been followed. In urgent cases, or where it is necessary to protect nature
conservation and/or cultural heritage preservation values, such enforcement or
prohibition orders may become operative immediately. Decisions to enforce
or prohibit certain action may be combined with financial penalties. Should a
person fail to comply with an enforcement order, the supervisory authority has
the right to order that the prescribed action be carried out at the expense of the
person at fault (Swedish Forest Agency, 1994)."

6.3.1 Mining and Environmental Protection

Finland and Sweden have been centers of mining for more than two millennia. Mine
production continues to thrive in these countries accompanied by ever-improving
environmental performance. Technological breakthroughs continue to be made by
metallurgical engineers in both nations. The most important of these was flash
smelting, which was developed in the late 1940s by the Finnish mining company
Outokumpu's engineers Petri Bryk and John Ryselin. Subsequent, patented evolution of the extraction process emphasized enclosed space. It captures heat potential of sulfide oxidation while it retains all emissions. It is regarded as the most
significant metallurgical breakthrough of the twentieth century (FundingUniverse,
2008). After only a few years of nonprofitablility in the 1970s, the company continued to diversify and grow. In the 1980s, it became one of the major mining
and metal-refining companies in the world, a position it still holds. Its technology
group recently became Outotec, an independent company "committed to developing technologies that help its customers to reduce the impact of their production on
the environment (Outotec, 2007)."

The Swedish iron mine at Kiruna, north of the Arctic Circle, is the largest and
most advanced underground iron mine in the world; it is largely mined remotely
with robotic equipment.

The length of time it takes from submission of a mining application to granting
of the permit varies widely around the world. Waiting is costly; many companies
cannot afford to wait. Sweden and Finland have bundled the permits required into
a single process, which significantly cuts the time, the cost, and the uncertainty of
permitting (Table 4.4).

Streamlining and increasing certainty in the mine planning process was arrived
at in Scandinavia through cooperation among corporations, mining jurisdictions, and government agencies; international competitiveness and environmental protection are high priorities for all sectors. In the United States, conflict
among the sectors has stalled progress toward streamlining, putting both US
mining corporations and international corporations mining in the United States

at a competitive disadvantage (E. Scott, 2003). Permitting delays in 2005 placed United States together with Zimbabwe at the bottom in permitting for all nations in the survey (TI, 2005).

The Swedish environmental law contains a comprehensive list of categories of activities that are given an administrative level corresponding to the level of hazardous materials involved in the activity. In the case of mining, for example, bulk mineral mining permits (e.g., sand and gravel, limestone) are in category C, handled by municipal boards. Normal mining exploration except uranium is in category B, corresponding to county boards. Full-scale metal mining requests are in category A, handled by one of the five regional environmental courts. These courts have advisory and oversight capabilities in addition to rendering legal decisions (Manheim, 2006).

> A regional administrator of the Swedish environmental code pointed out that Sweden had abandoned detailed regulations for mining because "it is difficult for a supervisory body with limited resources and detailed regulations to achieve better safety than the operator who takes active responsibility." The Chief Inspector for the national mining bureau within the Swedish Geological Survey stated, "In Sweden ... we should not have such regulations that the mining industry is forced to operate in countries that are less scrupulous ... we want a mining industry that operates under modern conditions ... that will also apply far into the future (cited in Manheim, 2006)."

6.3.2 Where Are the Regulations?

As part of preparation for an earlier lead article that included a review of the Norwegian offshore oil industry (Manheim, 2004), I searched for nearly a year for published Norwegian offshore oilfield regulations. Visitors to a Norwegian Embassy-sponsored symposium in Washington from the Norwegian Petroleum Directorate and a Norwegian consulting company tried to help me – but found nothing. I finally realized that there *were no* published regulations.

Unlike the United States, there are no uniform, fixed, and published regulations for the offshore oil industry in Norway, or forestry practice or mining in Sweden or Finland, as is US practice.[186] Rather, preparing an operating plan to obtain a license for timber removal or an environmental permit is the responsibility of the user/applicant, and is adapted to each case. Guidelines are provided by the supervising agency, for example, the mining bureau in the case of mining in Sweden, or the Swedish Forest Agency for forestry. But these are not fixed, but allow for negotiation and special provisions.

The environmental plan is approved by the relevant political unit, i.e., town committee, if the operation is small. It may be reviewed by the county (*län*) if larger, and one of the five regional environmental courts may deal with environmental management plans in cases of very large operations or where special hazards are involved. The courts are qualified to provide guidance. All jurisdictions draw on research and standards recommended by the Swedish Environmental Protection agency, which does not enforce laws as is the case with the US EPA.

The leasing system can yield rapid approvals by American standards. Both Sweden and Finland rank in the top tier of international competitiveness ratings for mineral investments (Table 4.4). A local mining permit (e.g., for a sand and gravel quarry) may be obtained in as little as six months (Manheim, 2006). But one should not assume that the flexibility of the system leads to corner cutting and careless practices. Unlike conditions in the United States, where local communities seek input through the federal regulatory agencies or by challenging operations through the environmental laws, local Swedish political units have full discretion about whether and under what conditions to grant the permit. So operators have an incentive to demonstrate their competence and responsiveness to local concerns.

The Mining Bureau is not obliged to accept lease requests from companies it deems unqualified to operate properly. It rarely needs to turn down requests because investing companies understand the system and only apply if they have confidence they can operate to satisfaction. The system places no ceiling on innovation. In fact, companies displaying potentially significant advances over standard technologies are likely to be given faster approvals, since this kind of innovation bespeaks a proactive, flexible organization that views the operation intelligently and holistically.

As in mining, rules for harvesting of wood depend on circumstances, as do potential environmental permits. For smaller owners or thinning, generalized guidelines may be followed without formal permits, but proper practice is expected – the Forest Agency provides help. For larger enterprises there are also no predetermined forms. Licenses are given by the Forest Agency and environmental permits are granted by the local jurisdiction(s) with special requirements according to the local conditions. And as noted in Section 35 of the Swedish Forest Act (Section 6.3) the Swedish Forestry Agency has full discretion to set special conditions for operations. Apparently, in some cases of the largest enterprises, long delays have been involved. Interestingly, these have led to formal editorials in forestry journals, complaining about the regulatory obstruction (Wibe & Carlén, 2002). In the strenuously egalitarian Swedish society, nobody is immune to criticism!

The development of environmental law in Finland and Sweden may seem lacking in rigor, but the results far surpass those in the United States (Esty, Levy, et al., 2008). Important reasons for this include the following:

- Citizens expect government agencies to carry out their duties expertly and efficiently, without political partisanship.
- Citizens expect competence and good performance from companies involved with natural resources.
- Representatives of political units (i.e., towns, counties) have decision-making powers over economic activities in their areas. At the same time, there is a tradition of allowing public access to private land under conditions of proper respect.
- Citizens have the right to formally question the laws but not to sue the government.
- A long historical tradition of mining, logging, and use of natural resources with proper stewardship.

6.3.3 Norway's Offshore Petroleum Industry: A Model for Advanced Technology and Environmental Policy

Norway joined oil producing nations in 1969 with the giant Ekofisk field in the Norwegian sector of the North Sea. Norway's national oil company, later called Statoil, shared offshore development with private partners along the entire Norwegian coast, including the North Sea, Norwegian Sea, and the Arctic Ocean (Barents Sea). In 1981 Gro Harlem Brundtland, former environmental minister and newly elected as the first female prime minister of Norway, faced a major decision. Should Norway produce its offshore oil and gas conservatively, or push drilling aggressively? Norway's socialist (Social Democratic) government chose to exploit the oil and gas resources at maximum speed consistent with efficient production practice.[187] Subsequently, Norway has become the world's third largest petroleum exporter (Fig. 6.3). Although its production from its existing 200-nautical-mile exclusive economic zone (EEZ) was expected to decline from peak reserves reached in 1995, Petroleum Minister Odd Enoksen announced in 2006 that Norway was claiming additional areas in the Arctic Ocean equivalent to half the land area of continental Norway (Norwegian Petroleum Directorate, 2006).

In addition to its oil production Norway has become a leading producer of offshore oil drilling platforms and engineering equipment, and the first nation to commit to zero toxic discharge, and sub-seafloor completions (i.e., connection of producing wells to pipelines) and pipeline systems as standard practice by 2006.

What is most significant from the environmental and policy perspective is a series of unusual developments on Norway's part.

- The decision was made to place the bulk of profits from petroleum exports in a foreign exchange fund, created in part to assure transition to other energy and industrial income after offshore resources were fully exploited. That fund stood at $392 billion as of May 2008.
- Beginning in 1991 Norway assessed a voluntary tax on its CO_2 emissions, of which offshore operations consumed the largest proportion (29%). The tax amounted to about $53/tonne of CO_2 in 2006 (Norwegian Petroleum Directorate, 2006).
- The first commercial-scale burial of CO_2 began beneath the offshore Sleipner field with a million tones per year in 1996, and continues to the present. CO_2 burial from the giant offshore gas field (Snøhvit) in the Arctic Ocean began in 2007, and a major burial operation in conjunction with an electrical power plant at Mongstad, is scheduled to begin demonstration operation in 2010.
- Norway's offshore petroleum operations take place in the same offshore ocean area as Norway's major fisheries.

The decision was made to place the bulk of profits from [Norway's] petroleum exports in a foreign exchange fund, created in part to assure transition to other energy and industrial income after offshore resources were fully exploited. Owing to rise in oil prices that fund stood at $392 billion as of May 2008.

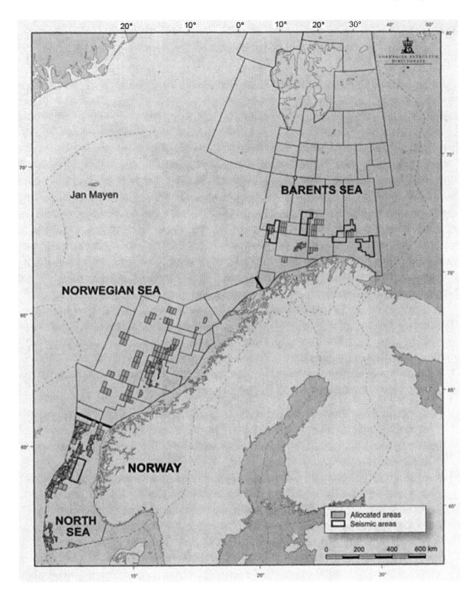

Fig. 6.3 Petroleum leasing areas along North Sea, Norwegian Sea, and Arctic Ocean sectors of the Norwegian EEZ
Source: Norwegian Petroleum Directorate

Norway's path to the present has not been trouble free. After a major blowout in 1977, it suffered the catastrophic capsizing of a "flotel" housing offshore oilfield workers, with more than 100 workers lost. Since then, however, mishaps have been minor.

Norway has become both an economic and environmental leader in the petroleum industry – using proceeds of petroleum exports to advance the carbon burial

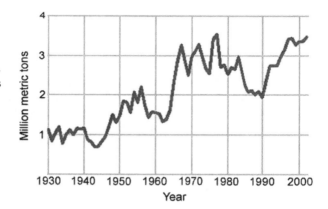

Fig. 6.4 Total production in Norwegian fisheries, 1930–2002; includes farmed fish
Source: Norwegian Directorate of Fisheries, and fisheries statistics, Statistics Norway

technologies designed to be exportable to China. It maintains fish productivity and sustains a major fish export industry in the same coastal area as its oil industry (Fig. 6.4). This is a picture that can be instructive for Americans.

> Norway maintains fish productivity and sustains a major fish export industry in the same coastal area as its oil industry.

In the United States, it is not even possible to hold rational discussions about offshore petroleum production outside the historic Gulf of Mexico and Alaska production areas. Petroleum leasing has never been permitted in the Georges Bank area off the northeastern United States – but that area's once world-class fishery has been devastated – by overfishing and poor management. Finally, though the United States has been a pioneer in the technology of returning gases to oilfield reservoirs for enhanced production, its only proposed CO_2 burial product (FutureGen) was "buried" in 2007 (the Department of Energy withdrew earlier funding commitments). The purpose of identifying problems like these plainly is not to assign blame, for no group's ideas or actions form in a vacuum. (Chapters 7–9 will discuss the origin of the problems.)

6.4 Discussion

This chapter may seem to show other nations' progress too favorably while seeing mainly faults and backwardness in US policies and developments. There may be a perception of stressing industry and economic activities more than environment. I have deliberately emphasized the difference in outcomes of policies abroad and in the United States, first because of America-centric tendencies in otherwise informative, multi-edition and widely-used textbooks like those by Walter Rosenbaum, Norman Vig, and Michael Kraft. Second, when American authors have examined European and other foreign environmental policy their studies have often tended to focus on specialized topics, for example, Dickson and Cooney (2005) and Axelrod, Downie & Vig (2004), missing or underemphasizing the big picture as I see it.

What is the "big picture"? One of the most striking differences between EU policy and US policy in general is how laws are made. The old joke that "there are two things, about which we are better off not inquiring: the making of sausage, and the making of laws" no longer applies in the EU. This chapter has summarized the multifaceted, careful way laws are prepared and analyzed for their potential effectiveness, fairness, and future impacts – before being presented for political decision making. Compare the United States, where hundreds of bills have been independently introduced to the 110th Congress on energy policy. Many of these bills obviously carry little expectations beyond registering their sponsor's interest or concern, with the outside possibility that some aspect may end up being included in more serious law preparation or amendments. More detailed, serious bills have been submitted without, or with only token bipartisan sponsorship – that have no reasonable expectation of passage – and only roil the political waters. Though more carefully developed than competing bills, the Lieberman–Warner Climate Security Act (S.2191) (Lieberman, 2007) was accompanied by no report detailing its potential impacts. The mandatory Congressional Budget Office analysis provides detailed projections of costs, revenues from sale of emissions permits, and predicted increases in costs to consumers and loss of tax revenue from business profits. But it offers no analysis of its larger impacts on American industry and the economy.

How can one expect such a serendipitous, dysfunctional lawmaking process to produce good guidance in one of the most critical policy areas for a nation of nearly 300 million that often regards itself as having world leadership responsibilities?

The laws and sociopolitical patterns they have created have been in place so long that troubles caused by them tend to be attributed to people and political groups, instead of being the consequence of the laws' imbalance and deliberate confrontational design. They are not corrected. Many knowledgeable scholars know about the problems, but assume that we are stuck with the present system, for example, "Assuming regulation will be with us, warts and all, for the foreseeable future . . . . . . (Nakamura & Church, 2003)."

If foreign experience reached US eyes and ears I believe that the alternatives could gain influence. However, "American exceptionalism" is not just a clever academic jargon term. Hubris and national preoccupation runs throughout the nation, from Congress to the man in the street to academia. Though I personally have spent an unusual amount of time, for an American, in study, research, and travel abroad, I did not realize the degree to which I too had been affected until I embarked on research for this book in 2003.

Consider the ability of an environmentally top-rated nation of 5 million population (Finland) to out produce the United States in cell phones – that were originally made possible by US inventions and technology. And that small nation has also made a high-tech, "clean" industry out of stainless and specialty steel production – a "smokestack industry" that many American economists and others assume that the

United States has "outgrown." The ability to simultaneously achieve effectiveness in environmental policy and industry is explained by Finnish management expert, Risto Harrisalo in a book section published in 2004 titled, Trust as Capital.

6.4.1 The Small, Homogeneous Society Explanation

An almost immediate instinct among Americans, whether experts or average persons, seems to be to dismiss noteworthy individual achievements by other nations, especially in Scandinavia, as the products of small, homogeneous societies. The implication is that such achievements cannot be expected in a large, diverse nation like the United States.

The *small* idea sounds plausible when applied to Sweden and Finland. But when details are examined, it fails to hold up. Consider the achievement of systematic cooperation and advancement within the EU as a whole. Do the 27 nations of the EU, comprising a population of 450 million, speaking more than 20 languages, and with highly diverse histories (including wars and hostilities) form a *small, homogeneous society*?

Next, what about US "diversity"? In point of fact, racially diverse or ethnic groups in the United States have minor roles in current political polarization and gridlock. The politicians or lawyers at each others' throats in Congress or in American courtrooms have quite similar racial and ethnic (mainly white) backgrounds. For example, studies of conflicting attitudes toward offshore drilling among citizens reveal partial differentiation based on economic factors (e.g., among low-income groups), but among people with strongly polarized views there is little difference in income, education, or ethnic background (Smith, 2002).

Finally, examination of US history prior to World War II, and also since that time, reveals many examples of American technological and policy leaps forward that required extensive social cooperation and coordination, and which resemble some of the Scandinavian advances discussed. US cities followed closely on developments in the origin of diseases, often building on each other's advances. The leadership of the Public Health Service helped create rapid advances in drinking water quality throughout the United States (Gurian & Tarr, 2001).

6.5 Alternative Energy in Europe

6.5.1 Wave and Tide Energy

France has long been the world leader in tidal energy, with 240 MW installed capacity at La Rance. Since its previously mentioned launch near the Orkney Islands, Ocean Power Delivery, Ltd. has set in motion or planned new Pelamis wave power installations off Portugal and Cornwall, South West England. Scotland and Spain have set ambitious goals for wave energy development. In contrast with the labyrinthine permitting processes, uncertainties, and lengthy delays for US ocean hydro developments, the UK ocean wave energy project, Seagen is currently being installed as a successor to the successful Seaflow offshore turbines that have operated continuously since 2003. Operated by Marine Current Turbines, Ltd., and supported jointly by the Department of Trade and Industry, United Kingdom, the EC, and the German Government, Seagen, rated at 1.1 MW was scheduled for installation in 2007. Power will flow into the UK grid. The next goal will be to install multiple Seagen units aggregating 300 MW along the coast of England, Wales, and Scotland, and possibly the French coast by 2010.

However, these projects would be dwarfed by a long-range Russian plan announced by Anatoly Chubais, CEO of United Energy Systems, and the huge Russian national hydroelectric combine. If experimental units now installed at Kislogubskaya in the Barents Sea achieve their promise, it has been claimed that giant tidal power plants with capacities of 10,000 MW will be built (Veletinsky, 2006).

What are the differences between European developments and those in America? The United States' tradition of entrepreneurship and innovation is reflected in the number of proposals received by FERC (over 50, as of 2007). A current listing of firms offering ocean energy services (Energy Source Guides, 2007) shows the United States leading in number of companies (19), followed by the United Kingdom (10), Australia and Canada (4), India (2), and a number of other nations with one company each.

However, the coordination of regulatory jurisdictions that has characterized United Kingdom's current leadership in ocean energy initiatives, as well as the cooperative support of ocean energy projects by other European governmental agencies, is missing in the United States. Preparations for the previously mentioned UK projects have included giving prior public information to local governmental units so that they and their local populations can make decisions *before* physical implementation begins. Barring unexpected developments, the extensive and often unpredictable permitting delays and confusion about regulatory jurisdictions that affect US ventures, are minimized.

There is also important structural difference between the US environmental regulatory system and those in EU nations. National environmental agencies of EU member nations subscribe to the principle of sustainable development enunciated in the Brundtland Commission report to the United Nations. If desirable technologies run into environmental problems, they are not simply left to sink or swim. The problems are addressed cooperatively with the goal of simultaneously seeking technological or economic viability while meeting or improving environmental performance standards. Regulatory agencies of leading European nations retain discretionary authority to this end, and often enjoy considerable public trust. Senior civil service policymakers as well as policies normally continue through changes in political administrations.

The holistic approach to regulation by European nations is typified in the recent "Prague Statement" issued by a meeting of 29 heads of European environmental agencies. The summary declared that besides achieving environmental goals, "a modern approach to regulation can reduce business risk and increase the confidence of the investment markets, assist competitive advantage, etc." RIA, or Regulatory

Impact Analysis, which gained importance after the Maastricht Treaty of 1992, has assisted in this process.

6.5.2 Biofuels

Rudolf Diesel, inventor of the diesel engine, already predicted in 1912 that diesel fuels from plant and animal feedstocks could be important in the future (Quick, 1989). Biodiesel gained an early start in Europe in part because of the high cost of petroleum products, and because of the wide use of diesel engines in both automobiles and trucks. It has been primarily produced from oil-containing plants like rapeseed, canola, sunflowers, and soybeans, as well as animal fats and tallow. Plant oils are esterified (by reacting with potassium hydroxide) which reduces viscosity and improves burning properties. Byproducts of oil extraction include protein (e.g., from soybeans) that can be used for animal feed and other purposes. The Austrian government has mandated use of biodiesel in areas of the high Alps because biodiesel's low sulfur content and rapid biodegradability of exhaust residues (a few weeks) in comparison with petroleum diesel residue (Weber, 1993).

6.5.2.1 European Development of Biofuels

D.K. Birur and coworkers at Purdue University (Birur, Hertel & Tyner, 2007) recently summarized world biofuel developments, pointing out that the EU has mandated a 5.75% share of biofuels on the liquid fuels market by 2010 (European Commission, 2003). Germany is the largest producer with 798 million gallons, or about 54% of the total for the 27-nation EU, followed by France (15%), Italy (9%), United Kingdom (4%), Austria (2.5%), and others. The spectacular growth in German production is the result of large exemptions from taxes for biofuels. Increasing emphasis is being placed on biofuels created by conversion of lignocellulose (woody wastes). However, in the light of the rise in world food prices, EU spokesmen have recently recommended against further increases in biofuel production.

German subsidies have made possible the production of electrical power from many small-scale biogas generators on farms. In this process whole corn plants (corn and stalk) from densely seeded crops are ground up, mixed with a little cow manure, and fermented. The gas produced by anaerobic fermentation is about 50% methane with the remainder mainly CO_2, which is separated by dissolving it in water under pressure. The methane is piped to the gas-fired electrical generation systems, whose power is transferred to the electrical grid. Fermentation residues are returned to the soil in order to in Germany minimize fertilizer supplementation.[188] Alternative energy production in Germany is shown in Fig. 5.1.

A typical aspect of the EU approach toward renewable energy as well as other future developments is to undertake detailed applied research with future projections relevant to the application of policy scenarios. An impressive example is the

European Environmental Agency technical report, *Estimating the environmentally compatible bioenergy potential from agriculture* (Petersen, Elbersen, et al. 2007).

The 105-page report offers a detailed assessment of many parameters of both conventional and biofuels crops, such as fertilizer, pesticide and water requirements, tendency for soil compaction, and compatibility for bird populations. The worst crops for environmental performance are oil rapeseed, potatoes, wheat, sugar beets, and corn (maize), while the best crops are clover-alfalfa, hemp, linseed, and mustard seed. The best yield per hectare is obtained from total maize cultivation. Biofuel production using cellulosic plants, like *Miscanthus*, switchgrass, reed canary grass, and giant reed, as well as entire wheat plants and double cropping (alternating corn with a more soil-improving crop in a single year) were evaluated.

The final matrix yielded by the detailed study shows the statistically optimum yields for various crops for each EU nation. The study also took into account future climate factors and environmental/ecological constraints. It excluded grassland-rich countries, such as Ireland, and densely populated countries like Belgium and the Netherlands. Most energy crops will be grown in seven nations: Germany, France, Finland, Poland, Spain, Lithuania, and Poland. Estimates of total hectares available for energy crops by various authors vary from 20 to 59 million hectares, with the current study arriving at 20 million hectares or 54 million acres potentially applicable to energy crops.

6.5.2.2 US Biofuels

In the United States the first major biofuel initiative came with the 1978 Energy Policy Act, which subsidized ethanol production – then almost exclusively from corn. This was followed by the 1990 Clean Air Act amendments, which required gasoline vendors to have as minimum oxygen percentage in fuel to allow gasoline to burn cleaner. A result was the development of methyl-tertiary-butyl-ether (MTBE), which was produced by oil companies from petroleum feedstocks. However, MTBE proved to have significant solubility in water and was found in ground water supplies, as a result of which it was banned in 20 states. Prior to passage of the Energy Policy Act of 2005 much debate occurred about whether energy companies should be given legal liability for lawsuits over MTBE issues. Eventually, Congress removed requirements for oxygen content of gasoline, leaving it to industry to meet clean-burning standards.[189]

The strong support for ethanol subsidies from corn-growing states plus the impetus to move toward alternative and neutral energy sources caused continued growth in subsidies for corn grown for fuel. In 2007, 25% of corn acreage was being dedicated to alcohol production. The diversion of US farm acreage from food crops to corn had long been criticized as having marginal or negative effect on net greenhouse gas emissions because its requirement of heavy nitrogen fertilization. Nitrogen fertilizer is currently extracted from air with the assistance of fossil fuel energy. By 2007, the adverse effect of devoting corn acreage to alcohol production reached serious proportions and is now strongly criticized by many scientists and persons concerned with world food supply.

Repeatedly since the 1970s idealistic Congressional initiatives launched with American fervor have resulted in failure or have had adverse results – as in the case of MTBE and corn alcohol. We can hardly expect outcomes to be different when a political body involves itself in detailed intervention in complex, specialized areas where it lacks expertise and has no ability to follow-up the results of new programs as a professional agency would. (Chapters 7 and 9 offer further discussions of this problem and suggestions for the future.)

Chapter 7
Reform Efforts and the Future: Where Do We Go from Here?

Governance – seeking good results, not just good methods

The tale of two US fisheries (Fig. 7.1) contrasts the good outcome – achieved with solid professional guidance and international cooperation – with the poor results achieved by the US regulatory system (see Section 8.2.1).[190] Figure 7.1 depicts fish landings for haddock in the US northeast. The haddock landings record the steep drop prior to 1976 owing to the heavy fishing of the foreign fleet, which included large Russian trawlers and factory ships.

The passage of the Marine Fisheries Act of 1976 (MFA) extended US control of fisheries to 200 nautical miles from shore – the Exclusive Economic Zone (EEZ). Foreign fleets were excluded from the EEZ, which led to the small surge of US fish landings in 1980 (Fig. 7.1). But the MFA failed to stop US overfishing and collapse of the ground fish populations on Georges Bank.

7.1 Introduction

Getting a balanced understanding of environmental regulatory policy from all shades of opinion is not easy. The subject touches visceral, controversial concerns in different groups. Arguments are frequently one sided. Overviews with extensive literature references are provided by multiedition texts on environmental policy, such as those by Kraft (2004), Rosenbaum (2008), and Vig and Kraft (2005).

Details must be sought in more specialized treatments in a literature expanded to nearly impossible levels (i.e., approximately 30,000 books are recovered on the query terms "Environmental policy" by the Amazon.com online bookseller web site or the Google Books search engine). Scholarly articles run into the millions.[191] Government documents and other kinds of data add to the retrieval and analysis problem.

Progressive environmentalists such as Christopher Schroeder and Rena Steinzor (2004) at the Center for Progressive Regulation tend to be satisfied with and protective of the existing regulatory system. They reject claims about inefficiency,

F.T. Manheim, *The Conflict Over Environmental Regulation in the United States*,
DOI 10.1007/978-0-387-75877-0_7, © Springer Science+Business Media, LLC 2009

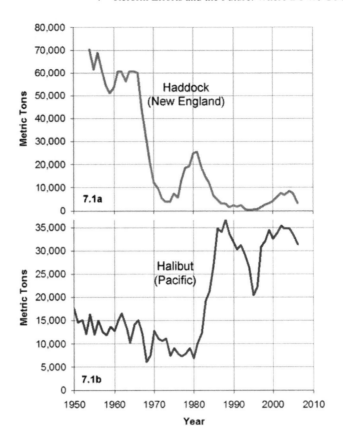

Fig. 7.1 Fish landings of haddock in New England (7.1a) and halibut in the Pacific (mainly Alaska) (7.1b), 1950–2006
Source: NOAA, National Marine Fisheries (National Oceanic and Atmospheric Administration, 2007)

arguing, for example, that the "best efforts" approach, i.e., the existing prescribed technology and standards-based methods, are *easy to implement and yield the safest results.* Many environmentalists' complaints about current laws involve inadequate funding for enforcement, or attempts by industrial companies to evade and avoid compliance (Natural Resources Defense Council, 2001). (The asymmetry of information sources was discussed in Chapter 2.)

Other environmentalists recognize outdated, counterproductive or self-defeating features in the "old" laws. According to Debra Knopman and Emily Fleschner (1999), for example,

"It would be sheer coincidence if the first generation of national environmental laws and regulations conceived 30 years ago would fit all of our needs at the start of the 21st century. Back then, most Americans lived in cities. The economy was dominated by heavy industry. Smokestack factories and big city sewers were Public Enemy No. 1" ...... Environmental problems the 70s "rules" cannot cope with are:

"3/5 of smog from nitrogen oxides comes from cars, trucks, railroads, planes and other non-industrial sources that are hard to control under Clean Air Act rules."

"Agricultural runoff, not included in the Clean Water Act is now the most extensive source of water pollution, affecting 70% of rivers and streams not meeting water quality standards. (Knopman & Fleschner, 1999)."

Recommendations by Knopman and Fleschner will be included among suggestions for reform (see Section 7.4).

Although the 1970s regulations created complaints and much antagonism from the business community when they were first implemented in the early 1970s, the nature and demands of business life are not conducive to in-depth policy study. Although the first libertarian-conservative think tank, the American Enterprise Institute, got started in scholarly analysis in the 1950s, in-depth analyses of regulatory policy from the conservative perspective did not emerge until 1981, in the form of 12 task force reports, summarized by Hinish (1981) in the massive guideline document prepared by the Heritage Foundation for the incoming Reagan administration.

Books from the industry perspective like that of Elizabeth Brossard (1983) are rare. The policy analysis gap from the industry-conservative point of view has been partly filled by the rise of conservative think tanks, notably the Heritage Foundation, the American Enterprise Institute, Competitive Enterprise Institute, and the Heartland Foundation. I propose that timely input should be gained by governmental coordinators for policy development by inviting input from affected industry groups or companies. EU policy does not rely on chance to inform policy preparation but seeks broad input.

In this chapter, I will emphasize concepts and quotes from some of the many thoughtful and nuanced studies of regulatory policy since 1980 that have tried to create light rather than heat. Special recognition is due *Resources for the Future* (RFF), a Washington think tank that is probably more closely associated than any other organization (with possible competition from the Rand Corporation) with in-depth analyses, influential ideas, and insightful publications about resource and environmental policy.

RFF has dedicated environmentalists as well as business leaders on its board of directors.[192] The publisher, Congressional Quarterly, is a prolific source of pertinent, nonpartisan assessments in the regulatory policy field. A recent release, *The New Regulations* (Fiorino, 2006) gets to the heart of the challenge of regulatory reform from one of the best available vantage points, the US Environmental Protection Agency (EPA). EPA's 18,000 employees handle most of the difficult environmental regulatory enforcement tasks in the United States, with the exception of the Endangered Species Act, in addition to extensive research, outreach, and maintenance of comprehensive online data. Daniel Fiorino is the Director of the Performance Track program at EPA. His program evaluates performance in a number of EPA's operational areas, makes awards to high performing industries and other organizations, and seeks special incentives and innovative approaches to better regulation. I draw especially on his book, which offers detailed and candid assessments of problems with current "old" regulations and suggestions for reform.

The remainder of this chapter summarizes critiques, reform movements since the establishment of the 1970s environmental laws up to 2000, analysis, recent proposals for reform, and an analysis of why major reform has been largely unsuccessful in the United States in the past. Finally, I review the experts' suggestions in the light of European and other foreign experience, offering my perceptions about what will *not* work (e.g., new laws and interventions by Congress) and suggesting concrete steps that could help build a base for a more cooperative and productive system.

The United States will probably more aggressively commit to reduce greenhouse gases beginning with a new administration in 2009, despite obstacles posed by our regulatory system and accompanying sociopolitical (NIMBY) barriers to innovation and cooperation (see Chapter 5). European nations have found that intensive technical and policy innovation and cooperation are needed to achieve progress in alternative energy strategies. Even if reforms in regulations and reduction in political polarization could be achieved in the United States (something about which many if not most experts from different points of view are skeptical) the road to "energy independence" and a green future would probably not be easy. *Without* significant changes to reduce conflict and change, the combination of problems in the United States, and its prevalent style of seeking short-range fixes for long-range problems, could well lead to steeply rising energy costs and shortages, impacts on industry and the economy, and difficulty in meeting looming demands for investment.

> Without significant changes to reduce conflict and change, the combination of problems in the United States, and its prevalent style of seeking short-range fixes for long-range problems, could well lead to steeply rising energy costs and shortages, impacts on industry and the economy, and difficulty in meeting looming demands for investment.

If further challenges beyond the present downturn do come, they could have at least one positive effect. When old paradigms fail and the consequences hurt, windows of opportunity for new approaches tend to open. The United States can move exceptionally quickly once opinion unifies around needs or goals. The EU offers valuable experience and models. The present book anticipates new windows of opportunity for change in the United States, if not in the near future, then when troubles mount.

> The United States can move exceptionally quickly once opinion unifies around the needs or goals. I anticipate new windows of opportunity opening, if not in the near future, then when troubles mount.

7.2 Selected Critiques and Problems with the Old Regulatory System

- *Lindall (1974).* "Environmental legislation, in a position similar to that of President Warren Harding, "has nothing to fear from its enemies – only its friends .... Unfortunately, the legislative approach has been piecemeal, and the effect of each new law on other laws or on the total legal structure has not been carefully evaluated." "There is much duplication, delay, potential conflict, and unnecessary and burdensome red tape."
- *Bardach and Kagan (1982).* For electric power, steel, foundries, copper smelting, electroplating, and auto industries, regulatory cost due to the Clean Air Amendments (CAA) dominated capital investment in the 1970s. It displaced capital needed for modernizing production facilities.

    - Rigid, cumbersome regulatory requirements (CAA and Clean Water Amendments – CWA) create antagonism and alienation.
    - Gross inefficiency, excessive compliance costs for OSHA fail to achieve objectives.
    - Permitting delays are created; they may extend from 5 to 10 years or longer.

- *Norton (1991).* Removal of discretion from Park Service staff to control fires, in conjunction with accumulated tinder and wood, created unusually intensive and destructive fires in National Park and Forest Service areas; 44% of Yellowstone Park burned. Radical changes in policies are undesirable.
- *Knopman and Fleschner (1999).* " ..... more than 2/3 of endangered species reside on private lands where the Endangered Species Act is not only least effective, but where the provisions or fears about the act may cause landowners to either conceal presence of such species, or even kill them so as not to be subjected to bureaucratic problems, or at worst, have their land placed under restrictive covenants or more."

    - " ...... of 3000 high-production chemicals in commercial use as feedstock, pesticides, or consumer products, the most basic information about 75% [of them] is lacking, a sign of dysfunction for the 23-year-old Toxic Substances Control Act."

    - "The first result of the Clean Water Act was to essentially shut down all innovation, because best available technology meant proven and available, not innovative and unproven."

    - Clinton Reinvention initiatives have attempted to deal peripherally with some of these issues, but were "neither broad nor deep enough to make a big difference in environmental quality."

    - "The truth is that uniform laws and standards, plus public information are the building blocks for lawsuits. Civic engagement is vital to solve place-based environmental problems."

- *Stroup and Meiners (2000)* cite " ..... the RAND study ..... that 88% of 13,000 Superfund claims were related to litigation and other transaction costs. Just 12% of total costs of claims went to clean up these sites."
- *Kagan (2003)*. "Lawyers for environmental advocacy groups used a blizzard of lawsuits in the early 1990s to bring logging to a virtual halt on extensive forest lands managed by the U.S. Forest Service and BLM in the Pacific Northwest."

  - "In the 1970s and 1980s the construction of nuclear power plants in the United States .... slowed to a crawl, [and was] then abandoned."[193]
  - Kagan also cites cases where industries used countersuits for their purposes.

- *Fiorino (2006)*. [U.S.] "Short-term technology-forcing strategy for end-of-pipe technologies sacrificed better technologies. In Scandinavia companies were given more time for adoption of production processes, which led to more significant improvements in technology."

  - "Environmental Law Institute's study of six industries found that 'technology-based emission limits and discharge standards, which are embedded in most of our pollution laws, play a key role in discouraging innovation.'"
  - [The] "compliance imperative makes firms reluctant to deviate from government-prescribed technologies .... Most firms fail to develop a culture of continuous environmental improvement necessary to sustain research, development and investment in innovation."
  - "So many specific provisions have been codified into law (Clean Air Act) that much of the effort devoted to maintaining compliance is delivering little environmental value to society."

7.3 Reform Efforts – History

7.3.1 Minor Reforms

Among these are amendments that sought to modify overly rigid or adversarial provisions that had aroused bad publicity or antagonism. For example, the Endangered Species Act's "stop the Tellico Dam" stringency was modified by amendment in 1978. At least five members of an eight-member cabinet-level contingency action committee can vote to make exceptions (e.g., allow an extinction) under special conditions.

7.3.2 Bubble Policy

In the 1970s the United States was rocked by two sudden oil shortages: the first Arab oil embargo of 1973, and later the Iran oil crisis of 1978–1981. In order to ease pressure on industry EPA introduced the bubble policy in 1976. It was rejected by the courts as not being consistent with the Clean Air Act (Tietenberg ,1999). During the 1978 Iran oil crisis, the requirements of the CAA on top of the severe economic downturn posed serious threats to major industrial firms, including the auto industries. On December 3, 1979, EPA Administrator Douglas Costle in the Carter administration announced the Bubble policy. It allowed industrial plant owners to draw a "bubble" over their entire plant. Within the bubble pollution controls for different emission sites could be adjusted to maximize reductions for the cheapest controllable emissions, providing the overall emissions did not increase. Although initially hampered by complications raised by state implementation plans (SIP), relaxation of the rules allowed the policy to become fully successful.

7.3.3 Reagan Counterrevolution

President Ronald Reagan maintained moderate environmental policies as Governor of California in the 1960s. However, Reagan's concern about the effect of the 1970s laws was added to by his anger at President Carter's "land grab." Carter sequestered 56 million acres of Alaskan land by declaring them National Monuments under the American Antiquities Act of 1906. Assisted by aides Jack Kemp, David Stockman, and others, the Reagan administration launched a multipronged regulatory rollback attempt (Durant, 1993). The administration apparently chose to bypass partly consultative reform approaches recommended by the Heritage Foundation's Mandate for Leadership (Hinish, 1981). These included early involvement of Congress in reform, special Commissions, and not entirely trivial amendments to the regulatory statutes.

Preplanned actions included instituting RIA (regulatory impact assessments) in the form of benefit/cost assessments for regulations, mandatory overview in OIRA, the Office of Information and Regulatory Affairs within the Office of Management and Budget, and delegating regulatory authority and money to the states. EPA Administrator Anne Gorsuch went beyond these measures and took a meat ax (40–50% cut) to the EPA enforcement budget. These and other egregiously confrontational actions that showed obliviousness to the concern about environmental regulation were guaranteed to inflame every last environmentalist as well as centrists. Was there a plan or a hope behind this kamikaze effort – i.e., that it might spontaneously inspire half the nation to rise up in support – or was Administrator Anne Gorsuch simply consumed by irrational fervor?

James Watt was a far more skilled administrator than Anne Gorsuch but just as zealous. He achieved some major changes (see Section 2.8.4.2) but competed with Gorsuch for bringing on disastrous backlash. William Ruckelshaus was called back to replace the resigned Gorsuch. Watt resigned under fire in late 1983. In the retaliatory wave by Congress, several environmental laws were reauthorized and strengthened: RCRA (1984), SDWA, and Superfund (1986). CWA (1987) was passed over a Presidential veto. Moratoria amendments attached to Interior appropriations bills beginning in 1983 shut off large parts of the Outer Continental Shelf (OCS) to oil and gas leasing (Kumins, 1997).

Reagan administration staff were not "climbers" or self serving bureaucrats. They believed in their mission, ideological zeal for which led to extremes. For example, Sea Grant, one of the most popular grant programs, was zeroed out in the budget 8 years in a row and the administrator was not allowed to report successes, although Congress restored Sea Grant's budget each year.

A detailed account of the first Reagan administration's environmental policies, including interviews with more than 400 persons was published by Jonathan Lash, now the President of the World Resources Institute in Washington, DC (Lash, Gillman & Sheridan, 1984).

One of the consequences of the furor over the actions of Gorsuch and Watt was that opportunities for legitimate reform that might have gained consideration in the typical "honeymoon" granted up to this time to new administrations were completely lost. Long-lasting distrust and hostilities were stoked.

7.3.4 G.H.W. Bush and the Clean Air Act Amendments

George Herbert Walker Bush declared during his presidential campaign that he would be "The Environmental President." William K. Reilly, who had served in the Council for Economic Quality in the White House under Russell Train, and was President of the World Wildlife Fund, was nominated as EPA Administrator. President Bush hoped to bridge the yawning gulf between environmentalists and industry and their Congressional supporters. In 1990 Bush, by presidential order, consolidated the offshore moratoria areas into an omnibus moratorium that covered about 85% of the EEZ in the lower 48 states. This act stopped the rise in membership for some leading environmental organizations like the Sierra Club, indicating the sensitivities over offshore oil leasing. After exhausting struggles, the Clean Water Act Amendments of 1990 were passed in November 1990. Getting it done required the combined efforts of Reilly, the formidable peacemaking skills of Senate Majority Leader Mitchell, Paul Portney of RFF, and Fred Krupp, President of Environmental Defense. The highlights of the 800-page Amendments were the provisions for emission trading to reduce sulfur dioxide emissions. This part of the law proved wildly successful, and was claimed by Krupp to reduce the cost of reducing emission by 80% compared with existing compliance models (Krupp & Horn, 2008). The breakthrough has an important symbolic importance by demonstrating the superiority of incentive models for regulation over command and control.

7.3.5 Clinton–Gore Reinventing Government Program

The National Partnership for Reinventing Government, under the leadership of Vice President Al Gore had massive scope (70+ reports) and the unquestionably committed intent to make the federal government smaller, simpler, more efficient, and helpful to citizens (Gore, 2001). How well it achieved its regulatory goals is more complicated to determine because of the difference between its claims and assessments of nonpartisan experts on regulatory matters. The overall reported achievements are so expansive that even if reduced by half or less, it is hard to doubt that a great deal in the way of cost savings, organizational reductions, and streamlining was achieved. Additional evidence confirms that administrative reforms[194] were achieved at the EPA, though there no major changes in the laws.

Speeches and statements as well as the tenor of relationships between EPA Administrator Carol Browner and her staff and the Office of the Environmental Secretary for Virginia, started with tough, confrontational attitudes (Dunlop, 2000). Descriptions from Nakamura and Church (2003), who followed the development of the CERCLA (Comprehensive Environmental Response, Compensation and Liability Act; Superfund) law for more than 10 years and who conclude their book by praising operations after a change, paint a stark picture of EPA's and Superfund's relationships to its clients during the early Clinton administration.

> "[Superfund] uses controversial liability principles to extract – under threats of treble damages – vast sums of money from entities ranging from corporate titans to cash-poor municipalities to mom-and-pop dry-cleaning establishments. These extractions typically have nothing to do with the current operations of these businesses or governmental units. Rather, compliance is commanded in a highly adversarial and quasi-prosecutorial forum in which the programs' targets – regardless of their previous or current actions – are treated more like criminals than responsible citizens and businesses (Nakamura & Church, 2003)."

Major changes took place in 1995. President Clinton had acknowledged in a State of the Union speech that the Superfund Act was "broken." The Republicans had taken over the House of Representatives in the 1994 elections and dressed down EPA administrators in acrimonious hearings. President Clinton announced that EPA would embrace a special "Reinventing Regulation" program. EPA program managers and enforcement staff – who previously did not communicate, now went on "road shows" together. These staffers as well as Department of Justice staff began regular communications. Instead of issuing peremptory demands, EPA employees began listening, dialoging, and negotiating with clients.

Environmentalists Knopman and Fleschner (1999), as noted earlier, felt that Reinventing Government had made mainly superficial changes with respect to the operation of the regulatory laws. But from my own experience and contacts with Superfund and other groups at EPA, as well as from participating in a recent EPA Environmental Partnership Summit in Baltimore, MD (May 19–21, 2008), there is clearly a new, more proactive outreach and helpful attitude at the agency. So perhaps both views have validity. The fundamental law structures have not changed but the attitude of the agency has moved to another level. In spite of highly publicized White House interference and controversies involving EPA during the George W. Bush administration, discussions with EPA employees at the Partnership Summit in Baltimore suggest that large parts of EPA activities were relatively unaffected.

7.3.6 The Republican Contract with America (1994)

After the House of Representatives was captured by the Republican tide in the US election of 1994, Speaker Newt Gingrich made the attention-getting announcement of the *Contract with America*, a bold and unusual statement of unified purpose by a political party. One of the key elements in the Contract was regulatory reform. Soon a flood of bills emerged in the House including many on regulatory subjects. Examples include H.R. 1080, "To authorize States to control movement of municipal solid wastes generated within their jurisdictions"; H.R. 603, "To authorize States to regulate certain solid waste"; H.R. 46, "To delay for two years [the] implementation date for enhanced vehicle inspection and maintenance under the Clean Air Act, and more", etc. Among the highest profile bills were H.R. 1022, "To provide regulatory reform and focus on greatest risks ..... " and H.R. 9, "To create jobs, enhance wages, strengthen property rights, maintain certain economic liberties, decentralize and reduce the power of the Federal Government with respect to the States, localities, and citizens of the United States, and to increase the accountability of Federal officials," cosponsored by 133 Representatives (Thomas, 1995).

Some of the goals espoused by the Contract had public appeal, and many passed the House. But the end of the Contract was as humbling as its opening announcement was confident. A single regulatory bill, *Unfunded Federal Mandates* (P.L. 104–4) became law, sponsored by then Senator Dirk Kempthorne of Idaho. In due course Speaker Gingrich himself was forced to leave Congress under a cloud. On inspection, most of the bills had failings that might seem obvious even to the nonexpert. In the more scholarly language of Robert Durant (1987), the bills' choice of subject, objective, and tactics lacked "*linkage*." Operational and tactical considerations failed to consider the views of Democratic senators who controlled the Senate and who would have to support House-passed bills before they could be approved.

7.3.7 Endangered Species Act and NEPA Reform

Chairman of the House Committee on Resources, Richard Pombo (R-CA) launched probably the most ambitious attempt at reauthorization and reform of two pillars of the environmental law pantheon since passage of the law (Zovod & SKaggs, 2005) (described in Chapter 2 and 8.4.4). Respect for the 86-page task force report was reflected in the passage of the ESA bill in the House, although both bills were ultimately defeated.

7.4 Proposals for Reform

Before listing some previous recommendations for reform, it is worthwhile pointing out that good ideas about breaking gridlock, like those of Anderson (1997) will not be enough to break through prejudice entrenched over time, suspicion, lack of communication, and other barriers. Increased trust will be needed among those who make, follow, and enforce regulations.

> Good ideas about breaking gridlock . . . . . will not be enough to break through prejudice entrenched over time, suspicion, lack of communication, and other barriers. Increased trust will be needed among those who make, follow, and enforce regulations.

I have mentioned the helpful and proactive attitude that has been cultivated by the professional civil service staff and contractor personnel at EPA. The old zero sum assumptions about economics versus environment have faded among EPA staff and an increasing number of operational level industry managers and workers. Fiorino (2006, p. 99) notes a greening trend that includes corporate goal setting and reporting. Much information from corporations is now shared voluntarily whereas in the past this would not have been the case. In 2005, 52% of *Fortune* global 250 firms issued separate corporate environmental reports. Many other features like Environmental Management Systems (EMS) are mentioned in Fiorino's book.

In my opinion, the ability of senior operational managers to speak and write candidly about the realities of regulation is a critical asset. Examples of insights that can be gained through such openness are presented in Fiorino's book (Fiorino, 2006). He reports that in the old "compliance" regulatory mode, everyone was treated the same, with poor performers being shamed or receiving penalties whereas good performers were of little interest until the next inspection. Interactions emphasized the negative, with few carrots and many sticks. Under the old system all facilities, good and bad, were subject to the same transaction costs: reporting, monitoring, record keeping, permitting – regardless of performance. States began to adopt performance tracks that recognized quality performance and might even offer reduced permitting fees for superior performance. EPA launched its own National Environmental Performance Track in 2000. Simple and logical as the relationship is – who would think of this if performance could not be discussed candidly with experts immersed in these kinds of relationships?

Remarkably, the Performance Track program was Administrator Carol Browner's final initiative at EPA! It has continued actively into the Bush administration. Environmental program continuity and predictability is standard in advanced European environmental agencies but has not been in the case in the United States. That brings us to elements of the "new environmental regulatory system" recommended by Fiorino, drawing on experience in the EPA Performance Track program and other sources.

7.4.1 Reflexive Law

Unlike other kinds of law, reflexive law "seeks to design self-regulating social systems through norms of organization and procedure." The law provides incentives and procedures that will induce people to assess their actions and adjust them to achieve socially desirable goals (Fiorino, 2006, citing Gunther Teubner).

7.4.2 Disclosure

Disclosure of performance or publication of toxic release (TRI) data, even though no mandatory action is required, creates conditions that promote change in desired directions. In fact, I mentioned earlier in discussion of the building of bridges and skyscrapers by early twentieth century engineers (Section 3.4.1) that the group identification and sense of responsibility promoted when quality-demanding work had high visibility yielded close to 100% performance. In contrast, a penalty-based system can never get below a certain percentage of failures.

7.4.3 Sociopolitical Governance

It would be unfeasible to efficiently operate a complex and diverse society like that in the EU in the kind of adversarial, distrustful environment that characterizes key parts of American society (Fiorino, 2006, p. 161) (see also Section 8.8, Guerrilla Warfare). While the United States is a volatile, ebullient, and competitive society, it was not adversarial in the past (see Chapter 3). EU techniques and insights can help us move toward cooperation that will be critical to meet the challenges of greening energy supply. Active cooperation by all will no longer be a luxury, but a necessity.

> It would be unfeasible to efficiently operate a complex, diverse, and dynamic society like that in the EU in the kind of adversarial, distrustful environment that characterizes key parts of American society (Fiorino, 2006, p. 161). EU techniques and insights can help us move toward cooperation that will be critical to meet the challenges of greening energy supply. *Active cooperation by all will no longer be a luxury, but a necessity.*

7.4.4 Innovation by States

During the 1990s, increasing impatience by states because of gridlock in the federal government led to agreements for joint innovative programs. To fully exploit states' capacity for innovation and independent decision making, changes will be needed in constraining federal laws; "second-generation laws proposed in the 1990s would have given EPA authority to make exceptions from statutes if regulated firms achieved better results (Fiorino, 2006, p. 209; Yandle, 1997)."

7.4.5   Other Features of the "New Regulation"

Adversarial relationships are replaced by communication; blame is replaced by seeking solutions. Decentralized partnerships recognize differentiated operations; ecosystem or ecological management is preferred over regulating through isolated parameters. Government maintains normative standards and applies pressure toward national goals, but avoids stipulating methods by which they are to be reached. Technologies and standards are not uniform above minimum standards, and innovation is continuous. Networks and colearning opportunities are fostered. Building mutual trust is a central objective. Integrated management can extend across media where advantages are gained.

7.4.6   From "Command and Control" to "Command and Covenant"

Evolution from a rigid, hierarchical system to a more flexible (covenant) system was proposed by Natalia Mirovitskaya and William Ascher (Mirovitskaya & Ascher, 2001). This seems similar to the "dual" or "competing" systems named by Powers and Chertow (1997), or to the XL concept adopted by EPA and used successfully in a major project with EXXON (funding for XL has been discontinued in the second Bush administration, according to EPA staff consulted in 2007). In other words, simultaneous options for a less flexible and a more flexible option exist. Though perhaps applied selectively at the outset, the goal would be for the entire system to move to the more flexible and efficient system.

7.5   Why Major Reforms of the 1970s Environmental Regulatory System Failed – and Lawmakers Are Deterred from Attempting Reform

The pollutant regulatory system of the 1970s was ingeniously designed for maximum political effectiveness (described in Chapter 4). Knopman and Fleschner (1999) pointed out that "uniform laws and standards, plus public information are the building blocks for lawsuits." In other words the combination of features of the laws based on specific pollutants form a system that allows even lay persons with minimal scientific training to observe failure to comply with standards, and with the assistance of a lawyer challenge a polluter or a lax enforcer in civil court.

NEPA's EIS provision (aided by the Administrative Procedures Act) added a powerful legal tool, unintended by NEPA's framers, to challenge new proposed activities before they get started. Provision for reimbursement of reasonable legal fees for successful suits reduces cost of litigation. The results of a regulatory system heavily

loaded in favor of private litigation as an enforcing mechanism are illustrated in
the *Friends of the Milwaukee River* case example (Section 8.4.2). A graduate of
the Northeastern University environmental law program, with volunteers, successfully challenged the Milwaukee Metropolitan Sewerage District and EPA for
"25 years of violations of the *Clean Water Act* by dumping raw sewage into the
river." A long string of newspaper articles in the *Milwaukee Sun-Times* portrayed
the lawyer, Karen Schapiro, as a hero and the agencies in question as ineffective
bureaucracies failing in their duties. As detailed in Section 8.4.2, my analysis of
this case reveals an example of the social, environmental, and economic damage
that disproportional weighting of environmental law to favor private enforcement through the *civil* courts has can do.

The "David vs Goliath" case was built on a limited understanding of waste water
treatment systems in mature cities. The "combined sewer overflow system" did
not have the capacity to retain all the water produced in extended rainstorms without disproportionate costs. It was also impacted by unanticipated growth of satellite
communities. In leading European nations like The Netherlands the complex issues
involved would have been examined – and solved – in appropriately wider societal
contexts with knowledgeable professionals and other sectors of society.

In the United States this is difficult because of compartmentalization of environmental management mandated by existing laws, regardless of the problems or
the best way to approach them. The CAA, CWA, and NEPA limit new conceptual
approaches to problems and organization. They govern even EPA's overall operating structure (divisions corresponding to water, air, enforcement, etc.). An ironic
nuance in this case is that Wisconsin (and the Wisconsin Department of Natural
Resources, involved in assisting the Milwaukee sewage treatment authorities) has
long been among national leaders in public health and wastewater treatment. It was
surely demoralizing for dedicated managers and engineers to see their efforts mis-
leadingly characterized by newspaper articles displaying only superficial grasp of
the issues involved.

The existing laws were brilliantly designed for political effectiveness, but they
violate criteria for good law and policy, and are ill suited for scientific effectiveness
(see Chapter 4). At the time of their framing that was not seen as important, for the
task at hand was to stop rampant pollution and societal obliviousness. "Industry"
that had only recently sent a man to the moon and brought him back, was widely
assumed to have unlimited power and resilience.

Any major change toward achieving flexibility and discretion desirable from a
scientific and efficiency point of view will potentially weaken a system that gives
citizen groups powerful legal levers. Herein lies the dilemma. Why should or would
the nation's 10- to 15-million-member strong environmental NGOs, the residents
of attractive environments, or progressively minded politicians agree to give up
existing, formidable powers to rely on the uncertain outcome of systems in which
politicians or powerful economic/social interests could exercise undue influence?
Are there not plenty of examples today of the risks that could entail?

Let us take a practical example. Let us take a proposal to integrate a city's environmental management across different media – air, water, toxic waste disposal, etc.

Projected money savings would be partially invested in new efforts for ecological management, including peripheral areas and waterways. This might be optimal from a scientific, environmental, and cost point of view. But loss of clear-cut emission or discharge standards and procedural checkpoints that now serve as simple indices for pollution and operational status would complicate local public interest organizations' or friendly national NGOs' ability to build a court case. Moreover, litigants would now have to have greater technical background to fully understand the principles, goals, and norms for the more complex management issues. Without more in-depth scientific background, it would be much harder to judge whether or not the new system might be simply a sophisticated device by the city and financial interests to put cost savings at the top of the priority list.

There is another problem that has headed off and is likely to defeat major reform efforts of the valuable kinds referred to in the previous section. Organizations with the greatest interest in simplification or relaxation of the laws, for example, industrial or commercial companies financially affected by regulations and their political supporters, have historically offered proposals designed primarily to weaken application of the laws. They did not show concern for environmental effectiveness. So environmentalists are and will probably continue to be justifiably suspicious of *any* reform effort.

There are only a few lawmakers in the Senate and House of Representatives with extensive scientific and environmental regulatory background knowledge. They, more than others, are likely to understand the suspicion and potential for opposition likely to be raised by even the most carefully drafted reform. With prospects for success so meager – most will probably conclude they have better things to do than tilt at windmills. So – a powerful system designed 35 years ago on assumptions of permanent and unavoidable divisions of interests in society continues to promote confrontation, lack of communication and cooperation, inefficiency, and worse.

7.6 Facing the Music

If we accept the reality in these relationships, isn't doing the best we can with the existing system, trying to improve performance around the edges so citizen watchdogs will have less reason to be suspicious and litigate the best we can do? This has been the pragmatic route for improvements in administrative operations and communications between EPA and corporations noted by Fiorino (2006).

But Fiorino and others make it clear that deeper and broader reforms are necessary and unavoidable. The existing system has led to mounting problems under the surface in the United States. One does not need to be an engineer or economic expert to recognize that the acknowledged deterioration of the United States' infrastructure means that something has been going wrong. $1.6+ trillion has been estimated as the cost to bring it into proper condition (American Society of Civil Engineers 2008). Nor does one need to be an environmental expert to recognize that a rating of US environmental performance of 28th–39th in the world suggests serious problems.

One does not need financial expertise to understand that record foreign exchange deficits and an unprecedented decline in the value of the US dollar against the euro (1.5:1.0) are not signs of the nation's fiscal strength or stability. Gasoline at $4.00/gallon for the first time really scares people and threatens the solvency of airlines, trucking companies, and other businesses. These signals remind us of how critical reliable and affordable energy are in high-GDP nations.

The returns on escalating transfer of US industrial operations and investments abroad may have temporarily sustained GDP, selected company profits, and stock market indexes. But how does hollowing out our technological, financial, and human resources square with the enormous challenges involved in proposed green transformation of our energy system? The *Lieberman-Warner climate security plan*[195] for reducing greenhouse gas emissions was recently defeated. However, we can use it as approximating potential future energy plans, since it resembles Obama and McCain energy policies released during the primary campaigns.

General Accounting Office (GAO) figures (General Accounting Office, 2004) show systematic declines in US productivity. These declines are understandable given the huge diversion of resources to subsidies for new technology development and compensations to population groups and affected national sectors. The GAO models assume gasoline prices of only $2.61 *in 2020*, whereas costs of fuel are already (in 2008) at $4 /gallon. The bold California plan to cut emissions (25% cut by 2020) created excitement and contains advanced and innovative concepts. But the reckoning was recently revealed in an unexpectedly high state deficit of $16 billion (Davis, 2008). Unlike the federal government, which can cover budget deficits by paper transactions, states cannot accumulate deficits, so California is a forerunner for and may be a predictor of realities that will be associated with new energy policies.

American politicians favor enthusiastic claims and visions, assignments of responsibility for problems, and simple fixes. ".... the U.S. economy is fundamentally robust, sound, etc.," ".....Bush's failed policies ...." "The current fiscal downturn is temporary," "Our foreign exchange problems are due to unfair exploitation of foreign workers" or "Chinese policies in maintaining a cheap Yuan ...." "The new green economy will produce millions of green collar jobs" .... "If the oil companies would only invest in renewable technologies we could be energy independent," ".... environmental fanatics are destroying the country." Colorful controversy is cultivated as an entertainment form. (See more on "green jobs" in Section 8.9)

But the reality is that in the past 40 years American political, economic, environmental, and also scientific and academic leaders have tended toward simplistic, short-range, or ideological approaches to problems, avoiding critical examination of long-term impacts or consequences. A division of society into camps like "environmental," "industry," and "don't know" makes intelligent solutions almost impossible. Could we get creative compromises between firemen and arsonists? Our current conflict over energy is much like that duality – with the each side in the debate assuming that they represent the firefighters and the other the arsonists.

> A division of society into camps like "environmental," "industry," and "don't
> know" makes intelligent solutions almost impossible. Could we get creative
> compromises between firemen and arsonists?

People are reluctant to address the problem of *Congress* in more than stereotypical ways because in one sense it represents the highest evocation and hope of our democratic system. Individual Congressional delegates do serve an almost universally appreciated and valuable function as ombudsman to government for their constituents. They are often brilliant, charismatic, and wonderfully articulate as individuals (they have to be to get elected). They may be much better informed and knowledgeable than people give them credit for – as not a few witnesses at hearings have discovered to their discomfiture. But the rating of Congress as a whole, Republican or Democratic, is abysmal – at an all time low. And this reflects dysfunction and reality, not just trivial disaffection. It's impossible not to conclude that the serendipitous whims and initiatives by Congress have, far more often than not, been unhelpful or damaging well back before World War II. But the precedents set in the 1970s environmental laws breached a wall. Former constraints against Congressional interference with the discretion of federal scientific and professional agencies to plan, develop, and adapt policies to emerging circumstances were abandoned. With the floodgates completely open, Congress was free to intervene as a micromanager in any area of society, large or small (like directing what labeling shall be placed on disposable batteries), where it had no inherent expertise or should have deferred to a responsible professional agency. Any temporary enthusiasm, concern about "doing something," or deeply held conviction that got enough votes could become law of the land.

Consider the 500 or more independently prepared, uncoordinated or trivial bills about energy policy that have been introduced into the Senate and House of Representatives during the 110th Congress. Although the standard report issued with bills approved by the standing committees gives descriptive information and history of the bills, and the Congressional Budget Office estimates their costs, beyond this there is no systematic procedure to coordinate bills or assess their short and long-term impacts and merit in the opinion of various stakeholder groups. In short, there is no process to lay out an impartial preliminary assessment of bills proposed to become law of the land, as is done in advanced EU nations. People may dismiss criticism of the prerogatives of Congress to run things the way it wishes. But given the stakes involved, it is hard to avoid the idea that such a haphazard way to make law of the land is an anachronism for an advanced nation of 300 million at a critical time.

7.7   Where Do We Go from Here?

I am not egotistical enough to assume that ideas like those described in Sections 7.4–7.6 will get immediate and respectful attention. What I do suggest is that near term Congressional reform efforts would not be wise or practical.

The divisions and problems are too great. However, the experience of the EU and the thoughtful ideas and knowledge of many insightful experts and observers in the United States do offer building blocks for consideration.

Meaningful cooperative or compromise solutions are not likely to be feasible until greater trust is established. A major step forward in building trust is to begin to depoliticize scientific and environmental policy.

The EPA in particular is a major asset in its repository of experience, staff, research, and connections to the industrial framework of America, as well as to academic and intellectual communities. It can be both an independent initiator of constructive ideas and a coordinator and mediator for communications between various groups in society.

- To depoliticize EPA, which is already an independent agency and thereby not part of the Executive Branch, it would only be necessary to give the EPA administrator a term of office long enough to bridge administrations – and the mission to steer a professional rather than a partisan course. The appointment system should include a bipartisan, professionally guided approach to selecting a candidate list. Perhaps the Comptroller General of the United States (head of the independent GAO) whose term of office is 15 years would be a good model. A nontrivial problem is the cost of living in the Washington, DC area – and competing salaries for an individual with the critical qualities needed in this job. Special provision for living costs, housing, and other ordinary expenses ought to be provided, if feasible, so that the salary becomes more discretionary. The same would apply to other leadership positions in critical federal agencies in which depoliticization is desired.
- A task of the new EPA Administrator should be to develop a unit within EPA to coordinate discussion and deliberations about reform and modernization of the United States' environmental – regulatory system, with full consultation from all stakeholders in the nation, and with vigorous contacts with the EU and other advanced nations.
- Regarding energy policy, with the consent of Congress, a high-level committee chaired by the EPA Administrator together with the leaders of scientific, engineering, or professional bureaus within relevant federal departments (DOE, Commerce (NOAA), Interior (USGS, etc.), Agriculture, Transportation) could develop policies for consideration by Congress.

In short, what seems most important is to begin more careful, thoughtful deliberation about the way the US plans its future, uses its resources, and solves problems. If those choices are solid and well prepared, they will gain public confidence, build trust, and begin to unlock the full energies and capabilities of a society with greater innovative capacity than any other. If greater trust is established, communication, knowledge, and desire to promote greater cooperation and efficiency may then facilitate a variety of more specific approaches to law and regulatory reform. (See also Harisalo & Stenvall, 2004).

Part II
Cases, Documentation, and Policy Analysis

Chapter 8
Case Studies and Examples

8.1 The Rise of German Science: Lessons Forgotten in US Science Policy After World War II

The rise of Germany as a major national power in the nineteenth century is a classic chapter in the use of science in national development. In comparison with Britain and France, which had become wealthy through their colonies and international trade, Germany in the early nineteenth century was fragmented into many principalities, impoverished, and backward. After the 1850s, Germany became the first country in the world to rise to the front rank of nations through scientific research and applications, creating the modern research university in the process. How it did so is relevant to environmental policy and effective use of science in US national life today.

In the eighteenth and early nineteenth century scientific research tended to be published in media that carried royal imprimatur and patronage, reflected in publications like the *Proceedings of the Royal Society*. The French Académie des Sciences was given its name in 1699 by King Louis XIV. Upon being reconstituted after the French Revolution in 1816, the Royal Academy, including France's premier scientific publication, *Comptes Rendus de l'Académie des Sciences* became part of the Institute of France, whose patron was the head of state.

In Great Britain and France leading scientists might come from backgrounds ranging from humble: Michael Faraday (son of a blacksmith), Humphrey Davy (son of a wood carver), Louis Pasteur (son of a tanner); to upper middle class or aristocratic: Henry Cavendish (eldest son of Lord Charles Cavendish), William Thomson (Baron Kelvin), and Marie Françoise Xavier Bichat (son of a French physician). Whatever their origin, British and French scientists appear to have absorbed something of the flair and romance of discovery that characterized premier scientific media in their countries.

In Germany most leading scientists came from middle-class backgrounds. They often displayed a dual motivation to excel at their work, and to contribute to the welfare of their nation. These tendencies, which were termed "The Protestant Work Ethic" by Max Weber (1904) had roots in the Reformation teachings of Martin Luther (1481–1546). Luther repeatedly emphasized the worth of all diligent work, as well as the duty to care for one's neighbor.[196]

From the beginning, German publication, especially in chemistry, tended to focus on practical documentation. *Chemisches Zentralblatt* became a comprehensive reference on chemical developments by 1830 (von Meyer, 1895). Major compendia began in 1837 with *Gmelins Handbuch der anorganische Chemie* later followed by the famous *Beilsteins Handbuch der organischen Chemie* (1887) and the *Landolt-Börnstein, Physikalisch-chemische Tabellen.*

Justus von Liebig (1803–1873), who became known as the father of soil and plant chemistry, came from a middle-class background. The aristocratic "von" was added when he was awarded a baronage by the King of Bavaria in 1845. Von Liebig not only made fundamental contributions to agricultural biology and chemistry. He founded a chemical journal, *Annalen der Chemie* (now *Liebigs Annalen der Chemie*) in 1832, and became a famous teacher who encouraged the dissemination of practical knowledge. Liebig even gave lectures to farmers on scientific farming and use of fertilizer. He also took interest in John Stuart Mill's *Logic*, arranging for its translation into German "in part because Mill promoted science as a means to social and political progress." Von Liebig also founded a company for producing meat extracts, which later trademarked the Oxo brand of bouillon cubes (Wikipedia, 2007).

In 1828 Friedrich Wöhler, a German student of the Swedish chemist, J.J. Berzelius, synthesized the first organic compound, urea. This demolished the "vitalism" theory that living organisms were required to make organic compounds, and opened the way to broader synthesis of organic compounds. A student of Justus von Liebig, August Wilhelm von Hofmann (1818–1892), laid the theoretical framework for the development of aniline dyes. In order to synthesize useful new organic and inorganic compounds, German chemists led in developing the new sciences of chemical thermodynamics and catalysis. Thermodynamics allowed prediction of temperature, pressure and other conditions favoring formation of new compounds, while catalysis involved surface-active compounds that dramatically speeded up chemical reactions.

Although William Henry Perkin, an English student of Hofmann, built a factory in England, and was the first to produce aniline purple commercially, cooperation between academic chemists and manufacturers helped Germany dominate the dye industry for nearly a hundred years (Murmann, 2003). Synthesis of organic compounds with medically useful properties became the basis of a new pharmaceutical industry, another giant field pioneered by German scientists. DDT (dichlorodiphenyltrichloroethane) was one of tens of thousands of new organic chemical compounds synthesized in the later nineteenth century in German laboratories. It would play a major role in post-World War II America, first as a "human-friendly, miracle pesticide." Subsequently, its misuse was attacked as an ecological hazard by Rachel Carson (1962).

German research developments, in which theory, applications, and systematized knowledge continuously interacted, helped Germany become the foremost nation in optics, microscopy, photography, and many other areas. New metal alloys were developed. The interaction between basic and applied science was facilitated by the fact that many famous German chemists were as comfortable with managers of factories as they were in the laboratory or the university classroom. The recruitment and education of gifted scientists was furthered by the positive image of science, the availability of good institutions of higher education, and demand for scientifically trained employees in industrial companies.

Beginning in the 1870s, Germany's polytechnic institutes were expanded along with the nation's network of social services by Otto von Bismarck. Bismarck, the "Iron Chancellor" who unified Germany and defeated France in the Franco Prussian War of 1870, was a brilliant and insightful leader. Fatefully, however, Bismarck's interests did not extend to developing greater participation of the population in representational forms of government. Thus, there were fewer political constraints on Kaiser Wilhelm I than on leaders of other major nations in the years leading up to World War I.

Germany's scientific prowess enabled it to overcome the effects of the Allied blockade during World War I in cutting off critical supplies of Chilean nitrate. Nobel prizewinning chemists, Fritz Haber and Wilhelm Ostwald, achieved the scaling up of nitrate production from electrical discharge in air in little more than a year and a half, supplying Germany all of its needs for gunpowder, explosives, and fertilizer. During World War II, Germany's scientific and technical capabilities remained formidable. Germany overcame loss or shortages of supplies of natural rubber and petroleum by synthesis from coal.

The purpose of highlighting German scientific and technical achievements is not to glorify Germany or scientific achievement for its own sake. The important issue for this book is that is that many theoretical breakthroughs achieved by German science emerged from scientists' search for new *practical* applications under conditions of active communication between academic research scientists and industrial organizations. This close-knit relationship benefited both partners, leading German firms like Bayer and I.G. Farben to lobby the German government to support construction and maintenance of first-rate educational institutions (Murmann, 2003).

In short, the processes that gave rise to Germany's achievements and industrial competitiveness in spite of disadvantage in access to raw materials and markets are opposite to central assumptions of Vannevar Bush and other leaders who forged a new US science policy after World War II.[197] They have closer affinities to models for centers of innovation recognized since the 1980s as "clusters" and networks where interactions of ever-increasing complexity take place (Rycroft & Kash, 1999).

8.2 Whatever Happened to the Blue Revolution?

In 1987 Dr. Judy Kildow, a science policy expert at Massachusetts Institute of Technology (MIT),[198] addressed a symposium on ocean policy at the Woods Hole Oceanographic Institution (WHOI) with a presentation called, "Whatever happened to the Blue Revolution?" "Blue Revolution" referred to a nearly universal expectation in the early 1950s of a bright future for America's ocean frontier. New equipment and biological knowledge would enhance fisheries and marine aquaculture. The

American offshore oil industry would find new petroleum discoveries off our coasts. The ingenuity that launched Liberty ships and aircraft carriers would advance the US merchant fleet. Engineers and planners visualized creative industrial and recreational use of artificial islands constructed off the US coast where dense populations limited developments on land. Robert Frosch, in a Seward Johnson Lecture at the Woods Hole Oceanographic Institution, explored unconventional values of the ocean – like flatness. This lent itself to airports adjoining the ocean in coastal cities – while it minimized noise impacts on populations. Coastal and ocean research would enhance knowledge that could aid in wise stewardship over our coastal areas and marine life.

During a leave of absence in 1988 and 1989, I pursued Judy Kildow's question in a research project with graduate students and faculty at the State University of New York (SUNY) Stonybrook, New York. In a message to Congress in March 1961, President John F. Kennedy had endorsed the National Science Foundation's Decade of Ocean Exploration, with the expectation that scientific exploration would usher in expanded ocean industry as well as environmental knowledge. Subsequently, US-led research on the world oceans indeed made many striking discoveries. From 1956 to 1975, ocean research centers increased from 16 to 134 in the United States (based on NSF data). Scientific research on plate tectonics helped explain earthquakes along the California coast, and hot submarine fluid discharge in active zones of the Pacific and Atlantic oceans. Knowledge of ocean geology, geophysics, biology and chemistry burgeoned. Some of the accumulated data is now playing a significant role in modeling the role of the oceans in global climate change.

However, in practical applications to national life the results were deeply disappointing. The competitive American ship building industry disappeared, except for military and other uses mandating US bottoms. The merchant fleet dwindled to the level of Yugoslavia's. The United States lagged far behind other nations with respect to deep ports for supertankers and cargo vessels – in spite of the fact that the United States had the world's largest international trade, 95% conducted by sea. US fisheries suffered steep declines. Marine aquaculture never had more than fitful, temporary success in spite of decades of liberal research funding. With a few exceptions such as Baltimore's waste island, even inquiries or proposals to initiate innovative offshore construction got nowhere.

The US offshore oil industry was a major technical and economic success – but crisis followed the Santa Barbara offshore oil spill. Ultimately 85% of the US Exclusive Economic Zone off the coasts of the lower 48 states came under moratorium for exploration and oil and gas or minerals production. The Coastal Zone Management Act and other legislation failed to halt loss of most of the nation's natural oyster beds, and environmental problems in Chesapeake Bay impacted the revered crab fishery. Unlike policy in most advanced nations, private shorefront ownership came to lock the general public out from more than a small fraction of the nation's shores, and excluded appropriate economic and social uses of ocean access. Only the US Navy and other military needs for ocean access were met.

What happened to the earlier expectations? How could so many things go wrong? A simplistic answer is neglect and preoccupation. Like a home, a nation has many components. If plumbing leaks are ignored water may cause rotting of wood supports; if utilities are not paid there will be no heat or lighting. A home and nation need balanced attention.

8.2.1 Postwar US Aquaculture History: Fish and Politics Don't Mix

Everything seemed ready for a blossoming of US aquaculture after World War II: the nation had a vast coastline ranging from tropical ocean to Arctic waters. It had scientific expertise in marine and biological research institutions that was liberally supported by grant funds to perform research on aquaculture-related problems for nearly 25 years. There was a broad demand and market for seafood – especially in tourist areas. Where wild species like oyster had dwindled, aquaculture offered a valuable supplementary industry that provided winter as well as summer employment to local residents. However, even the writing of special aquaculture statutes by Florida and Massachusetts failed to achieve more than isolated and short-lived results. Since the 1970s mariculture, or marine aquaculture, has achieved no more than low levels of activity, whereas freshwater aquaculture in interior areas of the country (catfish, trout, and tilapia) has enjoyed robust growth. Catfish farming grew by 500% from 1984 to 1999 (Harvey, 2000).

The big reason for the failure of mariculture is clear, though often obscured by details. Terraculture (land farming) has always enjoyed special status and protections, and states like South and North Carolina are similarly supportive of catfish farming enterprises privately and through state agencies. For example, North Carolina recently arranged to obtain Department of Agriculture grants to offset drought conditions for catfish raisers (Parker, 2008). In contrast, mariculture (sea farming) in the United States has been treated as an ordinary business – free to sink or swim in a competitive world. A National Research Council report in 1992 offered an assessment typical of many formal scholarly or governmental publications.

The writers of the report appear to recognize the realities but are satisfied to conclude the report in upbeat officialese, leaving it for "someone else" to develop specific real-world assessments and recommendations for needed changes:

> "The prospects of this emerging enterprise are for healthy and vigorous growth, given a fair share of support for the development of an advanced scientific and engineering base, along with a reasonable and predictable regulatory framework. On this basis, the environmental problems that presently constrain marine aquaculture are likely to be resolved so that it can contribute to the continued vitality of the nation's living marine resources (National Academy of Sciences/National Research Council, 1992)."

The plain facts are that competition at the ocean interface offers poor odds for the raising of marine species. Mariculture faces shore owners, yachtsmen, wild fishermen, boating and public recreation, citizens who for one reason or another raise objections, and local or regional politicians who lack relevant scientific background or knowledge. If operations reach viable commercial levels, regulatory problems offer new potential challenges. In Maine, a truckload of pan-size farmed salmon, whose development was aided by one state agency, was confiscated by another state agency for being under size. Small mariculture operators operating under everybody's eyes may be faced with requirements to deal with real or exaggerated concerns about environmental issues On the other hand, a shellfish operation in Virginia was ruined by outfall from an improperly operating municipal sewage treatment plant but got no help to cover losses. An early attempt at shrimp farming in the mid-Florida West Coast in the 1970s had just reached the harvest stage when scuba-diving vandals cut shrimp net enclosures, ruining years of planning and development – with no help or recourse.

Salmon farming, first perfected in Norway, has been banned in Alaska and there is only a small industry in Washington State. Most media and scientific reports in the United States focus on the potential problems of salmon growing, like lice infestations, and escape of farmed salmon to interbreed with wild species. Meanwhile, farmed salmon imports to the United States dominate supplies.

Although many factors combine to impede development of mariculture, one should not underrate the importance of the federal environmental regulations. Through their emphasis on the primary importance of environment, excluding all other values, they set up a legal and psychological framework in which the raising of commercial marine species is difficult. In sharp contrast, intensive culture of mussels in European coastal waters produces 50% of the world supply (Smaal, 2002). An additional benefit of this production is that filter-feeding mussels remove organic detritus from the water, helping keep waters clean and clear. Similar cultivation of filter-feeding oysters in Chesapeake Bay could have helped reduce the excess organic matter that has clogged bottom sediments and degraded the environment. Oyster cultivation has increased in recent years, but in order to make possible bold initiatives of meaningful scale, fundamental changes are needed. The changes involve reestablishment of public trust, depoliticization of regulatory policies, and willingness of coastal communities to accept activity balance that includes commercial cultivation of marine fish species.

8.2.2   Effect of US Policies on Innovative Approaches to Waste Management and Inhibition of Creative Enterprise

During my work in Woods Hole, I observed examples of the effect of the new regulatory regime on creative approaches to waste and coastal management. In the Falmouth-Woods Hole area (MA), experiments with spray irrigation of secondary-treated sewage effluent at the nearby Otis Air Force Base were discontinued when plant products from sprayed areas could no longer be marketed owing to regulatory problems or uncertainties.

A pioneering and highly promising project involving sewage effluent-fed aquaculture was developed by a leading biological oceanographer at the Hole Oceanographic Institution, John Ryther (Ryther, Dunstan, et al., 1975). The system was able to purify wastewater to levels cleaner than the product of conventional filtration methods. The project was later accepted for funding by the NSF (National Science Foundation) RANN (Research Applied to National Needs) applied research program, but numerous problems emerged. On the regulatory side the project required some ten different permits, even though it was experimental and involved no sale or disposal of seafood products. Whereas NSF treated its regular research awardees like "gentlemen and gentleladies," RANN managers handled scientists leading their projects as though they were contractors. So much paperwork was required (e.g., quarterly progress reports) that Dr. Ryther asked for – and received $20,000 additional funding just to handle the administrative tasks! Ultimately, problems with program management caused Ryther to voluntarily withdraw from the program.[199]

In 1988 an incident made the local newspapers around Long Island. Four trash bags of waste had been dumped into the coast off Long Island and were found on the beach. Some of the contents were medical waste, which on investigation, contained syringes that were contaminated with the AIDS virus – at that time especially feared. A local Congressman rushed a new law through Congress, the Ocean Dumping Ban Act of 1988, which permanently prohibited disposal of all sewage sludge or industrial waste into the ocean. The consequences would become costly for New York City, which had few other reasonable alternatives to disposing of municipal sludge well offshore into deepwater areas. This new law was almost uniformly regarded by policy experts as unwise overreaction, foreclosing potentially valid and useful disposal options for the foreseeable future around the entire United States because of a few bags containing medical waste (which the new law would have had little effect on in any event).

New York harbor was also the site of decades of fruitless attempts to gain approval for construction of a waste island that would have solved the perennial problem of disposal of dredged spoil, as well as ultimately creating valuable usable space.

A recurring issue is opposition by local Congressional delegates in the New Jersey area and local environmental groups to lease sales for offshore sand and gravel in federal waters. Such offshore supply would provide much-needed aggregate as supplies for roads and building construction in the greater metropolitan area. At times suppliers have gone as far as Indiana to obtain these materials. An irony is that MMS (Minerals Management Service) can authorize use of the offshore sand by federal agencies such as the Corps of Engineers to replenish the beaches of affluent shorefront owners – but the offshore sand cannot be used to build hospitals or roads for poorer people in the Metropolitan area. In contrast, offshore sand and gravel has been mined around the United Kingdom since 1948. This work is done in conjunction with cultural and fisheries agencies (DEFRA, 2007).

8.3 US Geological Survey

The US Geological Survey (USGS), founded in 1879, was among the United States' first four federal scientific agencies and was the first bureau-level agency. Its early achievements in research and service related to earth and natural resources have been regarded by thoughtful observers as among the most outstanding of any scientific organization in US history, public or private. USGS history is not without times of problematic scientific and administrative policies. Two of these periods led to major layoffs (reductions in force – RIFS). The period leading up to the most recent traumatic RIF has some features in common with those of a RIF a hundred years earlier. The diverse experience and activities of the agency, availability of detailed historical studies and my own 35 years of service in USGS led to choice of USGS for an historical vignette focusing on science and operational policy in a major federal agency.

8.3.1 Early History

After the Civil War, Congress shifted attention to the need for better understanding of the physical setting and vast natural resources of the reunited nation. The late USGS historian, Mary Rabbitt wrote: "On March 2, 1867, Congress for the first time authorized western explorations in which geology would be the principal objective. The leader of that survey, Clarence King, remarked that 'Eighteen sixty-seven marks a turning point, when the science ceased to be dragged in the dust of rapid exploration and took a commanding position in the professional work of the country' (Rabbitt, 1989)."[200]

USGS's founding came during a fortunate time window. President Rutherford B. Hayes (1877–1881) was elected in one of the closest and most angrily contested elections in US history.[201] The results of the election set in motion correctives for some of the rampant governmental corruption of the "Gilded Age." Hayes had as one of his three major goals the reform of the United States' civil service.[202] To free himself from the political distraction and conflicts of reelection politics, Hayes announced ahead of time that he would limit himself to one term. Hayes' choice for Secretary of the Interior was Carl Schurz, who had emigrated from Germany in the turbulent aftermath of the 1848 revolutions in Europe. He was a passionate progressive and advocate of civil service reform.[203]

Replacing three ongoing Western surveys under the leadership of F.V. Hayden, Lt. George Wheeler of the USACE, and John Wesley Powell, the USGS was created in 1879 as a last-minute compromise between the Democratic House and the Republican Senate and their choices of competing surveys. Powell had lobbied for creation of the USGS and therefore withdrew his name from consideration as leader of the new organization. Secretary Schurz chose as the Director for the new organization Clarence King, the leader of the 1867–79 survey for the Army engineers (Yochelson, 2006). The judgment with which King chose the initial employees of the USGS is partly indicated by the fact that though his objectives were practical, 9 of the 30 initial scientific and engineering appointees were elected to the National Academy of Sciences.[204]

Clarence King and J.W. Powell, who succeeded King as Director of the US Geological Survey, were both first-rate scientists, though their administrative policies were very different (Rabbitt, 1980). King gave the new Geological Survey a mission orientation, planned the goals, but gave his staff freedom to choose methods in order to achieve the goals. King resigned as Director in March 1881. The early separation was partly based on his cordial relationship with Interior Secretary Schurz, who refused to serve with James Garfield, elected President in 1880 (Rabbitt, 1989).[205]

Powell, a partly self-taught geographer who had a wide range of interests, gained large increases in staff and funding from Congress for the ambitious project of initiating geological mapping of the United States, undergirded by comprehensive topographic mapping. Like King, he gave his staff freedom to choose working methods. However, he also gave his scientists and engineers wide latitude to choose their projects (a practice that reemerged after World War II). Powell insisted that that basic geologic and paleontologic science precede more applied work, a goal that had questionable practicality, given expectations and needs for services, and USGS capabilities at this early time. He could be abrasive and tended to ignore the concerns of Congress and constituents when they conflicted with personal goals. He created irritation in Congress by lobbying for increased financial support. When he shifted effort from missions such as mining geology (e.g., mapping mineral resources in support of mining), and the Irrigation Survey, to topographic mapping and surface hydrology, and the results of his policies decreased service and delayed publication of reports to constituencies, he incurred the displeasure of powerful Congressmen. In 1892 Congress and President Benjamin Harrison drastically cut the USGS budget, forcing Powell to initiate a major reduction in force (RIF) in 1892. Although Congress could not fire Powell, it slashed his salary and ultimately achieved his resignation in 1894. For a time, the fate of USGS as an agency hung in the balance.

Powell's holistic view of the interrelationships between nature and people (he was the first to create ethnological maps of Indian cultures in North America), and philosophies like advocacy of public ownership of water supplies, have led contemporary writers like Wallace Stegner, Powell biographer Donald Worster (2001), and environmentally motivated politicians, including Bruce Babbitt (Secretary of the Interior under President Bill Clinton), to regard Powell as a visionary scientist and leader, whose ideas were ahead of their time. However, evaluation of Powell's leadership by USGS historians Mary Rabbitt and Clifford Nelson point to major failings as an administrator: "he had opportunities, made choices, received repeated warnings, refused to change, and nearly brought the agency down with him (Nelson, 2008)."

Powell was replaced by Charles Doolittle Walcott, who, as Geologist-in-Charge of Geology and Paleontology, had already assumed some of the Director's duties before being formally approved in 1894. Walcott, who became renowned as a paleontologist and stratigrapher, was elected while Director to the US Academy of Science. He served as President of the committee that founded the Carnegie Institution of Washington, subsequently becoming Secretary. Walcott added to his duties as USGS Director the Directorship of the Reclamation Service in 1902. After appointment as Secretary of the Smithsonian Institution of Washington Walcott resigned from USGS. He later served as President of the American Association for the Advancement of Science (1923) (Yochelson, 2006).

Of greatest importance for USGS were Walcott's administrative skills and alertness to scientific needs. USGS achievements during Walcott's 13-year tenure as Director, according to historian Rabbitt (1989), include fundamental studies in the genesis of ore deposits, paleontology and stratigraphy, glacial geology, petrography, stream flow measurements, volcanic and geyser action, rock composition, and structure. The geologic time scale was revised, definitions for rock classes were developed, and the first geologic folios were published. A Bulletin marking the 25th anniversary in 1904 noted that "USGS had grown from an organization of 38 employees at the end of its first year to one with 491 employees [and another 187 in the adjunct Reclamation Service] in 1904. Its first appropriation had been $106,000; the appropriation for the fiscal year that ended on June 30, 1904, was $1.4 million." Among achievements were preparation of topographic maps of 929,850 square miles of the United States (26% of the country including Alaska) published as 1,327 atlas sheets; geologic mapping of 171,000 square miles and publication of 106 geologic folios; studies of irrigable lands and detailed examination and classification of 110,000 square miles of the forest reserves.

USGS hydrologists supported use of water supplies for growing communities by gaging stream flow, testing pumping and other equipment, and publishing guides to operational equipment and techniques. In 1902, the Division of Hydro-Economics was added in order to assess the quality of water supplies and publish standards and methodologies for analyzing water.[206]

Besides meeting commitments to Congress and the Survey's regional constituencies, Walcott had USGS staff take independent action to fill needs for scientific information, branching out into unfamiliar fields if necessary. Congress showed its approval of new projects such as conducting soils surveys for the Department of Agriculture and classifying forest areas by increasing the USGS budget. These research areas were later spun off as independent units associated with USDA and Department of the Interior. Reporting of mineral production data was turned over to the US Bureau of Mines when that agency was founded in 1910 after a catastrophic mine disaster (see Fig. 3.2).

Walcott expected his scientists to operate independently in formulating strategies to accomplish their tasks, and report fully and promptly on their results. Once a new man was hired he was given a task to fulfill. If he did his job well he would be given greater opportunities and responsibilities. If he failed to come through satisfactorily, he was let go to seek work elsewhere (Yochelson, 1997). Having accomplished their primary responsibilities, staff were allowed and encouraged to push their studies to the frontier of knowledge.

Walcott's commitment to public service was aggressively adopted by a growing Water Resources Branch, led by Chief Hydrographer, F.H. Newell. A powerful statement of principles designed to simultaneously promote effective service by a governmental science agency and to assure its survival, is contained in a set of recommendations prepared by a three-man committee and presented by Marshall O. Leighton, who succeeded F.H. Newell as Chief Hydrographer of the USGS to a national meeting of district engineers for Water Resources offices in 1908 (Follansbee, 1994):

"As an official organization supported by appropriation from the national treasury, it is the duty of the Geological Survey and the Water Resources Branch to make available to the public in the most prompt and effective ways possible the results of its investigations. Failure to do this is failure in duty and it results in failure of public support. It therefore becomes a double duty – a duty to our organization and to the people."

The recommendations continued with what we now would term a multimedia approach to information outreach. It was recommended that:

> ' "Press bulletins be continued and no effort be spared to make them effective and secure for them a general circulation in the technical and popular press.' 'District Offices are best situated to respond to local needs, and offices should maintain lists of potential users, and respond promptly to local inquiries.' "

Engineers or other representatives were encouraged to join local organizations and become familiar with community needs, and to present lectures on relevant topics to audiences, and to seek to be helpful in every way.

There were also warnings about "Don'ts." Addresses or lectures for the sake of advertising work in programs where they were out of place were to be avoided. Lobbying and solicitations for support, and giving individuals undue prominence rather than placing emphasis on the work itself were also undesirable. Care was to be taken in correspondence to avoid "offensive, brusque, and unreasonable attitudes" (Follansbee, 1994).

8.3.2 From 1907 to 1971

Charles Walcott's successor, George Otis Smith (1907–1930) emphasized practical or useful service even more strongly than did Walcott.[207] This did not prevent USGS scientists, like the earlier G.K. Gilbert in earthquake science and geological processes, from becoming international scientific leaders in their fields. For example, Waldemar Lindgren wrote a pathbreaking book on ore deposits in 1913, and new editions came out until 1932. The volume still remains an important reference in earth science libraries. Oscar E. Meinzer led USGS to world leadership in the theory and practice of groundwater hydrology. F.A. Clarke, founder of the US Geological Survey's analytical laboratories, published a pioneering compendium on the composition of rocks, minerals, and waters, entitled, *The Data of Geochemistry*. First published in 1908, the compendium went through new editions until 1924, remaining in active use until World War II. The respect given Clarke's work led distinguished Russian geochemists to create the term "Clarke" as a unit describing the abundance of chemical elements in the Earth's crust. A number of geologists left the USGS during this period to work for companies.

USGS topographic maps provided products that were state of the art in accuracy and production technology, and USGS geological maps became indispensable tools for mining exploration, as well as for city and highway planners. But in the hard times of the Depression, under Director W.C. Mendenhall (1930–43), meager and fluctuating funding was often a severe trial partly mitigated by New Deal funds. Upon the outbreak of World War II USGS devoted most of its energies to the war effort. With the cooperation of university staff, it pursued search for strategic materials and metals, developed airborne magnetic surveying methods with the US Navy, helped prepare terrain reports for war fronts, and performed important war-related tasks like providing water supplies for troops and aerial cartography.

At the end of the war only half of the United States was covered by topographic maps, and only 10% by geological maps adequate to appraise mineral resources, susceptibility to landslides and other hazards, and irrigation potential (Rabbitt 1989). Directors William E. Wrather (1943–1956) and Thomas B. Nolan (1956–1965) placed increasing emphasis on basic research, while seeking substantial increases in funding to cover large gaps in coverage of topographic and geologic mapping, and groundwater surveys. President Roosevelt's science advisor, Vannevar Bush's influential blueprint for expansion of basic research after World War II, *Science The Endless Frontier* (Bush, 1945) affected the USGS as well as academic institutions.

With the Korean War strategic minerals again became a major concern, this time focusing on uranium and radioactive raw materials. Urban geology programs were created to aid in civil defense. New research fields included studies related to atomic energy, disposal of radioactive wastes, geologic hazards, and coastal geology, including undersea exploration of the Atlantic, Pacific, Alaskan, and Hawaiian island margins.[208] In water resources USGS initiated annual measurements of water properties in the Mississippi River from the 1950s, charted the water composition and discharge of US rivers and streams, and expanded knowledge of the capacities and storage capabilities of groundwater aquifers throughout the nation. Lowering of ground levels in coastal Louisiana was shown to be related to withdrawal of petroleum from reservoirs along the Gulf coast. During Director Nolan's tenure USGS operating funds expanded from $49 million to $104 million.

The actions of one of most effective leaders, Director William T. Pecora (1965–1971), during the offshore oil crisis in 1969 were described in Section 2.7. Under Pecora's leadership a national search for gold and precious metals was undertaken in cooperation with the U.S. Bureau of Mines. Appraisals of mineral resources of the nation and world were expanded, and the use of Earth-orbiting satellites to gain information about the earth and its resources was initiated. Geologic hazards in developing countries as well as the United States were analyzed, and astronauts trained by the new USGS Astrophysical Branch under the leadership of Eugene Shoemaker landed on the Moon in 1969. The USGS prepared the groundwork for the United States' underground nuclear tests on Amchitka Island, in the Aleutian Islands (1965–1971), and monitored the area for radioactive leakage after the largest event (1971).

8.3.3 Vincent E. McKelvey (1971–1978)

Director Vincent McKelvey enjoyed a worldwide reputation as an economic geologist. He had strong personal ethics, and cherished the USGS's reputation as an organization that could be relied on for authoritative, impartial science. He required scientific and administrative staff to divest themselves of securities involving mineral or energy resources after allegations of confict of interest emerged in the Senate Interior Committee in 1973–1974.

During McKelvey's tenure flagship in-house publications, the *Professional Paper* series, that often reported research requiring teams of scientists and other technical staff, large geographic areas, and might require years to complete, began to decline in number published. So did the focused *Bulletin* series. A part of newer geological research focused on "process-oriented" studies, localized earth science observations, relationships, and contributions to theoretical explanations of earth processes. Such research lent itself to publication in the rapidly growing peer-reviewed scientific journals favored by scientists in university departments and research centers. In the Geological Division promotion criteria favored productivity in such peer-reviewed publications while more specific USGS-oriented work was "pushed out the door" quickly in open file reports (Fig. 8.1).[209] In at least one major research group the Thomson-ISI Science Citation Index was used to rate scientists' publication performance.

Nevertheless, there were important practical achievements during the McKelvey period. The first Earth Resources Technology Satellite (later Landsat1) was launched in 1972. The National Atlas (1973), a product led by USGS with cooperation of ten other federal agencies, became indispensable in every US library. The first comprehensive assessment of the Nation's coal resources was published in 1975. The Atlantic and Pacific offshore areas were systematically mapped. In 1976, the USGS was given the responsibility for administering the exploration of Naval Petroleum Reserve No. 4 (redesignated as the National

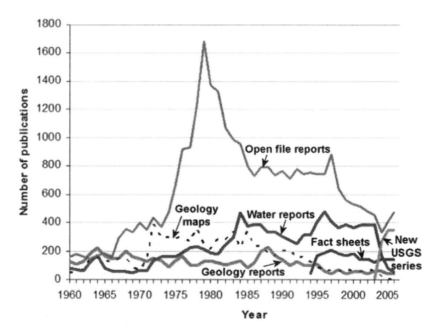

Fig. 8.1 Selected USGS publications, 1960–2006
Source: Compiled by Lucy Manheim from USGS Publications Warehouse (online)

Petroleum Reserve in Alaska) and the operation on the South Barrow Gas Field in Alaska (Rabbitt, 1989). Service-oriented research remained a strong emphasis in the Water Resources Division, in part because of cost-sharing cooperative programs with the states and local communities that were begun during the directorship of Charles Walcott and Hydrographer F. H. Newell from 1900 to 1906.

McKelvey was not a forceful leader, tending to defer to others' opinions. His policy of keeping USGS away from the hurly burly of politics was so strong as to become counterproductive. For example, in the aftermath of the Santa Barbara offshore oil spill (early 1970s), a local Congressman alleged that the well that USGS was drilling on Long Island for groundwater monitoring purposes was really to prospect for oil. This claim was totally unrealistic. Long Island had metamorphic bedrock at depths less than 1,500 feet, and no sediment layers with the remotest possibility for oil. Yet, intimidated by publicity in news media, McKelvey ordered the drilling rig removed. This action may have confirmed the opinion of the Congressman and conspiratorially minded but scientifically unsophisticated persons that USGS had been "caught in the act."

Under far more serious circumstances in the Santa Barbara oil spill, where there had been real fault on the part of the USGS, Director Pecora's response to public and political pressure was unflinching. He personally appeared in local public hearings at Santa Barbara, released all facts surrounding the incident, and accepted responsibility where due. He also rebutted with facts and vigor exaggerated claims or accusations. Moreover, against an outburst of public protest he ordered resumption of oil production in offshore wells seaward of the one that had blown out. He did this in order to reduce pressure on the leaking well so that it could be brought under control. Paradoxically, Pecora's willingness to take purposeful action in the face of political risks led to partial repair of USGS's image. Pecora himself went on to be appointed as undersecretary of the Interior Department. Director McKelvey's policies and ultimate fate were directly opposite to those of Pecora.

Up to this time, the USGS directorships had traditionally been treated as scientific and professional appointments, unlike many presidential appointments. McKelvey's background in the area of economic and mineral deposits did not sit well with incoming President Carter, an ardent environmentalist. He was said to be only deterred from immediately replacing the Director in his new administration by the intercession of Frank Press, Carter's science advisor and a noted geophysicist (Shapley, 1977).

Vincent McKelvey set in motion action on the part of President Carter in 1977, when he released a favorable report on offshore oil and gas resources. This brought him into direct conflict with Carter's conviction that "the energy crisis is the economic equivalent of war," as he expressed it in a famous TV address to the Nation (Carter, 1977). Physically weakened by the Lou Gehrig's disease that would end heroic final years, McKelvey became the first USGS director to be removed by the President, though the action was handled without overt publicity.

8.3.4 H. William Menard (1978–1981)

Incoming Director H. William Menard was personally selected by Joan M. Davenport, Assistant Secretary for Energy and Minerals in the Department of the Interior under Secretary Cecil Andrus.[210] Menard was an oceanographer at the Scripps Institute of Oceanography, and an expert on gedogic structure in the Pacific Ocean. He had shown avocational scientometric interests, authoring a book that charted the growth of academic science and scientific publications (Menard, 1971). The book included the budgets of federal government science agencies and very briefly reviewed their functions. The book gave hypothetical support for a federal Department of Science and Technology, while rather cynically dismissing its realistic chances. He asked, rhetorically: "Who will support it with determination and muscle? Not the unorganized scientists who come hat in hand to Washington; not the engineers and technicians who will be increasingly unemployed." Nor were there good candidates among "environmental departments of the federal government [who are] losers."

In his book, Menard lauded the early scientific achievements, but criticized recent USGS bureaucracy and delays in publication. As USGS chief, Menard's first official act was to ban smoking in USGS buildings, but he showed little interest in the traditional tasks and operations of USGS or its staff.[211] There was widespread relief among USGS employees when incoming President Ronald Reagan appointed USGS Chief Geologist, Dallas Peck, Director in 1981.

8.3.5 Dallas L. Peck (1981–1993)

Dallas Peck, a career USGS geologist (volcanologist) who had risen through the ranks, had a hearty persona and good people skills. The Committee on Earth Sciences (CES) under the Federal Coordinating Council was charged by William Graham, President Ronald Reagan's Science Advisor, with coordinating the nation's global change research activities. Allan Bromley, science advisor to President George H.W. Bush (1989–1993), credited Peck with handling the Committee in a manner that served as a model for later committees. Earlier, in the absence of interest by President Reagan, agencies had simply refused to cooperate (Bromley, 1994).

Peck's service as Director has been less favorably assessed by other observers, including USGS senior geologist, Robert Wallace in reminiscences of his association with six former directors (Wallace, 1996). Early in his tenure, Peck presided over loss of the USGS Conservation Division, whose most important responsibility was to manage offshore oil and gas leasing and production. Conservation Division had gained an unusually qualified Chief in Don E. Kash, a leading academic expert on energy and technology policy in 1978. However, a special problem for the Conservation Division was that it had a poor image among the Geological Division's influential leading scientists. This was partly due to the fact that maintaining proprietary data from oil companies inhibited publication by its staff in the open scientific literature.[212]

After a controversial offshore lease sale for the Baltimore Canyon trough area in 1981 and an equivocal response from the Director's office, James Watt, Interior Secretary in the first Reagan administration, proposed transfer of regulatory responsibilities from USGS to a new agency. According to William S. Radlinski, former Associate Director of USGS, who testified in Congress *against* the transfer after retirement from USGS, Director Peck testified in favor of divestiture,[213] which took place in 1982. Peck was said to be influenced by the Geological Division, many of whose leaders and scientists cheered when the transfer was announced.

User-oriented products and services continued in some areas of the broad range of activities, but many scientists' and projects' themes increasingly resembled – and in many cases cooperated or competed with – basic research programs at universities. Under the permissive leadership of Dallas Peck small projects blossomed and many new internal report series grew. Although USGS, especially the Water Resources and Mapping Divisions, maintained service oriented activities, the external impression of a "university-like" research culture at USGS contributed to subsequent events.

8.3.6 The Reckoning Approaches

The growth of discretionary and uncoordinated research activities within USGS did not escape the notice of budget-minded Congressmen. In 1993 a precursor development to the Republican Contract with America of 1994 took place. Representative Timothy J. Penny (D-MN) and John Kasich (R-IL) submitted an extraordinary set of amendments to H.R. 3400, the Government Reform and Savings Act of 1993. The collection of amendments, labeled the *Common Cents Deficit Reduction Act of 1993* by Congressman Penny incorporated an array of draconian budget cuts estimated to aggregate $90 billion. A significant proportion of the cuts were to take place through the creation of a new consolidated, Cabinet-level Department of Science, Space, Energy, and Technology. Transfers to the new agency include functions of the Department of Energy, NASA, the NSF, and parts of a dismantled Department of Commerce (which included the National Oceanic and Atmospheric Administration, and the National Institute of Standards and Technology). All bureaus of the Department of the Interior were pointedly missing from mention of transfer except the Bureau of Reclamation. The Bureau of Reclamation was to be merged with a trimmed U.S. Army Corps of Engineers. It was clearly understood that USGS and Bureau of Mines would be wiped out or privatized as a part of the massive reduction of 252,000 full-time equivalent federal positions called for in the amendment. US Geological Survey's total employees aggregated fewer than 7,000 employees.

In spite of vigorous opposition by President Clinton and other leading political figures, as well as outcry from affected agencies and functions and their supporters, at 9:14 p.m. on November 22, the *Common Cents Reduction Act* amendment came to a vote in the House of Representatives. It lost by only six votes – 213 for and 219 against (Penny, 1993). The bill containing amendments that would have dismantled USGS would have probably been vetoed by President Clinton even if it had passed the Senate and gained agreement with the House. But the message sent by the Penny–Kasich bill was not just a fluke created by erratic Congressmen. Even some members who voted against it suggested that a consolidated science agency and other features of the bill had potential merit. Traditionally liberal groups that could normally be counted on to support governmental services against conservative attack gave hints that they might look the other way if governmental science establishments were judiciously pruned.

8.3.7 Gordon P. Eaton (1994–1997) – Traumatic Years

USGS got only a temporary reprieve. A new USGS Director, Gordon Eaton, who had been a USGS scientist before occupying top administrative positions in major US universities and other academic organizations, was appointed by President Clinton in 1994. Implications of the Penny–Kasich bill were on his agenda. In August 1995 Eaton initiated the largest RIF since 1892. The action initially affected some 500 employees. Areas like Water Resources that maintained strong service commitments and had longstanding cooperation with local communities or users lost no staff in the RIF.[214] Hardest hit were those parts of USGS that had embraced more academically oriented research and emphasized publication in peer reviewed journals. This especially affected the Geological Division – the same division that had traditionally dominated leadership and had most aggressively launched discretionary research projects. Questionable methods, including targeting of older scientists, devastated morale. They created widespread disillusion, anger, and fear, and led to major class-action lawsuits lodged, after appeals to the US Merit Protection Board ("Adam et al. v. Babbitt," 1996), with federal courts in Denver and Menlo Park. A number of employees were reinstated, but the RIF action was upheld. Director Eaton announced to USGS leaders and staff that the organization had to reorient itself to foster internal and external cooperation toward serving public needs.[215]

In 1997, less than 2 years after the RIF, Chief Geologist P. Patrick Leahy announced a plan to halve new acquisitions for the USGS library system, the largest earth science collection in the world. A storm of protest in the United States and abroad and Congressional action forced withdrawal of the plan. It turned out that most earth science departments and organizations, including state geological surveys and also private companies were dependent on the USGS libraries' comprehensive and unique resources for loans and other services, and foreign institutions launched protests as well (Geoscience Information Society, 2007). In September, 1997, only a few months after the library fracas, Director Gordon Eaton, who had "lived by the sword," as a colleague put it, "died by the sword."[216] A report by *Science* magazine related his abrupt resignation and departure to a conflict with Interior Secretary Bruce Babbitt over government office sites in the San Francisco Bay area (Kaiser, 1997).

8.3.8 From Eaton to the Present

During 1996, incoming personnel from two additional groups compensated for USGS staff losses through the RIF and buyouts. One was employees from the dismantled US Bureau of Mines who had compiled mineral production statistics. This was a service highly valued by industry, and which involved holding proprietary data. The other group was the US Biological Service. It became the Biological Resources Division in USGS. The new influx of biologists, toxicologists, and other scientific expertise added to staff strength for dealing with environmental issues.[217] Diverse scientific skills equipped the USGS to partner with federal agencies that were overloaded with tasks but had too few qualified people to deal with them. The US Environmental Protection Agency, for example, was (and is) universally acknowledged to be overwhelmed with research and investigative tasks assigned by Congress. My experience suggested that USGS was not always open to opportunities to serve or cooperate on larger scales with other agencies, where cooperation was not fostered by emergencies like the Katrina Hurricane, or fit into existing programs.[218]

Incoming Director Charles Groat (1997–2005) announced a goal of reducing "stovepipes" (isolation of units) and fostering better cooperation among USGS scientific groups. But changing the names of the *divisions* (Water Resources, Mapping, etc.) to *disciplines,* moving staff to the Director's Office, and reorganizations did not produce visible results. Upon Groat's sudden resignation in June 2005, amid mounting problems including bad blood between DOE and USGS, and action to further cut already bare bones library staff, Associate Director Patrick Leahy (former Chief Geologist) became Acting Director. Groat and Leahy had declared commitment to service for societal needs. But abandonment of further USGS involvement in producing or upgrading topographic maps, the USGS's best-known and probably most widely used product, in favor of "more interpretive mapping research," made it clear that "service" would be rendered according to USGS leadership's interpretations, not necessarily what users wanted or needed. The former mapping discipline has been replaced by a "Geography" discipline.

On September 16, 2006, Mark Myers, who resigned as Alaska State Geologist along with other state employees in ethical protests over actions of Alaska state political leaders, was confirmed by the Senate as the Director of USGS. In a change from the pattern in place since the middle 1990s, Myers promised on arrival to hold off on near term reorganizations. He has maintained a lower-key profile than his predecessors. He has placed special emphasis on global climate change studies, in which USGS will have responsibility for three out of 13 current US research tasks, one being investigating the potential impacts of rising sea levels. Myers is said to have made thoughtful appointments. Individuals who have had private conversations with him noted his observations about the risk-averse character of the USGS.

8.3.9 Interpretation and Discussion

• *Authoritativeness.* USGS has prided itself on its tradition of producing authoritative and objective research and service. That tradition remains largely intact. Both pro and con factions in the debate over oil and gas leasing in the ANWR area of Alaska cite USGS resource estimates. The USGS inventory of trace constituents in US rivers is widely cited, as are USGS identification of the epicenters and strength of earthquakes. This credibility may be a key reason why, during years when the George.W. Bush administration proposed severe cuts in the USGS budget, that the Congressional Appropriations Committee (usually with bipartisan support) made restorations.

• *Morale and esprit.* In the past, morale and esprit of USGS staff were among the highest in federal service (the USGS was compared in this respect with the earlier FBI). The annual "Pick and Hammer" show (discontinued along with employee newsletters by Gordon Eaton) had as a highlight skits and songs that deftly skewered the peccadilloes of higher managers. Internal newsletters had featured letters raising honest critiques and concerns in a Geologic Division column aptly named "The Fumarole."

Morale suffered a devastating blow in the RIF of 1995, not merely because of staff separations and the implications that USGS was not pulling its weight among federal science agencies. The devious way the RIF was executed tore the fabric of the organization and shook employee confidence in management. It has been suggested that although evidence that OPM regulations had been violated was strong, judges may have been reluctant to formally overturn the RIF because of the chaos this might cause (e.g., insufficient funds to reinstate employees) and the precedent that could be set.

• *The damaging effect of academicized research from the 1960s to the '1990s.* During my service in USGS the postwar paradigm shift toward "curiosity-driven" research in the United States reached a peak. It affected not only the university environment, but also came to selectively affect USGS and other federal agencies through scientists' associations and through its influence on scientific leaders and administrators. The problem was not the quality of research. Nor did valuable service functions stop. But the new ethos had a number of negative effects on USGS service functions in US society:

  – Efforts became fragmented in semi-isolated miniprojects.
  – Initiatives based on arbitrary judgments, stimulated by buzzwords and fashionable concepts rather than careful evaluation of needs and goals grew.
  – Poorly coordinated or integrated open file reports and "outside publications" that were difficult to access proliferated. This trend is noted in Fig. 8.1. Open-File Reports tended to be issued serendipitously, without plans or larger goals. Outside peer-reviewed publications scattered the products of USGS research in media primarily circulated among research scientists.
  – The fragmentation of research and publication distracted from creation and completion of larger cooperative products with local, regional, or other partners,

or requested by Congress or other political organizations. A former Chief of the Office of Energy and Marine Geology once reported the opinion of OMB staff that USGS was "bright, slow, and irrelevant."

A subtle but perhaps the most damaging aspect of basic research emphasis that grew from the 1960s was the arbitrary division between more theoretically oriented research and applied or service functions. Where the two are separated, the "best and brightest" talents will tend to gravitate toward and compete more effectively for discretionary research opportunities. This adds additional prestige to the more independent research jobs, while it reduces the prestige and potential influence of applied activities. The diversion of talent may lead to the reverse of what is most needed. Talented people interfacing with "the real world" of local communities, state and federal agencies, private industry, etc., have almost unlimited potential to serve, see needs, build bridges and collaborative programs. On the other hand, attraction of the brightest talents to more academic and theoretical working environments limits their contributions to professional circles where their ability has less influence on agency outreach and service.

My point is not to criticize or suggest inhibition of discretionary publications by scientists. It is rather emphasizing that the primacy of "bread and butter" publication and service to society as an honorable and desirable activity lost ground during the post-World War II period as did the anchoring of the source of that service in the public's mind. I often think of the "Coast and Geodetic Survey" (now part of NOAA as the Coast Survey) whose name and identification with mapping became so strongly imprinted in coastal residents' minds that they often confused it with USGS. The "double duty" ethic driven home in the 1907 address by Chief Hydrologist Marshall Leighton in 1907, referred to earlier, eloquently describes desirable prioritization for government science, with a remarkably modern flavor and potential applicability.[219]

Directors after Vincent McKelvey held biases against more applied functions. While perhaps not obvious, these directly or indirectly affected decisions and relationships throughout the USGS. The most disastrous result is the loss of the Conservation Division. Some of the scientists who cheered its loss because such applied functions did not fit the "higher calling" of USGS science lost their jobs in the RIF of 1995. Ironically, no Water Resources scientists doing more service-oriented work were affected.

8.3.10 Conclusions

The conditions under which USGS achieved its best results have similarities to those in other peaks of scientific and technological achievement. These include the classic nineteenth century German university–industry cooperation that brought economic prosperity to Germany, while it created breakthroughs in many scientific fields; the Manhattan Project; achievements in US industrial laboratories in the early post-World War II period (e.g., IBM, AT&T, Polaroid, Xerox); the Apollo space program that placed a man on the moon; some of the United States' best biomedical research programs; and most recently, the Apple and Google companies.

Economist Michael Porter (Porter, 1990) in his book, *The Competitive Advantage of Nations* described academic-industry-government research clusters as highly productive generators of technical innovation. These environments seek breakthroughs to significant product or other concrete goals, and spin off creative side developments and new products as they do so. A similar orientation toward significant product creation characterized USGS's periods of greatest effectiveness, especially from 1896 to 1932, and during and immediately after World War II. Well-qualified scientists were allowed to choose the methods to achieve results. If they completed the "bread and butter" work, researchers were given the freedom to extend their research in discretionary ways. Currently, the Google Co. gives its researchers 20% of their time to pursue personal research. In such environments as well as in the German Siemens Corporation, people with different skills work together, and active communication between scientific staff and users of products is important, because needs often drive product creation.

Contrary to the assumptions of Vannevar Bush in his famous book, *Science The Endless Frontier*, applied research conducted by talented scientists is more productive of basic research breakthroughs, as well as more useful for society, than the reverse. This has been demonstrated by USGS history as well as in other productive research environments.

- Effective research to meet societal needs cuts across disciplinary and other arbitrary boundaries. Investigators are exposed to new or unpredictable conditions. If they are talented and open minded, new insights can lead to basic science breakthroughs.
- Work that benefits the public or produces "hard" results usually has less of a struggle getting funded – and therefore has a greater chance of continuity and stability. Interest in useful products of large scope builds as the goal or product is approached. In basic research the excitement is at the beginning or in early stages of a competitive grant or project, in confirming hypotheses or discovering variants along the way.
- Although leaders with vision enhance productivity in any area, their influence on activities that serve society more directly gain disproportionatly greater impact.
- The hallmark of effective research environments of the past and present is product creation as the *purpose*, not the byproduct of research strategy. The integration of data dissemination and/or product goals with research may be a key to future effective research environments at USGS.

It may be that a "marriage" or new, close cooperation between USGS and MMS could provide a buffer of qualified scientific information and reduce conflict or uncertainty in future proposals regarding economic use of the United States' vast offshore exclusive economic zone.

> The hallmark of effective research environments of the past and present is product creation as the purpose, not the byproduct of research strategy.

USGS (like other federal science agencies) had generally competent, service-oriented leadership, dedicated and qualified staff, and good productivity prior to the 1960s. Performance was adversely affected by science policy and political developments after the 1960s but may be in an early recovery trend.

8.4 Environmental Laws and Cases

8.4.1 California's Environmental Policy Act of 1970: Consequences Around San Francisco Bay (summarized from Frieden, 1979)

8.4.1.1 Background

The intended goals of NEPA, expressed by Lynton Caldwell, a senior advisor to Senator Henry M. Jackson, were to promote beneficial uses of natural resources and the environment while minimizing risks to health and safety, maximally preserving historic, cultural, and natural aspects of the national heritage, and fulfilling responsibilities of each generation to future generations. The law was intended to provide ethical and practical guidance to practical action, while forcing attention to environmental principles by all governmental agencies. Ideally, the principles would be internalized by the general public (Caldwell, 1998). However, Caldwell acknowledged that soon after enactment, "environmental nongovernmental organizations.....showed interest in NEPA primarily as the environmental impact requirement that enabled them to stop or delay specific government programs or projects to which they objected."

The California Environmental Policy Act of 1970 (CEPA) was modeled after the National Environmental Policy Act (NEPA) of 1969. The California statute was soon extended to apply to private as well as public actions. Its unintended consequences to the residential developments around San Francisco Bay are summarized in this vignette.

The almost unlimited scope of issues in CEPA impact reports (NEPA uses the term *environmental impact statement*, whereas the California Act speaks of *environmental reports*) offered similarly unlimited grounds for challenging their adequacy in court. Outcomes depended on a host of factors, including the issue being litigated, the strategies of litigants and skills of attorneys, serendipitous external circumstances, and the personal viewpoints of judges. In the described cases the original goals of both NEPA and the CEPA faded to near irrelevance as the Environmental Impact provision became the primary tool for legal challenge.[220]

8.4.1.2 Alameda and Oakland Airport Expansion Housing Problems

The passage of CEPA was part of political movements that also created new agencies to oversee land use policies. It fostered a new climate of litigiousness, as exemplified by conflicts over housing development in Alameda and the Oakland Airport expansion from 1972 to 1976. In 1969, Alameda was a city with a population of 70,000 on the eastern shore of San Francisco Bay, across the water from San Francisco. In that year a proposal for a residential development of 11,000 housing units "met a storm of opposition from Alameda residents" and was shelved.

A new proposal began with a joint development venture called Harbor Bay Isle Associates (HBIA) in 1972. This set aside 1/5 of the area for open space, lagoons, a marina, a recreation strip, two parks, and three school sites. A proposed second bridge across San Francisco Bay was to have provided direct access to downtown San Francisco. In a surprise move, voters in a special referendum voted down the bridge proposal.

In 1972 a new proposal with no bridge and with reduced density was offered. Opponents to a compromise by city planners introduced a city charter amendment prohibiting new apartment construction in Alameda, which was approved by voters in 1973.

A proposal with further reduced scope and no apartments was presented, but public resistance now focused on noise from increased air traffic from Oakland Airport, south of Alameda, which might have been encouraged by increasing the population of Alameda. This sentiment grew in spite of federal and state regulations that had resulted in introduction of new, quieter jet planes and noise abatement at the airport.

Both the developer and the airport were blocked from dealing with noise problems. If noise over Harbor Bay Isle reached certain levels, state regulations permitted only apartment buildings with acoustical provisions to reduce interior noise levels. But the amendment to Alameda's charter had prohibited new apartment buildings. Oakland Airport could have handled more traffic without adding to noise over Harbor Bay Isle by building a new runway allowing planes to take off over the bay. But this would require filling some land in the Bay. The recently established Bay Conservation and Development Commission had gone on record opposing any landfill used to extend airport runways.

In March 1974 the Alameda City Council approved rezoning for still lower density home construction. Oakland Airport authorities anticipated that new, more affluent residents occupying homes with increased prices in Alameda would bring pressure on the airport to limit flights. They therefore brought a lawsuit against HBIA association charging that its environmental impact report was inadequate. HBIA in turn filed a lawsuit against the Oakland Airport's expansion plans, charging that new flights would create impacts on Alameda. Oakland Airport won its suite against HBIA in 1975. Frieden (1979) indicates that one of the problems was the difficulty for Alameda to maintain an orderly review process under conditions where the developer was forced to constantly revise plans.

In April 1976, the San Francisco Superior court found Oakland Airport's Environmental Impact Report inadequate, and questioned the Airport's statistics about the need for expansion.

HBIA now prepared a new plan, reducing density of residential development still further, and commissioned a 3rd environmental impact report.

When the airport spokesmen tried to have the Court set still tougher conditions for developments, the Court finally lost patience and intervened with both litigants. In July 1976 Oakland Airport, the city of Alameda, and Harbor Bay Isle Associates reached agreement to drop all lawsuits under compromise agreements.

Under the May 1972 plan Bay area families would have gained more than 9,000 condominium units at prices ranging from $21,000 to $37,000. The April 1976 agreement reduced the number of permissible new dwellings to 3,000 while prices for the properties now ranged from $55,000 to $165,000.

To summarize: The new environmental regulations and accompanying creation of localized or fragmented authorities to oversee land uses, greatly expanded opportunities for any litigants to block undesired activities, regardless of the merits of proposals and larger needs of the areas in question. Technical studies increasingly began to be shaded in the direction of desired legal outcomes, rather than aiming at balance and accuracy. Siting new housing in Alameda became more difficult because it faced opposition by both existing residents and Oakland Airport at different times and under different circumstances. The changes in law only favored blocking actions; they weakened creative, balanced and coordinated planning and action.

The net effect of environmental regulation was to greatly increase prices of new homes in Alameda, favor more affluent residents, and force movement of lower-cost home development to areas more distant from urban centers. The utilization of naturally favorable areas for airport runways, building partially into bay areas, was blocked.

8.4.2   The Federal Water Pollution Control Act (1948, 2006; and Milwaukee Sewerage Case)

8.4.2.1   The Federal Water Pollution Control Act of 1948 (The original act)

The concise Water Pollution Control Act of 1948 established the general framework for control of water pollution; the details of implementation were left to the relevant agencies. In 1972, the Clean Water Act Amendments vastly expanded the clean water regulations, mainly because Congress chose to spell out the implementation details that were formerly delegated to agencies with professional expertise (Table 8.1).

The US Code for the Water Pollution Control Act of 1948, Public Law 845, from 62 Stat. 80th Congress, 2nd Session, Chapter 758, June 1948 (pp. 1155–1161):

– *To provide for water pollution control activities in the Public Health Service of the Federal Security Agency and in the Federal Works Agency, and for other purposes.*

- Sec. 2. (a) The Surgeon General shall investigate and, in cooperation with other Federal agencies, State water pollution agencies, interstate and municipal agencies, and industries, adopt comprehensive programs for eliminating or reducing the pollution of interstate waters. (b) The Surgeon General shall encourage cooperative activities to prevent pollution, enactment of uniform State laws, and encourage compacts between states. (c) Congress consents to two or more States entering agreements and establishing joint agencies. (d) Pollution of interstate waters which endangers health or welfare is declared to be a public nuisance and subject to abatement as described below.
- Sec. 3. The Surgeon General may, upon request of any State agency, conduct investigations and research concerning any specific problems of water pollution confronting any State, interstate agency, community, municipality, or industrial plant, with a view to recommending a solution.
- Sec. 4. The Surgeon General shall prepare and publish reports of surveys, studies, investigations, research and experiments, as well as recommendations.
- Sec. 5. The Federal Works Administrator is authorized to make loans to any State, municipality, or interstate agency for the construction of treatment works.
- Sec. 6. The Surgeon General and the Federal Works Administrator, shall conduct reviews of all reports, research and surveys pursuant to provisions of this Act and all applications for loans. A Water Pollution Control Advisory Board is established in the Public Health Service. The Board will have representation by the Departments of Army, Interior, Agriculture, and the Federal Works Agency, as well as six persons to be appointed annually by the President.
- Sec. 7. For each of five fiscal years from July 1, 1948, and ending June 30, 1953, a sum not to exceed $22.5 million is appropriated for loans under Sec. 5.
- Sec. 8. For the above period $1 million is allotted to the States for conduct of investigations, research, surveys, and studies, plus additional funds for other purposes.
- Sec. 9. Five officers may be appointed to grades in the Regular Corps of the Public Health Service above that of senior assistant. The Federal Works Administrator is authorized to hold, refund, or sell any bonds or obligations evidencing loans made u under this Act. Moneys received from sales shall go to the Treasury.
- Sec. 10. Definitions.
- Sec. 11. This Act shall not supersede or limit the functions under any other law, of the Surgeon General, or of the Public Health Service, or any other officer or agency of the Umited States, nor the Oil Pollution Act, 1924.
- Sec. 12. If any provision of this Act is held invalid in application to specific persons or circumstances, applications to other persons and circumstances shall not be affected.
- Sec. 13. This Act may be cited as the Water Pollution Control Act.

8.4.2.2 The Federal Water Pollution Control Act, as of 2006

Table 8.1 A small sample of the 607 US Code Sections (about 360 pages) of the Clean Water Act (Amendments of the Federal Water Pollution Control Act, as of 2006; codified generally as 33 U.S.C. 1251–1387)

Subchapter III -	Standards and enforcement	Section
1311	Effluent limitations	sec. 301
1312	Water quality-related effluent limitations	sec. 302
1313	Water quality standards and implementation plans	sec. 303
1314	Information and guidelines	sec. 304
1315	State reports on water quality	sec. 305
1316	National standards of performance	sec. 306
1317	Toxic and pretreatment effluent standards	sec. 307
1318	Records and reports, inspections	sec. 308
1319	Enforcement	sec. 309
1320	International pollution abatement	sec. 310
1321	Oil and hazardous substance liability	sec. 311
1322	Marine sanitation devices	sec. 312
1323	Federal facility pollution control	sec. 313
1324	Clean lakes	sec. 314
1325	National study commission	sec. 315
1326	Thermal discharges	sec. 316
1327	Omitted (alternative financing)	sec. 317
1328	Aquaculture	sec. 318
1329	Nonpoint source management program	sec. 319
1330	National estuary study	sec. 320

Source: EPA and US Code

Note the detailed specifications and prescriptions for programs, scope of activities, and procedures for arriving at standards and types and frequency of reports. Other sections, including naming of specific geographic areas, sites, etc., are equally detailed. Since 1987, Congress has been unable to reach agreement on significant amendments or updates to this law

8.4.2.3 Friends of Milwaukee's Rivers v. Milwaukee Metro Sewerage District and the US Environmental Protection Agency

During a trip to Chicago, I noted a *Chicago Sun Times* story of April 8, 2005, head-lining a victory achieved by Karen Schapiro, a lawyer working with volunteer assistance and minimal funding, in her suit against the Milwaukee Metropolitan Sewerage District (MMSD).

"CRUSADER FOR CLEAN LAKE GETS HIGH-LEVEL LEGAL HELP"

Milwaukee Lawyer fights to keep city's sewage out of water

By Gary Wisby

"Working for a small law office, a woman takes on an agency with a sizable in-house legal staff and an even larger outside law firm in a pollution-fighting case that goes all the way to the U.S. Supreme Court."

"Erin Brockovich? No, it's *Karen Schapiro*, a Milwaukee lawyer who is being honored for her long – and thus far unpaid – battle to keep the city's raw sewage out of Lake Michigan and its tributaries."

Details in the article conflicted with what I knew of Wisconsin's progressive traditions in respect to public health issues. I dug for more background. A long series of stories in the *Milwaukee Journal-Sentinel* by reporters Marie Rohde and Steve Schultze featuring the heroic public service legal efforts of the graduate of the environmental law program at Northwestern Law School turned up.

A three-judge panel of the federal appeals court had reversed the federal district court's dismissals of a suit by Schapiro's *Friends of Milwaukee's Rivers and Lake Michigan* dating back to 2001. The suit charged that the MMSD had been illegally "dumping raw sewage into the Milwaukee River for 25 years." These actions were alleged to be violations of the Clean Water Act of 1972, which stipulated that all effluent discharge from publicly operated wastewater treatment plants into public waterways must receive secondary treatment by 1977. According to the newspaper article "neither the Wisconsin Department of Natural Resources (DNR), nor EPA did anything."

However, a 46-page report (March 15, 2001) to the Wisconsin Natural Resources Board brought out a more complex picture than an environmentalist David taking on a bureaucratic Goliath. Sewer overflow discharges are allowed to take place under certain conditions – for example, regular maintenance, special emergencies, and especially heavy rains – which can fill total sewer storage capacity. Next, Milwaukee's system has systematically had to accept sewage from expanding satellite towns and communities whose sewer lines are not under the control of the City. According information from an official of the Wisconsin DNR, 300 miles of sewers are under control of the MMSD, whereas 3,000 miles are under the jurisdiction of surrounding suburban areas.

Wisconsin's Attorney General and the DNR argued that Schapiro's Friends of Milwaukee's River charges were false, that Milwaukee's outfall system is one of the best among the outfall systems for cities over 1 million population, and that a $900 million project to improve operations was hardly a "sweetheart deal," as was charged. A check with the web site of the *National Association of Clean Water Agencies* confirmed that in 2006 the MMSD was given the Platinum (highest) award for 9 previous years of "100% compliance with EPA's NPDES (discharge) permits (NACWA, 2006)."

A longtime official of the Wisconsin DNR added details. He reported private discussions with the *Friends of Milwaukee's Rivers* in which he asked what the group really wanted MMSD to do. There were few specifics except to "move faster." Reporters Rohde and Schultz had been provided extensive background data. But their newspaper reports continued to feature a crusader who had received the *Wisconsin Law Journal* award as a "Leader in the Law." Finally, he said, the editor of the *Journal-Sentinel* removed Rohde and Schultz from the MMSD case, replacing them with an experienced environmental reporter. However, he pointed out that the impressions made by Rohde and Schultz's many articles had wider impact than just on the general public. After the Appeals Court decision the three-judge panel was said to have been quoted as acknowledging that they got much of their background from coverage in the *Milwaukee Journal-Sentinel*.[221]

8.4.2.3.1 Discussion

Without getting into more specific details of the Milwaukee case, several conclusions can be drawn.

1. First, the MMSD is a technically qualified, award-winning agency with a strong tradition of good management. In recent years it has faced increases in sewerage inflows from satellite towns beyond its jurisdiction. It has proposed and executed systematic upgrades to operating capacities, in part by adding holding tunnels beneath the city. However, under severe rains no feasible sewer system short of unrealistic costs can contain all the incoming water. In big rains the system will continue to discharge diluted raw sewage to Lake Michigan, conforming to exceptions in interpretations of the CWA by EPA. Past tests show that during such periods 93–97% of near shore lake waters have met safe quality standards, due to dilution and rapid bacterial breakdown of organic matter.
2. The Milwaukee case shows the powerful and disproportionate influence that a single local lawyer with modest support can generate, tying up the activities and resources of major public institutions, even where grounds for suit are questionable. The desires of newspaper reporters for high-impact human-interest coverage created a shallow and distorted picture of complex issues that became disseminated to a large regional public – extended to the Chicago area from the originating Milwaukee paper. The newspaper coverage also appears to have disproportionately influenced the decision of a federal appeals court. I could not discern any specific or concrete improvement either sought or achieved by 6 years of litigation.
3. A rarely considered but important issue involves the impact of litigiousness promoted by current laws (e.g., CWA) on morale and effectiveness of public service organizations like the MMSD. Seeing their agency portrayed as incompetent or badly motivated through accusations or lawsuits given colorful coverage in local papers can hardly fail to affect employee or leadership morale. Defensive activities divert energy from the job, constructive initiatives and cooperative outreach. They may adversely affect even a good organization. A flawed organization operating in a hostile environment is less likely to attract the leadership and staff talent needed to improve its operations.
4. These issues inhibit the development of public trust, which is essential to allow agencies to be willing to consider bold, innovative ideas and actions that solve large problems more efficiently, and to overcome problems in the most effective rather than politically safest ways.

8.4.3 CERCLA (Superfund Act)

One of the prime examples of polarization and gridlock is the continuing controversy and dissatisfaction over the Superfund, created under the Comprehensive

Environmental Response, Compensation and Liability Act (CERCLA) (1980), to clean up hazardous wastes sites in the United States. One of the building blocks for CERCLA was the Water Quality Act of 1970 (Leon Billings, personal communication, 2008; Billings was Staff Director, Senate Committee on Public Works' Subcommittee on Air & Water Pollution, 1966–1978). The Water Quality Act, passed in the aftermath of the Santa Barbara oil spill, contains provisions for absolute liability for clean up of oil spills. The National Academy issued a lengthy report on the strengths and weaknesses of the Superfund activity in the Coeur d'Alene River mining area (National Research Council, 2005).

Among the criticisms in this report and cited references was that the fund had spent a large part of its money studying or administering sites or engaging in litigation, and left significant health risks for residents in more than a thousand sites remaining on the Superfund list. One of the pervasive concerns about Superfund sites was their potential effect on local groundwater (Lautenberg, 1997).[222] A reverse criticism was that the Superfund program lacked a cogent strategy to remediate existing sites and to identify as yet unidentified sites.

The reduction of support for the Superfund Trust Fund left Federally funded cleanup operations to annual appropriations of between 1 and 2 billion. The general population and writers in newspapers covering communities in the vicinity of known hazardous sites were left with a general perception of decline in effort and effectiveness on the part of EPA's management of CERCLA. This has been particularly true of the industrial corridor frequently termed "Cancer Alley," adjoining Baton Rouge, LA (e.g., Cox, 2003). Recognition that residents close to present or former industrial centers of activity often have lower incomes or larger minority populations, a new movement concerned with "environmental justice," which has become elaborated in the social science and law literature that emerged in the middle 1980s (Cole, 1994; Schlosberg, 1999).

Some earlier critics of the Superfund system, like Frank Lautenberg, have modified their stand to give it good marks with reservations (Lautenberg, 1997). Nakamura and Church analyzed the evolution of the superfund program and concluded that administrative changes in 1995 were the keys to successful modification. "It is our judgment that three factors accounted for the success of the 1995 reforms: the *leadership* at the top of the agency, the *management* of the reforms throughout the agency, and *attention to individual incentives* at both headquarters and regional levels" (Nakamura & Church, 2003). Conservative critics have a different view. The new cooperativeness on the part of EPA after the Republican tide of 1994 and the "Reinventing Superfund" initiative in the Clinton administration marked a major and welcome change in EPA's interface with users. But it does not change the inflexible, one size-fits-all mechanisms imposed on EPA by Superfund's Congressional mandates.

Superfund's fundamental identification with cleanup of dangerous toxic wastes has in the past always struck a responsive chord with both the public and Congress (Greve, Smith, & Wilson, 1992). Past attempts at reform have therefore not succeeded, in spite of the program's problematic history, as well as criticisms that EPA's interpretations of Superfund fail to deliver environmental justice. It is also probably true that past reformers were dominated by industrial or business interests whose main goal was to reduce economic impacts.

In the context of the huge costs potentially associated with coming energy policy changes and massive infrastructural modernization a blunt weapon like CERCLA is not only outdated but poses future risk. Its restrictive nature and legal complexities may block needed innovative and integrated plans in areas wherever newly found hazardous waste sites, or existing brownfields and other sites under CERCLA's jurisdiction are located.

8.4.4 Case Example of the Endangered Species Act (ESA): The San Francisco Bay Delta Smelt Petition

8.4.4.1 Background of the ESA

The ESA law was deliberately designed to protect endangered species against competition from economic activities. The law's designers reasoned that if competing considerations were allowed to influence decisions on endangered species, especially those that were not charismatic or had symbolic significance like the bald eagle, the endangered species would nearly always lose. Loss of priceless biological legacies would become inevitable. In subsequent years, controversies over species with insignificant-sounding names like the snail darter, the Furbish lousewort, and the Prebbles meadow jumping mouse would demonstrate just how powerful the ESA was. Litigation over the more charismatic spotted owl shut down a significant part of former western lumbering industries. ESA became a symbol of almost continuous conflict in the western states although supporters claimed that it had resulted in only a few cases of land taking.

In the 1960s, Rachel Carson's book on the impact of pesticides "heightened general public awareness of environmental issues, mobilized action groups, and stimulated development of..... preservation-oriented groups. These groups dramatically changed the nature of wildlife policymaking (Yaffee, 1982)."

The 1966 Endangered Species *Preservation* Act, stimulated by the Bureau of Sport Fisheries and Wildlife scientists, followed by the Endangered Species *Conservation* Act of 1969, outlined a program for aqueous habitat, regulated taking, and required interagency cooperation. A key provision in this bill gave the Secretary of the Interior authority to consider scientific evidence and make decisions on species listings and operational actions. Another was allowance for reimbursement of landowners when their property or activities were restricted through implementation of endangered species protections. The latter provision created options that take a step beyond the habitat management approach adopted by the Department of the Interior for biodiversity protection. This "biodiversity approach, adopted by Interior, allowed landowners who have endangered species on their property and agree to a habitat conservation plan to avoid having to take additional steps to protect a listed species (WEF, 2006)."

Stimulus for the far more drastic Endangered Species Act of 1973 was provided in President Nixon's February 1972 speech on environment. An initial draft of the act by Fish and Wildlife Service scientists was modified by powerful activist groups in the bill introduced by Senator Mark Hatfield (R-OR) and Representative John Dingell (D-MI). Most important, provision for citizen litigation was added. "The extremely strong, comprehensive and prohibitory statement" in the Conference-approved bill passed by a vote of 355 to 4 in the House, because "it was not clear that anyone would be hurt by it (Yaffee, 1982)."

8.4.4.2 The Delta Smelt Case

The Delta smelt provides a classic example of the collision between ESA and human needs. On March 8, 2006, the Center for Biodiversity, the Bay Institute and the Natural Resources Defense Council (NRDC) submitted an urgent petition to the FWS regarding the potential loss of the Delta smelt (*Hypomesus transpacificus*). Unusually large exports (usage) of water were held by petitioners to FWS to have sharply decreased the 2 parts per thousand salinity area in the delta, and caused major loss of the smelt. The petitioners asked that the smelt be given endangered status under ESA provisions that trigger mandatory action on the part of FWS. ESA was invoked to support increasing water supply to maintain the Sacramento-San Joaquin Delta, which formed part of the life cycle of the fragile pelagic fish community. If approved, it could lead to rapid promulgation of an emergency rule invoking special protections under the Endangered Species Act.

The habitat of the smelt is the Sacramento-San Joaquin Delta, part of San Francisco Bay where freshwater outflow of the combined Sacramento and San Joaquin rivers dilutes salty Bay water to create a delta characterized by brackish waters. The problem was stated to have been created by increased export of freshwater from the rivers to land uses. The petition claimed that some 12 of the 29 pelagic fish species endemic to the Delta area had already been lost, and the smelt was a signal for the imminent collapse of the entire brackish water ecosystem (Center for Biological Diversity, 2006).

The Water Education Foundation's brief on the delta controversy points out that the issue has been in contention for decades among farming interests, cities, and environmentalists. A complex and detailed agreement (CAFED) was concluded in 1994. It attempted to conserve water and institute actions that could help serve all the concerned interests. The river systems in question supply drinking water to about two thirds of the State of California, as well as local agricultural and industrial needs.

The fish in question is a pelagic (free-swimming) species that spawns in fresh or slightly brackish water and spends most of its life in a brackish zone where freshwater from the confluence of the Sacramento and San Joaquin Rivers (Delta area) mixes with salty San Francisco Bay water. The smelt's life and reproductive cycle is interfered with by human activities like "sand and gravel mining, diking, dredging, and levee or bank protection and maintenance and flow disruption (e.g., water diversions that result in entrainment and in-channel barriers or tidal gates)." But most important was maintaining enough freshwater outflow into the Bay. Upstream water quality was also important.

The petitioners offered what appears to be a strong case that the 2–3 in., semi-transparent Delta smelt qualifies for upgrading from *threatened* to *endangered status* from an earlier 1993 action by FWS, listing it among "threatened" species. FWS also determined that the smelt had a critical habitat. A similar "threatened" designation had been made by the California Division of Fish and Game, which was among the agencies conducting regular samplings.

The petitioners cited studies (Bennett, 2005) which show the smelt population had fallen to 2.4% of its 1993 population, well below "effective size", below which there would be inbreeding and genetic drift. It was now reported to have a probability above 20% that it would become extinct in 20 years. The 20%/20-year criterion was applied by the International Union for Conservation of Nature and Natural Resources to delineate an endangered species. A multidecade record of population sampling for the smelt showed irregular decreases in population along with low dips in population that correlated with increased water exports from the Sacramento-San Joaquin river system to land uses.

The water shortage in California, whose population is continually increasing and currently numbers about 23 million people poses a difficult battle to the petitioners to FWS. Though the Sacramento Chief of FWS' district office promised expedited evaluation, he served in a Bush administration whose policies favored land needs.

Although the President's influence was weakened by low ratings on public polls, if the petitioners' positions were accepted, water shortages in California could result in backlash against the environmental organizations and bad publicity for ESA, even in environmentally conscious California. Given this situation, it was perhaps understandable that when Ted Sommer, Chief of the California Department of Water Resources Aquatic Ecology Division, was asked when his group would finish its report, he said, "We still have a lot of questions that we need to answer over the next couple of years." According to a special web site on the delta smelt (http://www.deltasmelt.com/) it is clear that the issues have not been resolved.

8.4.4.3 Outcomes – an attempt at reform of ESA

The Delta represents a "zero sum game" confrontation in which straightforward decisions favoring either side involve losses to the other – with the values in question being asymmetric – i.e., having completely different character.

The Delta water and environmental issues also gave rise to political frustration and dissatisfaction (Thompson, 2006). Representative Richard Pombo (R), Chairman of the House Resources Committee, had his Congressional District #11 in Central Valley. Pombo, a critic of the Endangered Species Act, was quoted as saying, [Water managers have spent 15 years] "spending literally hundreds of millions of dollars, and billions of dollars in lost economic activity, and none of that has worked."

Dennis Cardoza (D), whose district is in the San Joaquin Valley, said "blame has centered variously on the pumps that divert water to farmers and cities, on power plants, on invasive species, on a decline in the food chain and on toxic contamination." He was disappointed by the lack of answers from numerous studies and conferences. George Miller (D), the former Chairman of the House Resources Committee and an activist on behalf of conservation, wildlife, and water issues criticized state and federal water managers "for proposing an increase in pumping without first studying whether it will further damage the delta (Pombo, 2006)".

Probably because of the protracted problems associated with the controversy, Representative Pombo, as Chairman of the House of Representatives Natural Resources Committee initiated an unprecedentedly detailed review of ESA. It culminated in a proposal to reauthorize and reform ESA in the form of TESRA (The Threatened and Endangered Species Recovery Act), approved in 2005 by the House Committee on Natural Resources. In this legislation an entirely new approach was taken. Opponents said that the Republican pollster and legislative strategist, Frank Luntz influenced the direction (American Civil Liberties Union, 2005).[223] A companion bill, S.2110 was subsequently introduced in the Senate by Senator Crapo (R) of Idaho. Many environmental organizations as well as the ACLU, which submitted a highly detailed brief (WEF, 2006), mobilized to defend ESA and reject TESRA.[224] The mobilization helped defeat S.2110, and went on to defeat Representative Pombo, who was weakened by associations with convicted lobbyist, Jack Abramoff, in the elections of 2006.

Environmentalists clearly feel that it is necessary to be uncompromising even at the cost of conflict with economic interests. Perhaps, if pressed about the economic impacts, they might argue that society can devise alternative solutions to meet water needs, but endangered species have no such flexibility. In the past, such impasses have invariably gone into the civil court system for resolution.

By its no-compromise focus on individual endangered species and excluding discretion by professional managers, the ESA raises each conflict or controversy over endangered species to the level of a fundamental choice guaranteed to generate conflict, winners, and losers but not thoughtful, balanced decisions.

8.4.5 Marine Fisheries Act (MFA) of 1976

Too civil for chaotic times, the MFA presided over the collapse of the Georges Bank and other American fisheries.

Fish landings for the northwestern Atlantic US fisheries were dominated by the great Georges Bank fishery after World War II. This fishery is notable for cod and ground fish like halibut, haddock, flounder, and pollock, as well as more migratory fish like herring, and valuable shellfish (sea scallops). Fishing pressure during the war had eased due to fishermen enlisted in the armed services and the absence of foreign fleets during the war. After the war fishing pressure increased rapidly. The Bureau of Commercial Fisheries was the US agency in charge of marine fisheries but it had no control of fisheries in international waters which then covered the productive areas. Fisheries matters were included in the International Council for Exploration of the Sea (ICES) meetings, and later the International Council for the Coordination of North Atlantic Fisheries (ICNAF). But the agreements were unable to control heavy fishing pressure from distant fleets (Pauly & MacLean, 2003). By 1976, fish landings had declined steeply from those of the 1950s.

The Marine Fisheries Act of 1976 was sponsored by Senator Warren Magnuson, an influential and highly respected Senator from Washington State. It began by declaring a 200 mile exclusive US fisheries management zone. With one stroke the law excluded fishing fleets from Spain, East Germany, Poland, and especially the Soviet Union, with its trawlers and huge factory ships, whose workers remained at sea much of the year, canning fish for Soviet consumers from Vladivostok to Leningrad. MFA was very different from the environmental regulatory laws of the 1970s. Instead of handing down detailed operational procedures and emphasizing compliance, MFA delegated regulatory responsibility to regional fisheries management councils. These were to have representation from all stakeholders, fishermen and fish processors, political and citizen representatives, and fishery scientists from the National Marine Fishery Service (NMFS), a division of the National Oceanic and Atmospheric Administration. Overall coordination was provided by the NMFS, which was made responsible for issuing fisheries regulations pursuant to the recommendations of the individual councils. There was praise for the law's balance, fairness, and provision for scientific guidance.

Figure 7.1 shows historical fish landings for haddock in the northeast Atlantic fisheries, and Pacific halibut landings for the same period. Haddock was a key commercial ground fish in the Georges Bank fishery. Having experienced a steep decline since 1967, haddock landings showed a bubble of increased landings after 1976. These represented stocks formerly taken by the foreign fleets, now exclusively accessible to American fishermen. However, within a few years the haddock landings crashed again. This time all ground fish populations in the once magnificent fishery were reduced to such low levels that even young gadoid (cod-family) fish were devoured by the now dominant elasmobranch (shark and skate) populations.

The entire bottom-feeding and mixed water-column-bottom feeding fish population including haddock, flounder, cod, and pollock on Georges Bank was devastated, in part by trawlers that not only captured bottom fish, but scraped virtually all life from the sea floor. Fishermen whose families had fished for three and four generations lost their livelihoods. Fish processors shut down, and seafood prices soared.

The Marine Fisheries Act handed over a huge fishery to US fishermen exclusively. The existing fishing boat fleet rushed to exploit the newly available resources within the exclusive 200-mile zone. NOAA–NMFS added to the fleet by subsidizing the building of new fishing boats. The ensuing overfishing caused the final collapse of the Georges Bank groundfish fisheries – one of the most spectacular fishery management disasters in history.

Progress in rebuilding the north Atlantic fisheries was slow and painful, except for rockfish (striped bass) populations. Recognizing that the original fisheries act had presided over decline or destruction of fisheries around most of the United States, except for the North Pacific Fisheries area, Congress amended the 1976 law to become the Magnuson–Stevens Fishery Conservation and Management Act of 1996. The original law's mandates to simultaneously foster maximum fish yields while preventing stocks from falling below sustainable limits had not put sufficient emphasis on maintaining fish stocks. The reauthorization Act strengthened these provisions and called for increased scientific research. However, after the law failed to produce significant improvements, it was amended yet again in 2006 to allow still tighter restriction on fishing (Federal Register, 2007).

A major problem with the Magnuson–Stevens Fisheries Act was that it assumed orderly and responsible functioning and self-discipline at all levels of the system. The regional fisheries Management Councils would produce wise fishing plans; NMFS would provide appropriate scientific oversight and regional coordination, including contacts with Canadian fisheries. However, these assumptions were all breached by real world factors and growing fragmentation of society. NMFS was not given the resources needed to make adequate scientific assessments of fish resources. The respect for scientific agencies that Senator Magnuson experienced in his earlier career broke down in the increasingly chaotic world of the 1970s. NMFS's rulings and recommendations – if "too stringent" – were overridden by officials in the Department of Commerce (Tillion, 2002).[225]

The Magnuson Act was not alone in affecting fisheries. These were subject to other federal and state laws, including the Marine Mammal Protection Act, the Endangered Species Act, the Coastal Zone Management Act, and the National Marine Sanctuaries Act, as well NEPA, the Administrative Procedure Act, Federal Advisory Committee Act, Paperwork Reduction Act, and the Regulatory Flexibility Act as well as international agreements. A major conference, sponsored by the North Pacific Fishery Management Council in 2005 (Oliver, 2006) concluded that:

"While each of these laws has a necessary purpose, a growing number of policy makers believe the manner in which they are individually applied in the process of marine fishery management results in incompatible regulatory requirements, management dilemmas and confusion among stakeholders. Conflicts among statutes have led to cumbersome and sometimes unnecessarily complex administrative procedures, on occasion resulting in lengthy delays in the approval of regulations. Some claim the interplay between statutes and conflicting mandates has resulted in jurisdictional battles and poor management decisions."

In other words, the overlap of legal processes, coupled with the outcome of various lawsuits complicated an unwieldy process for implementing fishery management changes. At the conference it was indicated that NEPA had become an overriding influence on fishery management. It was seen to pose insuperable difficulties in conducting efficient regulation.

In contrast, as noted, the International halibut agreements largely took management of this fish out of the realm of the United States' internal regulatory environment. Scientists and experienced fishery managers had the main role in developing management strategies. The results tell an important story.

The most effective approach to fisheries management has been a system that involves communication with fishermen and fisheries industries, but gives ultimate authority to professional managers guided by fishery scientists and research, as in Norway and the Pacific Fishery Management Council in the US northwest (McIsaac & Hansen, 2005).

8.5 Corporate Scandals

On April 26, 2005, The Washington Post reported an extraordinary list of recent corporate ethics scandals from its recent issues. Subsequently, the subprime mortgage problems revealed an unprecedented loss of judgment in important segments of the banking and finance community.

- *"Spitzer's Charges Face a Challenge*: NEW YORK – Theodore C. Sihpol III is fighting back. Former Bank of America Corp. securities broker Sihpol goes to trial today on 40 criminal charges for his role in helping a hedge fund place after-hours trades that Spitzer says are illegal."[226]
- *"Adelphia*: US settles for $715 million: founder, family to compensate shareholders (Post, April 26, 2005)."
- *"FBR Nears Settlement with SEC*: Firm has set aside $7.5 million related to CompuDyne Deal (Post, April 26, 2005)."
- *"Wal-Mart Says Inquiry Names Ousted Official*: A federal grand jury in Arkansas is investigating allegations that the former head of U.S. operations at Wal-Mart Stores Inc. misused up to $500,000 in corporate funds, the giant retailer said yesterday (Post, April 23, 2005)."
- *"Tax Shelter Leaders Get Jail Time*: Must pay restitution (April 23, 2005)."
- *"Analyst Fired for Personal Trading*: FBR says Kalla failed to disclose stock deals (Post, April 22, 2005)."
- *"BearingPoint Cites SEC Inquiry*: Warns investors (April 21, 2005)." "[Enron] Prosecutors Want Lay to Be Tried Soon (By Carrie Johnson, Page E02)."
- *"Dynegy Inc., a US power producer that nearly went bankrupt after Enron Corp. collapsed*, will pay $468 million to settle shareholder claims that it misled investors by disguising loans as energy trades in 2001 (Bradley Keoun, P. E02)."
- *"Tyco Case Juror Told Judge of Her Fears*: Phone call, letter upset holdout (By Carrie Johnson, P. E03)."
- *"9 U.S. Foodservice Suppliers Charged*: Prosecutors say defendants conspired to help columbia firm inflate earnings (by Brooke A. Masters, P. E03)."

- *"J.P. Morgan helped underwrite major WorldCom bond offerings....*: largest bankruptcy case in U.S. history in 2002. Plaintiffs..... claim J.P. Morgan and other WorldCom advisers and directors should have discovered the company's $11 billion accounting fraud and disclosed it to investors."

- *Term paper exchange site*: "Free Term Papers on Impact of Recent Accounting and Financial Scandals (OPPapers.com)." Adding an ironic twist, this web site helping students cheat offers essays on the financial scandals.

A Harris poll of November 1, 2007 showed that confidence in the ethics of a number of industries had fallen to the lowest levels ever recorded. Only 3% of all adults "believe that statements made by oil companies are generally honest and trustworthy" (HarrisPoll, 2007). Given the fact that the US oil industry plays a major role in national economic life, and will be critically involved in future energy policy, this abysmal situation has critical significance. It should not be dismissed with a "well, that's the way things are" attitude.

Following disclosure of heedless profit seeking in banking sectors, a barrage of criticism erupted against advocate's of banking deregulation.

8.6 Campaigns of Environmental Activist Organizations

These quotes are not comparable with the corporate scandals in terms of illegal and criminal acts. However, the campaigns of environmental organizations are often characterized by loaded language, exaggerations, distortions, and single-issue pursuit of ideological causes. Backed by skilled lawyers, these campaigns confuse and aggravate hostilities and have hardened political opposition to any environmentalist-supported energy policy initiatives. They are, therefore, ultimately counterproductive.

- *Environment New Hampshire* (environmentnewhampshire.org). "Offshore drilling creates a heavy burden on the oceans. Even when there are no large spills, the drilling and production process routinely releases hundreds of thousands of gallons of water and mud tainted with mercury, carcinogens and poisons into fragile ecosystems. And the risk of catastrophic disaster, like the 1969 oil spill that ravaged Santa Barbara, CA, is always present."
- *Environment Florida* (Stop offshore drilling – environmentflorida.org). "The oil lobby would like us to believe that after Katrina and Rita, we can drill our way out of our nation's energy problems. We're not buying it. Opening our shores to drilling would only put our beaches and coastal waters at great risk for a small, short-term supply of oil and gas. Routine drilling operations dump thousands of pounds of toxic chemicals into the marine environment, and a catastrophic spill – one that could spoil the ecology and economic value of Florida beaches for generations – is a real possibility. We can do better."
- *Environment America* (environmentamerica.org). "Offshore oil rigs pose catastrophic risks to coastal communities, ocean ecosystems and marine life, such as

sea birds, sea turtles, and marine mammals. For 25 years, Environment America's staff has been fighting successfully to keep offshore drilling out of the Atlantic and Pacific oceans."

- *Sierra Club Legal Defense Fund.* Promotional letter from Fredric P. Sutherland, President of the Sierra Club Legal Defense Fund in 1990, before the name change indicated in the next item; it listed 82 active lawsuits in 7 areas of activity. "Our world is drowning in filth. Garbage covers the land, and toxins seep into our groundwater and our homes."
- *Earthjustice* Promotional letter 2006 or 2007. "*Earthjustice* was founded in 1971 as the *Sierra Club Legal Defense Fund*; in 1997 we changed our name to *Earthjustice* to better reflect our role as a legal advocate for the environment (because, like our tagline says, the Earth needs a good lawyer)."
- *Defenders of Wildlife.* Filed suit with Sierra Club in Supreme Court to prevent *Dept. of Homeland Security* from erecting fence at the border with Mexico in the southwestern states, claiming "the fence will disrupt the migrations of various species, including imperiled ones such as jaguars (Eilperin, 2008)."
- *Natural Resources Defense Council.* Fundraising letter from Frances Beineke, NRDC President, 2008. "Only NRDC combines grassroots power with the policy clout of 300 attorneys, scientists, and other professionals – making us the nation's most effective environmental group. According to the national online news magazine *Salon.com,* NRDC boasts a more powerful climate arm than any other environmental group."
- *Natural Resources Defense Council.* Fundraising letter for NRDC from Robert Redford, 2007. "The Administration recently created a massive new loophole in the Clean Air Act that will allow 17,000 of the nation's worst polluters to spew more toxic chemicals into our air and harm the health of millions of Americans..... It has proposed new rules that will open all 155 of our national forests to logging, drilling, and mining..... and it has intervened to block a vitally needed program that would stem the flow of raw sewage into America's oceans, rivers, and other waterways."

Bush administration initiatives that are disingenuously named to sound like *environmental* initiatives (*Clear Skies* and *Healthy Forests*) anger environmentalists and convince them of the administration's bad motives.[227] On the other hand, I have to go back to the Gorsuch–Watt era to find such supreme assurance in the rightness of ideological causes as is expressed in suits declaring that the security of the US national borders is less important than animal migration paths (Eilperin, 2008).

8.7 Virginia Offshore Oil and Gas Issue

In 2005 and 2006, interest grew in Congressional legislation that would reopen areas under moratorium to oil and gas leasing where the adjacent states voluntarily accepted such action. Florida, North Carolina, and New Jersey had a record of political opposition to new development, but Virginia's position had been more favorable. Although the state had only a relatively narrow slice of the US EEZ within its purview, a modest potential for natural gas existed (Fahrenthold & Mufson, 2007). In the tradition of the administration of Governor Mark Warner, a pragmatic Democrat who had been successful in bringing the parties together in achieving concrete results, the Virginia General Assembly commissioned a detailed report on the merits of offshore oil and gas leasing, the House Joint Resolution #625. Coordinated by a former aide to Governor Warner, the Report found positive potential and recommended going ahead with experimental leasing. One of the attractions was that an augmented supply of natural gas would be a major asset to Virginia, which was faced with the possibility of new coal-fired power plants. Even if no hydrocarbons were found, the front-end bonus bids for offshore leasing could yield revenues to Virginia that could be invested in renewable energy development, including attracting relevant industries to the State.

Incoming Governor Timothy Kaine (D) stripped out the offshore component of the policy report and sent it back to the legislature (Town, 2006). Kaine offered the softening but largely meaningless note that he would not object to exploration of potential offshore energy resources more than 50 miles from the Virginia coast, provided it was funded with federal dollars.

This would have been the first major opportunity for realistic dialogs about energy policy since the 1970s. During the past 25 years, debates about offshore drilling have mainly featured spokespersons addressing their own constituencies or trying to influence decisions – never pursuing meaningful discussions. Pro-development speakers point out the need and opportunity for energy recovery as though opposition and deep concerns on the other side did not exist or that any reasonable person should agree. Opponents use slogans like *we can't drill our way to energy independence!* and offer arguments designed to raise emotions and evoke prejudice against offshore activities.

In a real dialog moderated by experienced mediators, the "lock them in a room until a breakthrough is reached" approach would probably show that offshore drilling was only the tip of an iceberg of misunderstandings on assessments of national energy policy issues. Previous US and EU precedents suggest that negotiations would yield arguments on some points, uncertainty on others that could be resolved by impartial assessments, and some remaining disagreements. Continuation of the process would expand the agreements and "solution area" and reduce the untouchable areas, while new conflicts would be reduced by the continuing dialog. Such an approach will involve leadership with recognition that high walls impeding realistic assessment must come down if policies and programs are to do more than paper over obstacles to progress.

Importation of gas is encumbered by another result of US gridlock. Of the dozens of proposed liquid natural gas terminals for the East Coast of the United States most have run into opposition citing fear of explosions or accidents. Thus, existing natural gas terminals are concentrated in the Gulf of Mexico area – necessitating long distance transport to users areas in the East and West Coasts.

8.8 Guerrilla Warfare

Chapters 1 and 2 reviewed the widening gap between the environmentalist community and "industry," taking "industry" as the larger industry, financial, and business circles. Antagonism began in the 1960s with antibusiness movements in the academic community, exacerbated by the Vietnam War. This discussion looks at the most polarized end members of the subsequent standoff. It is not my favorite area of the book, but is needed because some of the consequences of the activities have been overlooked.

Diverse opinions and beliefs, and criticism of concepts, behaviors, or actions are essential in democracies. They are a strength and help create robust policy as long as there is communication. Once groups become isolated and estranged from each other – i.e., groups get their information and reinforcement largely from their own or like-minded sources, radical voices tend to exercise increasing influence – which further increases alienation. Exaggerated or distorted categorization or labeling of the "out" group or ideas increases mutual antagonism and ideological distance. In some cases alienated groups spring up *de novo.* In other cases they may get increasingly radicalized owing to prolonged conflict.

Guerrilla warfare is used in this section to describe radical antagonisms that express themselves by vilifying, attacking, or demonizing targeted individuals or groups, or use associations to stigmatize other individuals or ideas as inherently bad or dishonorable.[228] A thesis of this book is that whereas group isolation of any kind leads to poor decision making, more extreme attack strategies, even where they achieved short-term success have harmed both sides and society as a whole in the long run. Those EU nations that have simultaneously achieved success in both environmental and economic areas, like the Scandinavian nations and Austria, have no conflict or gridlock.

8.8.1 At the Extreme End

The web site of the environmental "group" Earth First! describes its philosophy frankly:

> "Earth First! was named in 1979 in response to a lethargic, compromising, and increasingly corporate environmental community. Earth First! takes a decidedly different tack towards environmental issues. We believe in using all the tools in the tool box, ranging from grassroots organizing and involvement in the legal process to civil disobedience and monkey wrenching" (http://www.earthfirst.org/about.htm).

It regards itself not as a *group* but a *movement.* Its logo includes a raised, shaking fist. Earth First! operates worldwide, from Canada to the Amazon basin, the Philippines, Australia, The Netherlands, Britain, and elsewhere. A conservative opposition tracking site (National Center, 1993) claims that:

"David Foreman, a former Washington lobbyist for the Wilderness Society, was the co-founder and board member of The Cenozoic Society and the founder of Earth

First! He is author of *A Field Guide to Monkey Wrenching* and *Ecodefense*, books that provide detailed instruction on how to sabotage equipment, industrial projects, roads and vehicles in the name of environmental protection. Foreman left Earth First!"

Going beyond Earth First!'s organized and publicized tactics, some individual and small-group eco-anarchists have branched out into car bombings and death threats that have effectively curtailed some medical research efforts in the United States, United Kingdom, and elsewhere. They typically remain underground, out of sight, until they claim credit for a high-profile illegal act as a warning to others they oppose.

I am not aware of any current corporate counterpart to Earth First! However, there is an extensive literature on violent hired strikebreakers used by some US companies from the 1870s to about the 1920s (Malmgren, 2005).

8.8.1.1 Angry Voices

Al Franken, a Harvard-educated comedian, political commentator and radio host, is an Emmy winner and a candidate for Senator from Minnesota. However, his visceral antipathy to President Bush is expressed so graphically that a continuation of the following passage, "Vast lagoons of pig feees: The Bush environmental record" (from Franken's book, *Lies and the Lying Liars who Tell Them*, Dutton, 2003) is not suitable for this book.

Ann Coulter. Best selling author, syndicated columnist, and outspoken conservative American commentator, Ann Coulter, mercilessly scores the left wing establishment, as is revealed in book titles like: *"How to Talk to a Liberal (If You Must): The World According to Ann Coulter," "Godless: The Church of Liberalism," "If Democrats Had Any Brains, They'd Be Republicans."*[229] She has captured the imagination and consolidated the beliefs of conservatives who may not have realized their principles and way of life was under attack.

Franken and Coulter's popularity with large constituencies demonstrates the tendency toward desensitization of the nation to vilification and ridicule of individuals or beliefs over the past 30 years. A mainstream manifestation of this trend is found in reality shows that exploit the putdown or exclusion of individuals from groups as popular entertainment. In the often tolerant style of the times, *Paul Starobin* (Starobin, 2004) looks at the "Angry American" from various political and other perspectives. He makes a valid point – that Americans would not be angry or support partisan causes "if they didn't think there was a payoff," i.e., they are not blasé and or apathetic. He finishes his article with the remark that perhaps the country needs even more anger. Since Starobin wrote his article in 2004, he may have gotten what he wished. Or, if he fell into the trap of thinking that more anger could let "his side" (whatever that might be) "win the battle," he is following a common pattern of thinking that has yielded ever greater dissatisfaction in the United States and no light at the end of the tunnel.

8.8.1.2 Intellectual Analysts and Opposition Trackers

Joel Bakan, a professor at the University of British Columbia who teaches constitutional law, is the author of a widely known book, *The Corporation: The Pathological Pursuit of Profit and Power* (Bakan, 2007). The alliterative title reflects Bakan's view that corporations are designed and operate only for profit and power, without regard to negative consequences; they are inherently pathological and criminal. The book was made into an award-winning film, evidently has a wide following, and was persuasive to the majority of 45+ Amazon.com reviewers. The point is that Bakan, who advocates strongly limiting but not banning corporations, articulates what a significant number of North Americans feel and think – that corporations are all bad and need to be curbed, regulated, and opposed at every turn. A Harris poll yields a rock-bottom 3% of all adults who think that "statements made by oil companies are honest and trustworthy."[230] (The point that scholarly justification for antagonisms intensifies US gridlock is explored further in Chapter 9.)

Sharon Beder. Australian political scientist and writer, Sharon Beder (2006a; 2006b) has waged a prolific campaign to characterize American corporate culture and free market policies as subverting democratic principles and values.

> "Corporate values emphasise mass conformity, subordination to authority, obedience and loyalty. Ironically, these values, which undermine individuality and freedom of expression, have been encouraged in the name of individuality and freedom. The market values of competition, salesmanship and deception have replaced the democratic ideals of truth and justice. Economic relationships have replaced social relationships. The power of the state has become subordinate to corporate interests. The realm of politics has increasingly narrowed as all major political parties are enrolled in the service of corporate interests (Beder, 2006a)."

Beder's books and articles point to the United States as the cradle of both the modern democratic state and the new "ideology of the free market missionaries." The idea that corporate public relations campaigns have created an oversimplified ideology that has had undesirable consequences deserves thoughtful consideration. The EU could hardly have reached its level of sophistication, cooperativeness, and effectiveness if member nations had mainly used US Advertising Council concepts as philosophical guidelines. The trouble is that Beder finds practically nothing to like in the United States during the last 100 years. More friendly, balanced evaluation might have had more constructive influence.

Ron Arnold and Alan Gottlieb. Fourteen years ago, a disillusioned Sierra Club environmentalist (Arnold) and an articulate conservative writer (Gottlieb) teamed up to expose what they saw as environmentalism's true impact. In *Trashing the Economy – How Runaway Environmentalism is Wrecking America* (Arnold & Gottlieb, 1994) Arnold and Gottlieb catalogued the organizational framework, links, and support systems of the environmental movement in a series of books. Arnold initiated a movement started in the late-1980s called "wise use" that combined concern for the natural environment with concern for human use of the environment. Critics of the movement call it anti-environmental, supported partly by the fact that the major funding has come from extractive industries. Wise use advocates speak of stewardship and question putting trees and animals ahead of humans. Arnold's work includes historical and scholarly research, and he documents his books intensively, as does Beder. What they share is an inclination to impute conspiratorial links that in a number of cases break down on examination.

Trackers. On both sides of the environmental-industry divide are tracking organizations and web sites. On the left is *The Sourcewatch Encyclopedia*, an activity of the Center for Media and Democracy that "Documents the PR and propaganda activities of public relations firms and public relations professionals engaged in managing and manipulating public perception, opinion and policy." SourceWatch also includes profiles on think tanks, industry-funded organizations, and industry-friendly experts that work to influence public opinion and public policy on behalf of corporations, governments, and special interests.[231]

Funding activities, scientists, and think tanks sponsored by the bête noir of the environmental establishment, the Exxon Corporation, are tracked by *ExxposeExxon. com* and by Greenpeace's *Exxon secrets*.[232] The activities of the Bush administration are tracked by BushGreenwatch.com.

On the political right the web site, *Activistcash.com* and *DiscoverTheNetwork. org* show profiles, histories, and funding sources for a long list of environmental organizations, foundations, and celebrity-activists. The Heartland Institute maintains *Polybot*, a massive database of articles, books, and other documentation on a wide variety of themes, compiled from some 350 conservative and free market sources. The George C. Marshall Institute provides supplementary material from the free market point of view.

Although not assured, in most cases the raw numbers and documentary information in the tracking organizations on both sides, listed above, seem reasonably accurate, although interpretive comments follow the persuasions of the sponsoring groups.[233]

8.9 Green Jobs, United States and Sweden

A comparison of responses retrieved for a query on "green jobs" was conducted for US and Swedish web sites through the Google search engine. The query for "green jobs" yielded 19,200,000 web sites. The query for the Swedish equivalent, "gröna jobb", yielded 259,000 hits. Responses on the first page of each search were reviewed and annotated in Table 8.2. The weighting systems built into the Google search engine (relevance to the query, frequency of hits, and other factors) identify national differences that are similar to other evidence discussed in Chapter 6. The sites in the English language query appeared to be from US sources, though several referred to overseas activities. All sites in the Swedish-language query were Swedish except one that was dominantly Finnish (Finland has a Swedish-speaking minority).

A large share of jobs offered on the American web sites are from environmental organizations. These and openings for attorneys, and managers for company regulatory

Table 8.2 Green job advertisements in the United States and Sweden

Response to "GREEN JOBS"	Response to "GRÖNA JOBB"
Green jobs – renewable energy jobs: senior solar product manager, three product marketing engineers, driver, solar crew member	Grönajobb.se: 262 openings for diverse jobs, many for work with animals, garden managers, landscape architects and engineers, mechanics, salespersons
Green jobs – cleantech: school program manager, corporation attorney, wetlands planner, two corporation regulatory affairs managers, solar installer, project managers and officers for three environmental organizations, two public administrators	Svart på vitt om gröna jobb: report on cooperative program between Forest Service and Employment Agency to place chronic unemployed workers in training positions as environmental and "culture" caretakers
Sustainable business: "green investing is hot," "worlds top sustainable stocks," three jobs identical to items in foregoing ad; distribution partner for jute manufacturer in India	Article in "To Work for Change," online magazine: "More green jobs with refuse." Sweden is a leader in export technology for incineration and material recovery
Greenbiz.com: bookkeeper, communications director sought by two environmental organizations; writer/PR director for reusable bag company, Chicago	Gröna jobb: Ympäristöprojekti: Green jobs in Finland, especially Lappland (border Sweden-Finland)
Green jobs: tree hugger: three lobbyists for Greenpeace, two finance managers, customer representative, "owner-manager," Rainforest Alliance, fundraising manager, Wholeness for Humanity, two summer marketing interns	Forestry Service Announcement: risk for continuation of project involving 500 workers and managers in Scania; work discussed in Svart pa vitt, above, involves care of parks and nature preserves and search for cultural remains in forests
Environmental job listings: provides links to various web sites	Cooperation, Forest Service and Office of History and Culture: documents program to hire workers
Green collar jobs: overview of green manual labor job types available in the San Francisco Bay area for "race, poverty, and the environment" organization	Norden: cooperation between Scandinavian countries. Article "Global Climate Crisis creates new jobs: In Denmark windmill industry increases jobs from 2,900 to 21,000; Norwegian firm leads in solar energy; 50,000 engineers will be needed in Sweden by 2010; New jobs as "energy inspector" created by new law requiring 600,000 homes be inspected for energy leaks
"Green jobs" to outweigh losses from climate change: This headline refers to an article about green jobs in Indonesia and Brazil, where 500,000 jobs have been created in the ethanol -sugar cane industry. [However, independent report indicates that most of the jobs involve hard manual labor at low wages]	

Source: Google.com, May 2008

affairs in US sites have no counterpart in Sweden. Sales clerks and communications managers round out the US scene, along with openings for two engineers and solar installers. Webzine articles discuss minority and poverty-related openings in the San Francisco Bay Area, and jobs harvesting biofuel crops in Indonesia and Brazil.

The Swedish responses to *gröna jobb* begin with a Swedish employment site offering hundreds of green jobs. Articles and reports involve cooperative programs of the Swedish Forestry Service and Employment Office, as well as Office of Historical and Cultural Affairs training opportunities in nature and cultural resource jobs for the chronically unemployed. Cooperation and advances achieved elsewhere in Scandinavia are reported in online magazine articles. A characteristic example of Swedish candor is provided by a warning from the Forest Service that one of the governmentally supported programs faces layoffs if fund shortages are not overcome. Is this a hint for readers to encourage funding for the agency project in question?

8.10 Alternative Energy Sources and Emission Reduction Technologies; Carbon Capture and Storage (CCS)

Among renewable energy sources, hydropower and ocean tide energy sources, as well as biofuels were discussed earlier. This supplementary discussion briefly reviews solar energy, geothermal energy, and other newer advanced technology approaches, and CCS, also referred to as carbon sequestration.

Table 8.3 provides an overview of potential technologies applicable to counter greenhouse gas emissions in various regions or types of areas in the United States, along with the technical and political (organizational) requirements to implement effective action. Some of the technologies were discussed in Chapter 5. Solar, geothermal, advanced "cleantech" systems, and carbon capture and storage are discussed here. Political, as well as technological advances will be needed.

8.10.1 Federal Solar Energy Programs and Management in the Past 25 Years: Erratic Policies Yielded a Poor Record

Solar energy development in the United States has a history that deserves careful study. The first Congressional laws and developments set in motion by the 1973 Arab oil crisis included the Solar Heating and Cooling Demonstration Act of 1974, the Energy Policy Conservation Act of 1975 and subsequent amendments; establishment of the Energy Research and Development Administration (ERDA) (1975), Solar Energy Research Institute (SERI), and Regional Solar Energy Centers (RSECS) (1977). The Department of Energy (DOE) was created under the Carter administration, with James Schlesinger as Secretary (1977). A further flurry of laws was passed from 1978 to 1980, followed by programs initiated after the "Second (Iran) oil shock" in the spring of 1979. Outlays for solar energy research and development peaked at about $1.6 billion in 1980 (in 2007 dollars). They dropped sharply in the Reagan administration, and remained low in the early Clinton administration (Beattie, 1997).

Although some good initiatives were set in motion, all eventually failed as a result of on-and-off federal support policies, chaotic tax policies, and lack of coordination with and among federal and state agencies and municipal and private utilities. Poignant consequences were noted for *Luz International Limited*, described

Table 8.3 Alternative energy strategies

Area/topic	Power source, emission reduction	Technical requirements	Organizational requirements	Transportation	Construction
Urban residential	Solar (distributed PV, CSP[1]), trash, biogas[2], heat pump	R&D/interconnection, metering	Organization, incentivation, financial support; policy and regulatory reform	Enhanced public transport/ automotive R&D	Insulation, appliances, appropriate design
Business, office, industrial sites	Solar, (distributed PV, CSP, solar collector, low-temperature geothermal, CCS[3]	R&D/interconnection, metering	Civic management and organization; policy and regulatory reform	Enhanced public transport	Green buildings
Farm, varied landscape	Biomass, limited biofuel, tree farming, wind, hydro, low-temperature geothermal, CCS	R&D/biotechnology/ interconnection/ metering	Interactive planning, management, incentivization	Rebuild innovative passenger rail	
Mountainous	Hydro, limited biomass, geothermal	Interconnection	Policy and regulatory reform and reconciliation	Variable, depending on area; extended rail	
Forest	Biomass (forest litter and selective removal)[4]	Monitoring, ecological and policy research	Policy and regulatory reform and reconciliation, interactive planning, management, incentivization	Minimally intrusive road system	
Coastal/ocean/ lake	Wind, ocean wave and tide, nutrient retrieval via algae; water recycling[5]	R&D development	Policy and regulatory reform and reconciliation, public interaction and education, interactive planning, management, incentivization	Passenger rail ocean transport	
Utilities	Cogeneration carbon burial, algal absorption, wind farms	Carbon sequestration, algaculture, coal technology	Political policy development, interactive planning, management, and incentivization	Variable/extend rail where not already served	Cogeneration, CCS

Source: Compiled by F.T. Manheim

[1] CSP: concentrated solar power includes concentrated solar photovoltaic (CPV) and concentrated solar thermal (CST)

[2] Pellets – trees, heavy brush

[3] Carbon capture and storage

[4] Dead, damaged, over mature; comprehensive removal after catastrophes, e.g., hurricane, flood

[5] Explore methods to reuse water discharged to ocean through large rivers

as a rare merger of business and social responsibility (Gozan, 1992). The company employed 1,800 people and constructed and operated nine solar electric generating systems (SEGs) in California. These had a rating of 354 MW, or 95% of the world's solar electricity at the time. Despite heroic efforts, the company declared bankruptcy in 1991, "an end deeply dismaying to solar advocates," but the systems are still operating.

In 2006 there were about 550 MW of installed solar/PV in the United States, with special interest being generated by new printable solar cells made of a copper-indium-gallium-diselenide semiconductor. Although some communities in the United States have not developed specific policies for residential solar, Feeney and Neumann (2004), report permits and fees as high as several thousand dollars in some parts of Southern California, not including switching and metering systems to connect to the electrical grid. Such regulatory issues are regarded as barriers to the development of otherwise promising solar photovoltaic systems not only in the United States but throughout the OECD (Philibert, 2006).

Beattie (1997) points out that after DOE assimilated previous solar-related programs in 1977 each subsequent administration reorganized solar activities. According to the author, the history of solar energy management by the US federal government is characterized by faulty models of resource base, demand growth, technology, and market capabilities for solar thermal and other systems, and wildly optimistic projections for future development. *The National Energy Plan* of 1977 projected 2.5 million solar homes by 1985 (Bezdek & Wendling, 2002). In his speech on June 20, 1979, recommitting to solar energy, President Jimmy Carter announced the goal of 20% of US energy needs to be met by solar resources by 2000 (Beattie, 1997).

Carter's policies share with other US presidents a pattern that one might call "inspirational," to distinguish it from the "three C's" prevalent in the EU policymaking: *consultation, coordination,* and *continuity.*[234] Carter's initiatives were often driven by short-term urgencies and personal convictions. Although he was trained as a nuclear engineer, was far better equipped to understand with the technological sides of energy issues than most members of Congress or other Presidents, and was skilled in political tactics, Carter made no intensive efforts to achieve consensus even among Democrats, and puckishly acknowledged that he had severely provoked Ronald Reagan, then an emerging Republican candidate for the election of 1980 (Carter, 2000). In the event Carter lost reelection he thereby undermined any chances for program continuity.

8.10.2 *"Cleantech" Power Projects*

In spite of reservations I expressed earlier about hype and hope, the push for green energy has clearly resulted in an explosion of advanced, impressive R&D developments in alternative energy. These include projects led by people famed for their IT credentials, and backed by discerning venture capital firms. The 15 leaders among the "25 Tech leaders who have jumped to Clean technology startup firms" (Fehrenbacher, 2008) are as follows:

- *Shai Agassi,* Founder, CEO Project Better Place.
- *Bob Metcalfe,* Partner, Polaris Venture Partners, CEO GreenFuel.
- *Vinod Khosla,* Founder, Khosla Ventures.
- *Sunil Paul,* Seed investor, early stage cleantech, Nanosolar, Oorja.
- *John Doerr,* Partner, Kleiner Perkins.
- *Elon Musk,* Chairman, Tesla, Chairman, CEO SolarCity
- *Steve Jürvetson,* Partner, Draper Fisher Jürvetson.
- *Bill Gross,* Founder, Idealab.
- *Ray Lane,* Partner, Kleiner Perkins.
- *Al Gore,* Chairman, Generation Investment Management, Partner, Kleiner Perkins.
- *Raj Atluru,* Partner, Draper Fisher Jürvetson.
- *Steve Westly,* Founder, The Westly Group.
- *Dan Whaley,* Founder, CEO Climos.
- *Martin Eberhard,* Founder, former CEO Tesla.
- *Martin Roscheisen,* Founder, CEO Nanosolar.

All individuals on the list have credentials. For example, Vinod Khosla, an immigrant from India who founded Sun Microsystems, "pioneered open systems and commercial RISC processors," became a leading venture capitalist and is now a cooperator in Representative Jay Inslee's *Apollo Project* (Inslee & Hendricks, 2008). Steven T. Jürvetson (b. 1967), son of Estonian immigrants, completed a Bachelor's degree in electrical engineering at Stanford in 2.5 years (#1 in his class) before continuing to M.Sc. and MBA degrees *en route* to becoming Managing Director of Draper Fisher Jürvetson, a global (33-nation) venture capital firm with $5.5 billion in investments. Al Gore, former Vice President and Nobel Prizewinner, needs no further comment. Martin Roscheisen's firm is a pioneer in printed solar PV panels, and received the largest Solar America award from the DOE in 2007.

New areas of biotechnology have burgeoned. One of the most interesting because of its potential for removal of CO_2 from coal-fired power plants is "algaculture," removal of CO_2 through photosynthesis by green algae in nutrient-rich solutions. Selected types of these algae grow at rates 7–20 times the rate of land plants and also produce a type of oil usable for biodiesel or plastics, and nutrient-rich residues. The diversity of current approaches to this process is indicated below, in a sampling from Fehrenbacher (2008):

- *Greenfuel technologies Corporation.* Bob Metcalfe. Uses recycled CO_2 to grow algae; backed by Polaris Ventures, Draper Fisher Jürvetson, and Access Private Equity.
- *Solazyme.* Uses biosynthesis and genetic engineering to produce algal strains for better biofuel yields. The San Francisco company grows algae in fermentation tanks without sunlight by feeding them sugar; it is supported by Chevron, Imperium Renewables, and Blue Crest Capital Finance.
- *Blue Marble Energy.* Seattle-based company finds algae-infested polluted water systems, and cleans up the environment, turning algae into biofuel. It

argues that ponds will never be large enough to support power plant CO_2 removal. Questions remain about control of wild algae.

- *Inventure Chemica.* Seattle firm seeks algae-to-jet fuel product; supported by Imperium Renewables and Cedar Grove Investments.
- *Aurora Biofuels.* Using genetically modified algae developed by Professor Tasio Melis at the University of California (Berkeley), the company claims to produce biodiesel fuel with yields up to 125 times greater than current production methods; backed by Gabriel Venture Partners, Noventi, and Oak Investment Partners.

All algal production methods would need huge areas and water supply to convert CO_2 produced by coal-fired electrical generating plants to biomass. An experimental algal production system has been employed on a large (750 MW) New Mexico power plant site. More than 8,000 acres yielded over 1,000,000 tons of algae per year, which represented 40% CO_2 capture, and recycled water (Sun & Hobbs, 2008). On the other hand, Briggs reported mean production of 50 tons biomass per hectare (2.7 acres) (Briggs, 2004). Clearly there is a much more intensive potential for algal growth under specialized conditions. Not mentioned in the former report are the nitrogen and lesser amounts of other nutrients needed for algal growth. Nutrients would be available from sewage, eliminating the need for synthetic nitrogen, which would require energy input.

Algaculture to remove CO_2 and useful byproducts will need (1) intensive research to develop scalable technologies – as well as creative cooperation with waste treatment systems, ideally in close proximity to the power plants; (2) reducing current barriers of regulations, permitting systems, governmental jurisdictions, zoning and public approval; (3) stable tax and incentive (subsidy) systems that favor good outcomes and are not corrupted by private greed or local officials' favoring special constituencies; (4) political cooperation – so that systems set up in one political administration or jurisdiction are not swept away or buried in the next, or otherwise destabilized, and (5) a more knowledgeable and concerned citizenry (without the latter the other political requirements can't be met). A tall order!

8.10.3 Geothermal Energy

Most existing geothermal energy comes from five states, listed in order of installed capacities (MW): California, 2,490 (beginning in the late 1970s); Nevada, 297; Hawaii, 35; Utah 26; and Alaska 0.4. California's geothermal energy production provides 5% of the state's electricity needs and is claimed to exceed installed capacity of every country in the world, with Indonesia a close second (Bertrani, 2007).

The largest facilities are on high-temperature sites that produce electrical energy by steam turbines, with others being intermediate-temperature sites that drive turbines using "binary cycle" (use of low-boiling point liquids) or "Rankine cycle" methods (Energy Efficiency and Renewable Energy 2008a).

A currently unused but related potential source of geothermal energy would be to make use of deeper petroleum producing wells that produce hot oil or water. This is similar to cogeneration of heat with electricity or industrial processes like steelmaking. About 7% of power plants or industrial facilities in the United States utilize cogeneration of heat, whereas higher levels of cogeneration are used in Europe and Japan.

Scattered facilities throughout the United States use low-temperature geothermal heat *via* heat pump systems to warm buildings, swimming pools, and other structures. In the 1970s the NSF RANN (applied research) program conducted valuable, relatively low-cost mapping and engineering and economic research on geothermal energy potentials in Middle Atlantic coastal areas. Heat-flow maps were paired with comprehensive locations of various commercial facilities whose requirements potentially matched heat sources. Applications ranging from heating of chicken farms to drying in automobile body shops and other commercial opportunities were identified (Hirst & Moyers, 1973). The RANN program was cancelled by the Carter administration in the late 1970s in large part because of its unpopularity with academic research leaders (Green & Lepkowski, 2006). A few of the innovative energy programs that were transferred to the DOE were terminated in the Reagan administration.

Potential geothermal sites have experienced regulatory problems or objections from local residents. Examples include a federal appeal court's siding with the Pit River Tribe's suit in a longstanding legal action against power producer Calpine Corp. for a facility near Medicine Lake, 30 miles east of Mount Shasta, OR. On the other hand, May 23, 2008 information released by northwest Geothermal reported that legal claims were finally dismissed, clearing the way for a 120 MW geothermal plant on the flank of Newberry volcano in eastern Oregon.

EIS-related permitting procedures applied to federal leasing or land acquisition for resource or energy recovery operations since the 1970s have frequently worked in such a way as to foster suspicion or negative attitudes in local residents, while creating delay and risks for operators. An unusually favorable opportunity to develop a new industry in Hawaii using geothermal energy in 1987 broke down over psychological factors related to the circumstances of hearings in Honolulu.[235]

8.10.4 Carbon Capture and Storage (CCS): A Critical Technology

For several years the leading Norwegian environmental organization, Bellona, has vigorously argued that carbon burial is the single most important technology needed to counter greenhouse gas emissions. Why? The leading developing nations, China and India, it reasons, will not be denied economic growth. China is building two coal-fired power plants a week and the equivalent of the UK power grid per year (Deutch & Moniz, 2007). The least expensive and most readily available source of energy for them is coal. Neither conservation, energy efficiency, nor renewables, even nuclear energy, can fill the gap. Moreover, many European nations and the United States will also remain dependent on coal for many decades. So as costly and difficult as it may appear, Bellona asserts that the CCS technology must be developed as rapidly as possible. Other nations must assist China and other developing nations to acquire and use the technology. Bellona estimates that coal-fired electrical power plants can be upgraded, where suitable substrates are available, for about $100 million each, and this action should be implemented as rapidly as possible (Bellona, 2005). Norway, Sweden, and other EU initiatives are responding to this need (Table 8.4).

The United States has about 500 coal-fired power plants in operation, on average, discharging about 3 million tons of CO_2 per year (Deutch & Moniz, 2007). Where does the United States, with more than a half-century of advanced technological expertise with recharging gases to producing oil reservoirs for enhanced oil recovery, stand with respect to CCS? The answer is more than embarrassing. DOE announced in early 2008 that it was dropping its commitment to the only US national project, *FutureGen* (Armistead, 2008). In contrast, ten other nations either already have commercial-scale or demonstration projects for CCS in progress or are moving toward implementation (Røkke, 2008).

Consider the leader, Norway, for example. This country of 5 million people began commercial-scale burial of 1 million tons of carbon per year already in 1996 in the Sleipner offshore oil field, after authorization in 1992 (Røkke, 2008). The United States has "talked about" and "planned for" the now sidelined demonstration project much longer than it required Norway to move from decision to commercial-scale burial of CO_2 produced from offshore wells.

Table 8.4 Carbon storage projects started or planned

Year started or projected	Name	Tonnes CO_2[1]	Country
1996	Sleipner	1 MT	Norway
2001	GPS Weyburn	2 MT	Canada/IEA
2004	BP Salah	1 MT	UK
2004	GdF K12B	30 kT	Holland
2006	Snøhvit	800 kT	Norway
2007	BGR Schwarze Pumpe-Vattenfall	30 kT	Germany/Sweden
2008	Hazelwood-IntPower	20 kT	Australia
2009	Total Lacq	30 kT	France
2009	ENEL Polar Brindisi	20 kT	Italy
2010	BP-Muller	400 kT	UK
2010	Mongstad-SINTEF demo	0.1 MT	Norway
2011	BP Carson	500 kT	UK
2011	EON IGCC	450 kT	UK
2012	Statoil/Shell Halten CO_2	850 kT	Norway/Netherlands/UK
2012	SaskPower	300 kT	Canada
2012	Karsto	400 kT	Norway
2012	FutureGen	250 kT[2]	(USA)
2012	Hypogen	400 kT	France
2014	Mongstad-SINTEF full	1.3 MT	Norway
2014	RWEZeigCCS	450 kT	Germany
2015	Vattenfall Oxyfull	1 MT	Sweden

Source: Røkke (2008). [1] Per year   [2] Support discontinued as of February 2008

One of the United States' leading experts on CCS reported that the United States is well endowed with geological storage space and the technologies for adaptation to new and existing power plants is available in most high-density power plant areas (Friedmann, 2007).

8.11 Positive Developments in the Environmental Community

Much of this book has detailed confusion, dysfunction, and internal conflict as a consequence of divisive developments in the 1960s and 1970s. There are obviously many constructive activities that have been overshadowed in this book by "what's wrong" explorations. The following vignettes are only isolated examples of the many constructive developments that would be enhanced by systematic search. I justify emphasis on the more problematic issues by their negative effect on cooperation and positive initiatives such as those mentioned below.

Fred Krupp and Environmental Defense Fund. A disproportionate number of the positive news items involving cooperative agreements or constructive developments relating to environmental developments in the past 20 years have been associated with the name *Fred Krupp*, President of Environmental Defense Fund (EDF), an organizations that once had the slogan "sue the bastards." His approach inclues talk (communication) and an open mind, backed by a strong organization and research. Feature articles (Schulte, 2007; Wessel, 2007) provide key details.

> "Krupp didn't invent the idea of teaming with business on pollution problems, but he proved perhaps the most daring in putting the theory to work. After taking the reins of Environmental Defense in 1984 at age 30, the New Jersey native with a law degree from the University of Michigan proclaimed his intent to use profit motives to achieve environmental ends in a Wall Street Journal op-ed. He said it was a fallacy that either the economy wins or the environment wins: 'The new environmentalism does not accept 'either-or' as inevitable.' Back then, the notion was 'revolutionary,' says Dan Esty, director of the Yale Center for Environmental Law and Policy. Other green groups 'thought it was their job to beat up the private sector,' Esty says. 'Fred understood that luring people into the dance was more likely to produce results (Schulte, 2007).'"

Krupp is enthusiastic about the cap and trade provisions of the 1990 Clean Air Act Amendments, in whose passage he had a role. The flexible and incentive-based system of controlling sulfur dioxide emissions into the air reduced by 80% the costs to industry and society that might have been incurred using the old methods. That figure should alert us to the costs of the vast areas of the remaining system, where flexible procedures are not in place.

Krupp was personally involved in extensive negotiations with and browbeating of the huge Wal-Mart Company, the world's largest retailer and company. As *Newsweek* indicated, EDF set up a special office in Bentonville, AR, near Wal-Mart's headquarters, which aided in achieving agreements over Wal-Mart selling energy-efficient flat-screen TVs instead of the older models.

EDF and Krupp were involved in negotiations over the agreements with TXU, a major Texas utility, which resulted in scrapping of 8 of 11 coal-fired power plants, with a part of the lost energy capacity to be provided by new wind farms in the panhandle of Texas.

Other achievements were agreements with McDonalds to replace former polystyrene boxes with paper, and slash waste, and improved fishing practices through cooperative agreements with fishing companies.

World Resources Institute. WRI differs from many environmental NGOs in that it is not membership-based, but operates primarily on grants and contracts, using professional staff. Some of its initiatives include:

(1) *Promoting the Greenhouse Gas Protocol* (GHG), a system of accounting to support global emissions reductions (WRI, 2007). WRI has been involved in cooperative discussions through which some of the 42 global companies have adopted GHG protocols, for example, Alcoa, BASF, Bayer, BP, Caterpillar, Daimler Chrysler, DuPont, Exelon, FedEx, Ford, General Electric, General Motors, Georgia-Pacific, International Paper, PG&E, Royal Dutch Shell, Unilever, UPS, Volkswagen, and Weyerhaeuser.

(2) Promoting development of CCS for use in coal-fired power plants.

(3) *Eradicating Poverty Through Profits.* A program promoting private sector approaches to reducing poverty and increasing sustainable development on a global level.

(4) *Exploring critical global issues involved in greenhouse gas emissions*, such as export of electronic waste from the United States (Nakagawa, 2006), tropical deforestation, and many other issues.

Jonathan Lash, President of WRI since 1993 has an exceptional background in interfacing with the United States and global financial and business community. "[He has served with] a broad range of national and international groups, including: the Organisation for Economic Co-operation and Development's Round Table on Sustainable Development; the DuPont Biotechnology Advisory Panel; the Tata Energy and Resources Institute (India); the Keidanren Committee on Nature Conservation (Japan); the China Council for International Cooperation on Environment and Development; the Pew Environmental Health Commission; and the Research Institute of Innovative Technology for the Earth (Japan) (Lash, 2008)." He and WRI are prominent among movements bringing the United States together in constructive developments rather than partisan conflict.

Jay Inslee and the Apollo Project. Inslee is a liberal Democratic member of the US House of Representatives (1st District, Washington State) and one of Congress's most charismatic enthusiasts for reducing greenhouse gas emissions. What distinguishes him from other politicians is that over the past several years he has led a dynamic team effort to explore innovative but realistic solutions toward achieving breakthroughs in new renewable energy technologies. He and the Apollo Team have gotten their hands dirty in undertakings like "reinventing the car," energy efficiency, wind energy, coal, and nuclear energy. An initiator of federal government bills to support developments, like the *New Apollo Energy Act* (H.R. 2809) (Inslee and Hendricks, 2008), he recognizes that government efforts can do little without private investment, technological input, and the longer-term developmental and operating role. Many past governmental initiatives subsidized new developments based on their initiators' concepts of what the nation needed but failed to incorporate in-depth business insights and perspectives and left behind them strings of stranded investments, failed bureaucratic projects and disillusion. Former President Bill Clinton, who wrote the forward to the book by Inslee and Hendricks (2008), became so enthused about the Apollo Project that he was scheduled to be a coauthor in a follow-up volume (Manheim, 2008).

8.11.1 Environmentalists' Movement Toward Realism in Energy Policies

A decade ago virtually all major environmental organizations did not want to hear about oil, coal, or nuclear power. They embraced policies that relied largely on conservation, increased efficiency in use of energy, and renewable energies – which would assure a bright new green future. Proposals for nuclear power were vigorously attacked. As major national and international policy development and in-depth analyses struggled to define the realities (IPCC, 2007), it has become clear that even the most optimistic scenarios retain coal in the global energy budget long into the future. The United States and many European nations are currently reliant on it for 50% or more of their electric power. The cost of oil is making people realize how much all advanced societies' economies and basic functioning depend on transportation. It's a serious issue – not to be dealt with through assumptions and hope. The more intensive European experience has shown that replacing coal and oil is hard-won and expensive. Sweden decided in a national referendum 20 years ago to dismantle its nuclear power plants. But, more than 40% dependent on nuclear energy for electrical power, the Swedes have suspended those plans, while Finland has moved ahead with its 5th reactor.

The *National Resources Defense Council* was the first environmental NGO to mention the word "coal" in talking about future energy policy. A recent report commissioned by the NRDC includes realistic assessment of all the key power sources: coal, natural gas, nuclear as well as renewables; it participated in the MIT coal studies (Deutch & Moniz, 2007). In potential scenarios *Environmental Defense Fund* has followed suit (Bradley, 2008). Inslee's Apollo Project accepts reality. These are encouraging developments. Removal of artificial barriers to discussion, exploration, and negotiation makes cooperation and progress possible.

Progress has a way to go. The *Sierra Club* still opposes nuclear power and coal as well as biotechnology and genetically modified crops. Solar energy proponent and venture capitalist, Vinod Khosla, recently noted:

> "For every nuclear plant that environmentalists avoided, they ended up causing two coal plants to be built. That's the history of the last 20 years (Fehrenbacher, 2008)."

Greenpeace and *Friends of the Earth* adamantly oppose technological development of CCS because it could encourage coal use. They have no answer when it is pointed out that, whatever we choose to do, China is building new coal-fired power plants on a weekly basis. My discussions with German Greenpeace energy leaders indicated that their concerns about nuclear energy were not just gut-level fear of radioactivity. They were primarily concerned about spread of nuclear weaponry – a reasonable issue. All NGOs currently oppose opening new exploration and leasing for oil and gas from the US offshore – although this reduces rather than increases resources available for renewables. As long as auto and aircraft transportation is to be maintained in the United States, a barrel of available oil not recovered from the U.S. offshore means importing it from abroad at $100+/barrel – while perhaps also exporting US capital and technology.

Chapter 9
Policy Analysis

9.1 Post-World War II Influences on US Academic Research and Policy Studies

9.1.1 Effect of the New Science Paradigm

A paradoxical change took place in US intellectual life after World War II. A country whose traditions had always tended (and in many respects still tends) to the pragmatic and practical side built a vast new complex of research universities, research specialties, and institutes.[236] The United States became the world leader for theoretical explorations in the natural and social sciences. In my earlier professional field this growth was marked by increase in the number of marine or oceanographic institutions from 16 in 1956 to 134 in 1976.

Federal stimulus for science after World War II was unbalanced in the direction of basic, discretionary research (see Chapter 2). This is documented by the fact that funding of applied research by the National Science Foundation was excluded until modified guidelines were signed by President Johnson in 1968. But by this time, the die was cast. The prestige accorded to basic research, and the lower status given to applied science and practical applications, though modified, continues to the present. The new paradigm set in motion trends unanticipated by Vannevar Bush, its main promoter, though he began to complain about them by the 1960s (Zachary, 1999).

Funding for basic research, vigorous advocacy by academic scientific leaders, and the heady freedom to pursue discretionary research in any desired direction produced a kind of special euphoria. This was especially true for younger scientists who got in on the action in the early decades when the ratio of funds to researchers provided generous support. Many senior scientific leaders today had their outlooks forged during the "golden era" for research in the 1960s–1970s. The attraction of the academic environment was heightened by the rewards, prestige, and excitement of research, the ego gratification provided by the respect and attention of students, the sense of accomplishment provided by peer-reviewed publication, and finally, the spice of (paid) travel to meetings to joust with specialist-colleagues. As it was designed to do, this system attracted many of the nation's "best and brightest" talents. The system was partially followed in other nations, but only in the United States and

F.T. Manheim, *The Conflict Over Environmental Regulation in the United States*,
DOI 10.1007/978-0-387-75877-0_9, © Springer Science+Business Media, LLC 2009

the former Soviet Union (L. R. Graham, 1993) was the dominance of basic research maintained past the 1960s (see Fig. 2.3).

The result was that a universe of research science became semidetached from society. Skilled researchers in disciplines within this universe often communicated more with each other – wherever they were located, which could include foreign nations, than with fellow faculty members in the same department. Subjects of research could be relevant to society, but were studied for the sake of knowledge rather than in the pursuit of applied goals.

9.1.2 Robert Maxwell and the New Model for Science Publication

Before the war most science publishers tended to favor books whose authoritative content and synthesis of important or broad fields would encourage wide distribution. This reduced cost per book, further favoring distribution to individual scientists, who were the main market. Content that would have long shelf life was favored by both buyers and publishers. The publications system in turn influenced the research products of scientists and engineers. Books that synthesized knowledge in broad fields also lent themselves to use as textbooks.

But in the early 1950s Robert Maxwell, a bold entrepreneur in scientific publications, realized the potential for an entirely new concept for scientific publication, which he launched through Pergamon Press in the United Kingdom.[237] Governmental encouragement of scientific research was fostering rapid growth of scientific departments and institutes. Maxwell realized that publications serving emerging scientific fields and disciplines provided dynamic new market opportunities. The number of potential institutional buyers (libraries) had reached the point where Maxwell could estimate the costs and income of new publications based primarily on institutional purchasers who were obliged to acquire new journals and books relevant to their sphere of activity. After Maxwell took leadership of the small Pergamon Press in the United Kingdom, he dramatically increased its publishing activity, which reached 100 journals and books per year by 1957. Elsevier Publishing Co. in The Netherlands soon adopted similar policies and became a leading world publisher.

The new publications system expanded into the social sciences and other fields. Former constraints on subject matter and numbers of publications decreased. Marketing of new journals was aided by "name" scientists on the masthead and requests by scientists to their institutions for acquisitions of journals. In the case of books the former emphasis on breadth, authoritativeness and comprehensiveness – attributes for long shelf life or multiple editions could now be replaced by specialized topics. Criteria for promotion and tenure based on publications credentials had created a "publish or perish" syndrome that further pushed publication in peer-reviewed media.

The ultimate effects and subsequent history are beyond the scope of this book. But one of the many byproducts of the system, redundant or duplicative publication, has become so serious in medical fields – where these practices can distort medical statistics, that editors are adopting stringent warnings and other measures (e.g., Stahel, Clavien, et al., 2008).

Offsetting the steep price rises for scientific books starting in the 1980s has been a trend toward inexpensive policy books issued by think tanks, university publishers, and electronic publications.

9.1.3 Social and Natural Sciences

A strong interest in theory had already characterized growth of the social sciences in Germany and Central Europe before World War II. Emphasis on theory increased along with a growing number of subdisciplines in the social sciences in the United States after the war. An example of the "theoretization" tendencies is offered by the treatment of "environmental responsibility," a central concern for environmental policy, by Hobson (2006). The article was derived from the Hobson's recent Ph.D. thesis:

> "Human geographers have explored at some length the discourses and subject positions implicated in the recent rise of 'environmental responsibility'. Assigning it either as an individual disposition enacted in various spaces, a performative 'othering' tool, and/or a form of ecological governmentality, these *debates* have said little about the role of research and researchers in encouraging environmental responsibility. Utilising arguments from William James' 'radical empiricism', I argue that exploring practices through a pragmatist lens enables a tentative re-envisioning of environmental responsibility." (FTM italics)

The academic jargon in this example is enough to keep nonspecialists who might be interested in the subject safely away. However, the larger barrier to wider utilization of insights that the author may have achieved lies in her primary interest in *intellectual debates among peers,* rather than the application of environmental responsibility in society.

Even where social scientists have important information and insights to offer general audiences and speak in clear or even colorful language, if their work is embedded in a mass of literature whose overriding focus is on internal theoretical debates, there may little chance that the work will come to the attention of those who could use it.

Theoretical research, especially "empirical studies," i.e., confirming or rejecting hypotheses or otherwise quantifying relationships through statistical analysis, has been much in vogue in the political and policy sciences, even though the assumptions or constraints of this kind of research limit its value for practical goals. In the early 1960s, Congress added social sciences to the research areas that the National Science Foundation was authorized to support. This may partly explain movement of social scientists toward practical involvement in societal affairs in the early 1960s. Phyllis Grodsky describes the results:

> "... social scientists assumed that their research methods were a technology for restructuring society, and were analogous to the technology that eventually placed a human being on the moon ... [i.e.] the process of replacing older knowledge by new, so that eventually the older data would become obsolete (Grodsky, 1982)."

The preferred research approaches lost sight of the need for historical perspective and the fact that even the nature of questions could change with time. Grodsky cites several authors to the effect that when investigators dealt with the interface between evaluations research and politics, many unforeseen problems emerged. "Decision-makers often appeared to the social scientists as irrational....Many [social science] investigators returned to their [traditional] research methods as... a harbor of safety in an uncontrollable sea."

One of the problems that I encountered in tracking the source of polarization and gridlock in US politics was the tendency among knowledgeable researchers to avoid dealing with emotional or other issues of obvious importance in preference for more quantifiable factors. Examples are incidence of divided or unified government (i.e., President and Congress divided among the parties), correlation between Party identification and issue positions, ticket splitting, majority seat advantage on standing committees, and Presidential vs. Congressional initiatives (e.g., Binder, 2003; or chapters in Bond & Fleisher, 2000). These and other relationships proposed by various authors as potential *causes* of polarization are more likely *outcomes* of the larger rift.

The natural sciences have not been immune to the academicization trend. In geochemistry, for example, special prestige in the postwar decades became attached to mathematical modeling, complex statistical distributions, new disciplines like deep earth crustal studies, properties associated with rare elements like germanium, and analytical measurements involving great sophistication like study of osmium isotope ratios. In the enthusiasm for theoretical and conceptual exploration and discovery, societal needs for applied science needs were relegated to low levels of prestige and importance by academic science leaders.

So great was the hubris and single mindedness among US marine scientists in the late 1970s that a poll was published that rated the quality of academic marine scientific work funded by various federal agencies (Table 9.1). Quite naturally, NSF (National Science Foundation)-supported research (which essentially excluded applications) rated highest, while science supported by federal agencies with successively more practical functions rated lower (among the lowest were the Corps of Engineers and Coast Guard). Note that the authors go so far as to egregiously create the opposite of "*High quality*" rating, labeled "*Substandard or not science at all*"! I knew some of the good scientists involved in organization of the poll. These pollsters were operating under widespread mindset that elevated pure discovery and intellectual inquiry to the highest and best (or only esteemed) use for science.

That the elite scientists were not thinking through the role of science in society beyond their personal pursuits is indicated by an informal poll I conducted at a national meeting of marine scientists in the early 1980s. I selected a random sample of 10 abstracts that described breakthroughs with potential relevance to societal concerns. I sought out the authors and asked them what they thought the "next steps" in their research might be, justifying the nation's investments. Who would use their work? None of the ten had given any thought to this question. Some dismissed it saying "that's not my concern," whereas a few replied with statements something like "that's an interesting point....."

Table 9.1 US academic marine scientists' assessment of the quality of academic marine scientific work funded by the federal agencies[1]

Agency	High quality (%)[2]	Mediocre (%)[3]	Substandard or NS (%)[4]	Rank	Number of respondents
National Science Foundation – Deep Sea Drilling Project	95.5	3.7	0.8	1	243
National Science Foundation – Oceanography Section	94.5	4.8	0.8	2	564
Health, Education and Welfare – National Institutes of Health	89.9	6.9	3.2	3	188
National Science Foundation – Environmental Biology	87.9	9.3	2.7	4	182
Office of Naval Research	86.3	12.3	1.5	5	562
National Science Foundation – International Decade of Ocean Exploration	79.6	17.3	3.1	6	515
National Science foundation – Polar Programs	76.1	17.2	6.7	7	267
US Geological Survey	70.4	24.9	4.6	8	281
Department of Energy	57.2	33.6	9.2	9	339
National Oceanic and Atmospheric Administration – National Marine Fisheries	42.1	42.9	15.0	10	273
Sea Grant	34.4	40.6	24.9	11	566
Environmental Protection Agency	32.1	42.7	25.2	12	396
US Army Corps of Engineers	24.0	42.7	33.3	13	342
US Coast Guard	23.0	43.6	33.3	14	204
Bureau of Land Management	13.1	33.4	53.4	15	419

[1]Source: This table, including the title and content, are from Ladd, Palmer, & Shannon (1978); F.T. Manheim simplified the column headings
[2]Percentage of respondents rating the agency as conducting high quality scientific investigations
[3]Percentage of respondents rating the agency as conducting mediocre quality scientific investigations
[4]Percentage of respondents rating the agency as conducting substandard scientific investigations or conducting work that was not science at all

It was not until later that I encountered the model of German science in the nineteenth and early twentieth century and took interest in other American examples of where the *best* scientific talents had been engaged with applied scientific goals. It appears that where the best talent is paired with critical national problems under suitable conditions, superior results are not only gained for the practical goals, but basic research breakthroughs are made as well (mentioned in Chapter 2; see also Chapter 8, German Science, and USGS).

Some reasons why applied research under proper conditions stimulates basic research discoveries – the reverse of the basic-to-applied research pattern – are discussed in the last section of the case study of the US Geological Survey (see Chapter 8). Basic research creates a temptation for researchers to elevate the importance of conceptual approaches within fields. Applied goals (like good environmental regulatory policy, for example) typically require crossing many scientific disciplines and areas of knowledge in changing frames of reference. For researchers open to them many opportunities to see new scientific relationships are available. This point deals largely with the motivation and general strategy of research. Stern, Porter & Furman (2007) describe a large set of factors that have been identified in technologically productive "clusters."

9.2 Shifts in Attitudes of Scientific, Industrial, and Governmental Leadership During and Since World War II

9.2.1 Pearl Harbor and Achievement of Unity During World War II

Nothing could have drawn together American society more effectively than the Japanese attack on Pearl Harbor. President Franklin D. Roosevelt quickly addressed Congress and the public by radio on December 7, 1941, "A date which will live in infamy (Roosevelt, 1941)" and asked Congress for a declaration of war. As a young boy, I remember block-long lines at military recruiting stations as men of all ages from teenagers to their 30s signed up. The nation was addicted to Hollywood movies – most people attended at least once a week. The newsreels that came before all movies showed recruits training with wooden rifles because of the shortage of all war equipment. Mobilization was rapid. A new spirit of solidarity united the nation. Families collected tallow from animal fat to help make nitroglycerin, newspapers, tin from toothpaste tubes, and cans. War bonds had brisk sales. The war front was carefully watched in school classes. News correspondents headed by Edward R. Murrow became known for their honesty and integrity.

Put aside were animosities that had grown between the second Roosevelt administration (1936–1940) and the business community ("The captains of industry and Wall Street barons hate me and I welcome their hate"[238]). Roosevelt called on the top business leaders in the nation, like William S. Knudsen of General Motors to help in the war effort. It had a galvanic effect on industrial mobilization for war (Braeman, 1994).

The Ford Motor Co. converted its River Rouge plant to making Pratt & Whitney aircraft engines, and its Willow Run facilities to making bombers. Chrysler made tanks. Willys made Jeeps. We should recall that, in comparison with the years of research and planning assumed to be needed to achieve carbon capture and storage, Ford's conversions took place in a single year! Everett DeGolyer, the nation's leading petroleum geologist and oil investor was asked to advise in organization of oil supply. Vannevar Bush, a brilliant, MIT-trained electrical engineer, pioneer in computer development and founder of a technology company, became Roosevelt's science advisor. He and MIT President Karl Compton headed the National Defense Research Committee. Industrialist Henry J. Kaiser, tapped by the Maritime Commission to build merchant and later military vessels, introduced radically new methods for building ships in seven shipyards around the country (Adams, 1997). Kaiser's yards built more than a quarter of the Maritime Commission's vessels (or 1,490 ships) including the famous Liberty ships, in what has been stated to be two-thirds of the time and at a quarter the cost of the average of other shipyards.

Recovery of the US economy had been slowed and delayed more than in other nations by flawed economic assumptions and policies. Bruce Smith provides a retrospective summary, especially as related to science and technology:

> "Excess production and expansion that was too rapid were commonly considered the causes of unemployment. The codes of the National Recovery Administration, for example, stressed production limits to be negotiated industry by industry. The idea of expansion through innovation was distinctly unpopular. Technological unemployment was a very common explanation for the Depression, and there were calls for a moratorium on scientific research and technological development (Smith, 1990)."

During the war, industrial activities, which had to be divided between war needs pushed by the military, and civilian needs, were placed under the overall direction of the War Production Board, led by Donald Nelson, a former Sears Roebuck executive (Tassava, 2008).[239] Enough "old" or "top-down, command-and-control thinking" remained to provide war parallels to the inefficiencies created by postwar detailed technical and scientific prescription of environmental regulation. For example, Bruce Smith cites reports that in aircraft manufacture, "development of engines was only fully satisfactory when the services gave firms all possible freedom in deciding on details of design and development (Smith, 1990)." The overall performance of American technical and industrial innovation was astounding, producing a ten-fold increase in production in planes, munitions and ships from 1940 to 1944, with 3-fold increase in aluminum production and 35% increase in steel production. Unemployment dropped to negligible levels while the combination of an income tax (the top tax bracket was 94% in 1945) and heavy war bond sales kept inflation down. Roosevelt, whose policies had been psychologically effective but deficient in economic terms during the Depression, proved to be a quick study and capable of good judgment in the choice of effective leaders in the war effort.

Many university scientists and engineers were brought into war-related technological developments, among the largest of which were MIT's famous Radiation Laboratories, which employed 4,000 workers, primarily in radar development. The wartime mission-oriented teams included scientists and engineers, industry, government and often the military – coordinating technical specifications and logistics, as well as government as overseer and paymaster. Team efforts across a variety of fields achieved goals, overcoming obstacles through innovation. Project budgeting and personnel management became an integral – not a separate – part of the efforts. After the war scientists often became successful in a variety of occupations: industrial companies, federal and state agencies, and university academic centers. In the latter centers they became especially adept at garnering grants from federal agencies among which the Office of Naval Research became the initial leader.

9.2.2 Postwar Developments and Talent-Leadership Shifts

The nations' economy surged under the pent-up demand for autos, consumer goods and housing, the dynamism and strong leadership in many of the more advanced industries, and the flow of talent from the nation's universities and returning veterans. Technology-rich companies like IBM and AT&T sponsored some of the world's most advanced research in their laboratories, attracting top scientists and engineers. The oil industry had attracted outstanding scientists like M. King Hubbert (now known for his Peak Oil concept), who later moved to USGS and also served as President of the Geological Society of America. Earth science, chemistry, and engineering students from Ivy League schools had no hesitation in entering careers in the oil industry and (hard rock) mining. Harvard University had the world famous soil mechanics and civil engineering expert, Arthur Casagrande, and leading physicists cooperated in geochemical – isotopic research, opening understanding of the Earth's crust.

Problems were concentrated in the steel, rubber, and especially coal mining industry, where unions had grown strong during the war and now had conflicts with management. The coal-mining industry in particular was avoided by young business and engineering talent because of its poor safety record, reputation and conflicts with the United Mine Workers. Another industry that was seriously impacted was the merchant shipping industry, whose vessels were pressed into service during the Korean War. The US merchant marine academies, such as the one in Buzzard's Bay, MA were first-rate, but failure to keep pace with technological development left the fleet at a competitive disadvantage.

Academicization trends in a segment of scientifically-and technologically motivated young people modestly increased through student deferment from military draft in the Korean War, though the effect was minor compared to the mass flight to academic deferments during the Vietnam War. The McCarthy era search for Communist agents and sympathizers (1950–1954) was closed by Senator Joseph McCarthy's censure by the Senate but left behind polarized feelings in both the academic community and Hollywood, as well as among conservatives.

9.2.3 Fast Forward to the 1960s – Alienation and Narrowing of Attitudes

A series of destabilizing events took place in rapid succession in the 1960s. Cuba and the Bay of Pigs fiasco shook confidence in the military-political professionalism of the new Kennedy administration. Rachel Carson's book in 1962 created alarm within environmentally concerned groups and aroused the ire of the powerful chemical industry. Then followed the traumas of the assassinations – John Kennedy in 1963, Malcom X in 1965, and Robert Kennedy and Martin Luther King in 1968. Unease about the growing engagement in the Vietnam War, radical movements (see Fig. 2.4) and antibusiness sentiment grew in academia.

Shifts in attitudes and leadership of the oil industry were highlighted for me through discussions in 1970 with Max Blumer,[240] a former Shell Oil Co. petroleum chemist and one of a parade of talented scientists who left oil industry research groups in the early 1960s. After the Santa Barbara spill in January 1969, a barge that ran aground in Buzzard's Bay, MA in 1969 spilled massive oil on the very doorstep of marine scientists in the Woods Hole, MA complex of research institutions (Reddy, 2007); Blumer reported on toxic components in the complex chemical makeup of crude oil and its products in the Buzzard's Bay spill. He catapulted to national prominence, testifying in Congress and participating in national meetings with the heads of many of the nation's leading oil companies. Engler cites many examples of diverse, freely expressed perspectives by oil industry leaders prior to the 1960s (Engler, 1961). A few were outspokenly self-critical of the business community.[241]

In 1969–71, mainly insensitive or obtuse attitudes of oil executives were cited in the newspapers and other media accounts. I asked Max, whom I knew well from my work in Woods Hole, "Were there any presidents and other oil executives with whom he met and spoke that showed open-mindedness about the concerns of scientists or the environmental community? Max reflected then shook his head. There were none, he said, except possibly the President of Atlantic Richfield (ARCO).[242] Oil companies stopped recruiting at Ivy League universities.

The rise in litigation and litigiousness in the 1970s (Fig. 4.2) brought new emphasis on defensive strategies for the oil and chemical industries. Challenges for companies in natural resource sectors tended to promote the advancement of people with legal skills, thicker skins and concern for the bottom line. The American Petroleum Institute became dominated by EXXON by the mid-1970s, and largely ceased outreach activity after 1980. Richard E. Ayres, a founder of the NRDC, who has been intensively involved with environmentalist efforts to control air pollution from the middle 1970s up to the present, recently related to me observations that leaders in American electrical power utilities seemed to represent a selection of engineering graduates who, as a group, tended to lack creativity, and lacked ability to identify with environmental concerns beyond formal compliance requirements.

Hardening and narrowing of attitudes on the environmentalist and political sides has mirrored trends among industry and business captains. Even the name, *The Union of Concerned Scientists,* which was founded in 1969, implies an organization with coherent advocacy positions on public policy issues.

Earlier environmental pioneers, however committed, tended to maintain the ability to see wider perspectives. Rachel Carson, though passionate in terms of concern over pesticides' effects, targeted causes and practices, not people or political movements. Unlike the singlemindedness of more recent polarized times, she demonstrated active and detailed interest in helping farmers replace chemical pesticides in controlling pests for the purpose of maintaining agricultural productivity.[243] Even Aldo Leopold, revered as a leading proponent for non-materialistic environmental values, recognized loss of the sense of husbandry and identification achieved by "game-cropping of the Scottish moors and the German forests" when responsibilities for environmental management were exclusively relegated to government in distant protected areas (Leopold, 1949).

Passionately devoted to the value and "beauty of the untouched forest," Arthur Carhart involved himself actively in creating opportunities for human interactions with the wilds so that people would gain familiarity and a sense of identification with nature (Carhart, 1959).

Former Vice President and Nobel Prize winner Al Gore's work suggests the effects of polarizing trends. When I read his earlier book, *Earth in the Balance* (Gore, 1992), I knew only about the political Senator Gore. Gore's knowledge, insights and original ideas about scientific data on climate cycles took me by surprise. I think that if the Vice President had been an academic researcher rather than a politician, he would have unquestionably made a good name for himself. Similar creativity as well as dynamic entrepreneurial qualities characterize significant parts of the Gore Reinvention programs (Gore, 2001). But when it came to applying his insights to politics Gore seemed to enter another realm. He had little more to offer than stereotypical criticisms of Republicans. Missing were bridging solutions, citations of political sins by both parties as a way to disarm opponents, or new, nonpartisan ideas for progress, such as bipartisan-backed task force studies.

Gore's celebrated documentary film, *An Inconvenient Truth* is a truly remarkable *tour de force* in scientific-policy filmmaking. The inclusion of extensive hard data, effective presentation both in visual imagery and in imaginative stagecraft could have put his film and book on a level with Rachel Carson's *Silent Spring* as a work of power and influence. But unlike Carson's apolitical focus on issues, Gore was apparently unable to resist mixing science with partisan politics. On page 269 of the book version of the film, Gore lists prominent scientists who "believe that President Bush and his administration are threatening the Earth's future."[244] In the final chapter, ironically entitled *The Politicization of Global Warming* (Gore, 2006, pp. 284–287), Gore pulls out all the stops in identifying the enemies to progress. He attributes resistance to recognition of the climate crisis as the result of people's reluctance to change habits, behaviors, or experience inconvenience, and attacks multinational companies with heavy financial stakes in the *status quo*. Gore seems unable to recognize that associating science, however valid, with a political message stamps it "political" as well. It virtually guarantees automatic rejection by those in the opposing ideological camp, and invites skeptical attitudes even among moderates or uncommitteds. The same kind of "guilt by association" is involved in the rejection by the environmental movement and its political supporters essentially sight unseen of any scientific or technical proposal with links to the Bush administration or Republican Party.

"End-member" polarization and partisanship was illustrated by the task force on energy policy, led by Vice President Cheney early in the George W. Bush administration (see Chapter 2). Later examples included harsh assessment of abuses in the Interior Department criticized by the Inspector General in 2006 (Devaney, 2006) and a series of complaints regarding interference with orderly functioning of scientific and regulatory employees. It is important to recognize that this hardening to perceptions of impropriety may be one of the products of prolonged conflict and distorted public accusations against industry.

Not comparable in scope, but similar in terms of single issue focus is a recent "Coastal Campaign" web site of the California branch of the Sierra Club. Celebrating victory against a proposal to build an LNG terminal in Cabrillo, CA, the site singled out a local businessman for praise.[245]

It had evidently not occurred to Sierra Club writers and web site managers that there was anything unusual about the individual's business. It is a multifaceted service helping American manufacturers to outsource operations in China. The company, besides serving as a bridge to 100 factories in China, offers accounting software that simplifies tracking of production and logistics overseas. Seen from the distress expressed throughout the United States during the past Presidential primary campaigns about loss of the United States' industrial base, the idea of an entrepreneur using well-developed business skills to encourage and facilitate further movement of US industry abroad is a matter of concern.

9.3 Mediator–Leaders: A Critical Need for Future Progress

Mediators have had a critical role in US history (see Chapter 3). These are individuals who seek to understand the views of all sides in conflicts, promote communication and recognition of common interests, and see potentials for creative problem resolution. Notable mediators within the past two decades include Senator George Mitchell, who served as an effective majority leader in the Senate, helped mediate the conflict in Northern Ireland, and who served in a critical capacity to help achieve passage of the difficult 1990 Clean Water Act Amendments. Lee Hamilton served on both the 9/11 Commission and as a coleader with former Secretary of States James Baker for the Iraq Study Group. Senator Richard Lugar (R-IN) and former Senator Sam Nunn (D-GA) provided exceptional leadership in the creation of an initiative to assist the former Soviet Union in decommissioning nuclear weapons during the window of opportunity after 1990.

President Fred Krupp of Environmental Defense Fund (EDF) first became visible when he and EDF defended use of a cap-and-trade system for controlling sulfur dioxide emissions in the 1990 CWA amendments. That system became one of the greatest successes in the history of US environmental regulation. Krupp has achieved other major successes by talking to, understanding, and negotiating with business organizations. This tactic has proved more effective than the earlier EDF slogan, "Sue the bastards."

A leader within EPA (Environmental Protection Agency) is Daniel Fiorino, head of EPA's Environmental Performance Tracking program. The program creates awards and specific benefits for high-performing industrial companies, and has fostered an active and well-regarded innovations office. Fiorino is also a leading author on environmental policy (Fiorino, 2006). Jonathan Lash, an earlier chronicler of the sins of the Reagan attempt to roll back regulations, has now become a leader in forging cooperative links to industry for green development.

9.4 US and EU Lawmaking

The administrations of the first six US presidents (ending with the single term of John Quincy Adams) emphasized qualifications and competence for federally appointed officials. The qualities were also reflected in Congressional lawmaking. Important laws were couched in clear language with sufficient but not excessive detail. It was evident that officials were expected to carry out their duties responsibly and independently. The influence of an educated stratum of the nation was sharply reduced with the election of Andrew Jackson. Subsequent changes in Congressional lawmaking patterns are reflected in Fig. 9.1.

Lawmaking in four time periods: First six Presidents—competence era; Gilded Age pre-1960s environmental lawmaking; and Congress as micromanager.

- **The 1796 Sale of Lands Act** governs surveying and recording of land distribution north of the Ohio and Kentucky rivers. The Act designates appointment of a Surveyor General, gives concise guidelines for surveying land, including observation of relevant features like streams, salt licks and springs, and the general "condition" of the land. Government control of waterways is specified. Multiple copies of the observations are designated for different officials, one to be accessible to the public. In short, authority and responsibility are linked. Duties as well as the writing are clear and straightforward and encourage land description beyond the minimum. Salt springs are unusual but valuable to animals and human populations, indicating that the framers were knowledgeable about the areas to be mapped.
- **The Agricultural College Act.** In 1884 the nation was in Mark Twain's "gilded age." Corruption was rampant. Seats in Congress were sometimes bought and sold, and fiscal abuses abounded. Confusing and verbose writing of the laws was combined with nitpicking detail because agencies and officials were not trusted to exercise independent judgment where money was involved. An awkward, ponderous style had evolved to lend legal and official weight to what were often trivial acts.
- **The Federal Water Pollution Control Act of 1948.** By the early years of the twentieth century Congressional respect for the competence of federal scientific and regulatory agencies was reflected in Congressional approaches to lawmaking. Congress created new or modified agencies, designated missions and goals, and provided funding, but generally left discretion to agencies to plan, create detailed policies, and implement goals. The influence often extended beyond formal or enforceable responsibilities of agencies, as can be seen in the development of drinking water statutes. Following the first standards for drinking water issued by the Public Health Service in 1914 (and applicable only to interstate commerce), further developments led to the United States becoming the first nation in the world where customers in urban restaurants anywhere in the nation automatically received a glass of safe drinking water.

FOURTH CONGRESS. Sess. 1. Ch. 29. 1796.

Chap. XXIX.--*An Act providing for the Sale of the Lands of the United States, in the territory northwest of the river Ohio, and above the mouth of Kentukey river.(a)*

Section 1. *Be it enacted by the Senate and House of Representatives of the United States of America in Congress assembled,* That a Surveyor General shall be appointed, whose duty it shall be to engage a sufficient number of skillful surveyors, as his deputies; whom he shall cause, without delay, to survey and mark the unascertained outlines of the lands lying northwest of the river Ohio, and .....

FORTY-EIGHTH CONGRESS. Sess. I Ch. 20. 1894

Sec. 3. That the State of Colorado, in selecting lands for agricultural-college purposes under the acts of July 2, 1864 and July 23, 1866, may select an amount of land equal to 30,000 acres for each Senator and Representative which said State is entitled to in Congress, from any public land in said State not double-minimum-priced land; or selections may be made from said double-minimum lands, but in the latter case the lands are to be computed at the maximum price and the number of acres proportionally diminshed; but no mineral lands shall be selected.
Approved. April 2d, 1884.

EIGHTIETH CONGRESS. 2nd Session. Ch. 758, June 1948

To provide for water pollution control activities in the Public Health Service of the Federal Security Agency and the Federal Works Agency, and for other purposes.

Sec. 2 (a) the Surgeon General shall investigate and, in cooperation with other Federal Agencies, State water pollution agencies......adopt comprehensive programs for eliminating or reducing the pollution of interstate waters. (b) the Surgeon General shall encourage cooperative activities......

ONE-HUNDREDTH CONGRESS.   1987.

TITLE 33--NAVIGATION AND NAVIGABLE WATERS
CHAPTER 26--WATER POLLUTION PREVENTION AND CONTROL
SUBCHAPTER III--STANDARDS AND ENFORCEMENT

Sec. 1316. National Standards of Performance
(b) Categories of sources; Federal standards of performance for new sources
(1)
(A) The Administrator shall, within ninety days after October 18 1972, publish (and from time to time shall revise) a list of categories of sources, which shall, at the minimum, include:
pulp and paper mill;
paperboard, builders paper and board mills;
meat product and rendering processing;
dairy product processing;

Fig. 9.1 U.S. lawmaking: Then and now
Source: U.S. Code and Government Printing Office archives

- **The Federal Water Pollution Control Act (FWPCA) (Clean Water Amendments, as of 1987).** Stresses that arose after World War II and failure of the Executive branch to act on urgent environment problems led to loss of confidence in government. In the 1970s a group of strongly motivated senators and environmentalists spearheaded enactment of sweeping, detailed environmental laws designed to be enforced in part by civil litigation through the civil court system. The detail of the sample sections (Fig. 9.1, 1987) is further amplified in the full text and even more in thousands of pages of regulations associated with the laws.

Accustomed to equating "environmental commitment" with command and control laws, Americans may see older statutes like the 1948 FWPCA as weak (Lazarus, 2004). What they fail to recognize is that the earlier system partly reflected professionalism, communication, and cooperation among public health administrators. That system placed responsibility on them to adapt operations to new guidelines and scientific standards as they became available. The system of cooperation and professionalism that laws of that era envisage has parallels in the evolution of environmental laws in advanced EU (European Union) nations.

While this system had partly broken down by the 1960s in both Europe and America, the Europeans gradually increased the effectiveness of their institutions. The United States abandoned reliance on leadership by the network of responsible scientific and professional agencies.

9.4.1 Lawmaking in Advanced EU Nations (see more detail in Chapter 6)

Concepts for major new legislation are submitted by the ruling party or coalition to the relevant ministry.

1. Task groups prepare background data and draft legislation that is subjected to assessment of short-term and long-term impacts, including exposure to a range of potentially affected groups. A revised draft goes to Parliamentary committees that check for consistency with previous laws and EU policies and prepare a final draft.
2. The proposed law or policy is submitted to Parliament for debate, amendment, and vote. [Variant – some nations operate on provisional policies subject to continuous modification for years, with laws only passed after much experience has been gained.]
3. Advanced EU nations have increasingly tended to combine and simplify their fundamental environmental laws, changes that are made possible by decentralization and trust in official agencies. Sweden has a single law of fewer than 200 pages.

9.4.2 US Lawmaking

Any Congressional member may submit a bill for a proposed law alone or cosponsored.

1. The bill is assigned to a relevant standing committee of the House of Representatives or Senate where it may or may not be scheduled on the calendar for action, which may include hearings.
2. If passed by committee, a report is prepared on the bill, which is received in the relevant chamber of Congress and placed on the legislative calendar at the discretion of the Speaker of the House or the Majority Leader of the Senate. The bill may be assigned to additional committees for comment.
3. If passed, the bill goes to the other chamber, and if passed there is reconciled and sent to the President for signature.
4. The process is almost exclusively political, avoiding any impact assessment beyond funding.

9.4.3 Discussion

The main difference between the EU and the United States is the serendipitous, poorly coordinated way in which US bills are handled. There is little constraint on the number or intrusiveness of bills, and no significant vetting or testing of laws' potential impacts. Since the constraints of previous lawmaking patterns were lifted by the detailed and highly intrusive laws of the 1970s, lawmakers feel free to respond to any emerging problem or concern with micromanaging, rapidly generated laws. The results have been poor.

Impulsive support for production of ethanol from food corn yielded unanticipated adverse effects. Such outcomes are typical rather than exceptional. The public does not understand the details but knows something is wrong, and gives Congress failing marks. Individual Congressional members usually give their constituencies good personal assistance, which helps account for the fact that the public may like its individual Congressional members while the system remains broken.

A set of guidelines for good laws by McGarity (1983) has been widely cited : (1) administrative feasibility, (2) survivability (under existing conditions of judicial and political review), (3) enforceability, (4) efficiency, (5) fairness and equity, and (6) ability to encourage technological advance. More universal guidelines used in OECD and UN discussions are shown in Fig. 9.2. Comparison of the United States and OECD/EU/UN models brings out critical missing elements of consensus, inclusiveness, benchmarking, and efficiency in the US adversarial system. Behaviors that are assumed by some to be unavoidable attributes of US industry may, in fact, be products of this adversarial system.

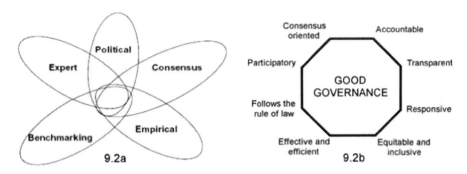

Fig. 9.2 Decisionmaking models
Source: 9.2a, OECD; 9.2b, UNESCAP

The existing matrix of environmental laws has virtually impregnable defenses and maximum survivability (see the McGarity model in Chapter 4 and the EU model in Chapter 6). But in the majority of criteria relating to scientific efficiency, ability to foster cooperation, innovation, and social and political harmony its failings help explain the systematic deterioration that has taken place in past decades.

The environmental movement has won most of the legal battles and virtually all of the legislative battles. Through successful tactics, 107 million acres of land have been shielded. That may be satisfying to activists. But the preemptive exclusionary policy has little resemblance to the kind of responsible and proactive interaction with nature practiced in Scandinavia. Continued forced compliance to rigid regulations, and demonization will not attract bright, responsible, and innovative leaders to industry. It is hard to see how it can enhance voluntary cooperation. Changes in the US system will likely be very difficult. Just as it took a Nixon to initiate communication with China and Clinton to change the earlier welfare system, it may be that new ideas and initiatives may have to come from the environmental side. (Some suggestions are reviewed in Conclusions, Section 9.6.)

9.5 Approach to the Present Research and Book

It might be assumed from these discussions that I mainly champion applied studies and downgrade the value of more theoretical approaches. Rather, I share with Alrøe and Kristensen (2002)[246] the idea of unitary rather than divided science, and that a policy analyst should interact with as many affected groups in society as possible during the learning process and then communicate the results in full rather than in compressed or analytical format. Many of the essential pieces of information and critical insights in this book have come from probing the social science literature, often through the suggestions of experts, and following up citations in the literature. Other insights surely remain to be found.

There are many techniques of the social sciences, catalogued in the exhaustive treatment by Mirovitskaya and Ascher (2001), that are relevant to aspects of environmental regulatory policy. However, for the present conflict and comparative study, more analytical approaches, i.e., systematizing or classifying policy phenomena in specific models or theoretical approaches to environmental conflict analysis are not suited to the emotion-laden, multidimensional problems of environmental regulations. Cheldelin, Druckman & Fast (2003) comment:

> "Scholars of regulatory politics have long puzzled over environmental regulation"…"Public choice theory has explained much of economic regulation as the product of concentrated interest group politics, but the theory has not furnished a convincing account of environmental regulation." …"Civic republican theory has offered an alternative hypothesis for the emergence of environmental regulation, but has often been more normative than positive" (i.e., has not offered statistical probability evidence). "Both the origin and content of environmental regulation remain enigmatic."

Different groups do not even agree about the nature of the problems with US environmental regulation. This type of problem could therefore be included among "wicked problems" (Rittel & Webber, 1973) or "social messes" (Horn, 2001) that involve complex interdependencies. However, such problems respond to in-depth historical analysis. In this case I believe that major factors, events, and time frames can be tracked even though the interactions become complicated and the trail for some effects gets diffuse as time goes on.

The starting point for my study was: what was the origin of the current conflict between US environmentalists and industry and when did major forcing events happen? I worked backward from the present, soon finding and bracketing two critical events: Rachel Carson's 1962 book, *Silent Spring*, and the Santa Barbara oil spill of January 1969, followed by passage of groundbreaking environmental laws.

I had acquired background about the fact that US history of health and environmental policy was more complicated and informative than just a dark age prior to the conservation and preservationist movements of the nineteenth century. Nor did enlightenment begin with Rachel Carson's book. Historical review beginning from the early history of the United States was clearly needed to work forward to meet the retrospective search in the 1960s. Successive sets of questions emerged about events, new laws, groups, and issues during these historical reviews.

I had the good luck to build on the insights of Don Kash about energy and environmental policy and federal government developments during the Carter and Reagan administrations. I drew on my own collection of documents and gray literature, literature searches, and Freedom of Information Act acquisitions of correspondence during the Carter and early Reagan administrations.

My location in Fairfax, VA, a suburb of Washington, DC, offered advantages for conferences, briefings, and for personal interviews and special library resources (e.g., Interior Department and US Geological Survey), as well as the excellent collections in the George Mason Library and Virginia university system collections. Over the course of five years I conferred by telephone or had interviews with more than 100 individuals, gaining special information from many distinguished and knowledgeable persons, the oldest of whom was 101-year-old Herbert Graham, a former Chief of the National Marine Fisheries Service in Woods Hole.

I published a pair of studies on offshore oil and mining and environmental policy in 2004 and 2006. These studies revealed important differences between Scandinavia and the United States. The Scandinavian nations had never had political polarization or gridlock over natural resource policy. The Norwegian Supreme Court (*Høyesterett*) had also rejected jurisdiction over offshore oilfield development because it felt it lacked competence. Unlike the US experience, litigation was not an issue. Lack of conflict, however, did not mean having no environmental problems, differences of opinion, or targeted exclusions from development.

Study of EU environmental policy included interviews with officials from seven nations: Austria, Finland, Germany, Hungary, Ireland, Sweden, and Ukraine, during three visits to Europe in 2003, 2005, and 2007. In addition, I have had extensive discussions with Norwegian environmental experts during conferences hosted by the Norwegian Embassy in Washington, DC. During the 2007 trip, European experts affirmed that the number one problem for their nations was global climate change.[247]

Before I began to compare US conditions with those in Europe, the intractability of the US impasse and its adverse influence on environmental and energy policy, as well as industrial – infrastructural development, left me discouraged. Almost no American from any perspective appeared to hold out much hope for real change. That is now more understandable in the light of European developments. The United States had a network of capable scientific and professional federal agencies. Because of failings in political governance, it was not able to deal effectively with environmental problems after World War II.

Instead of correcting governance and building a more effective cooperative system, the science and professional agencies were relegated to the role of technical functionaries as the political system assumed total responsibility.

Meanwhile, European nations moved from a roughly similar point to systematically strengthen their professional and governance systems. They decentralized environmental management and simplified laws through extensive experience and with good communication. If the United States opens its eyes to foreign experience, constructive change – though late – can begin.

9.6 Conclusions

The present book offers what may be a first (limited) attempt to integrate understanding of US environmental policy and political polarization problems after World War II with the history of political governance since the nation's founding.

1. During the administrations of the first six presidents, the United States maintained exceptionally high quality of governance. But failure to exclude slavery from the Constitution – because alternatives would have risked the unity of the 13 former colonies – created underlying instability.
2. The governance flaw due to the slavery provision led to political dominance by slave states and the inevitability of the Civil War. Later consequences were weak or no land use policies for the West, governmental corruption, and unstable lawmaking processes in Congress.
3. Isolated, high-quality federal agencies had existed (e.g., Coast Survey) since the early nineteenth century, but the Pendleton Civil Service Act of 1883 first

permitted the broader development of effective federal scientific and professional agencies and improved operation of Executive Departments. This helped implement progressive forest and conservation policies around the turn of the century.

4. The growth of competent federal agencies prior to World War II in turn encouraged better policy development and improved Congressional lawmaking, but enough arbitrariness remained to undercut evolution of stable and integrated systems and more effective guidance for the Executive Branch.

5. President F. D. Roosevelt was a charismatic and skilled political leader. However, discouragement of technical innovation and research under flawed economic assumptions and the Roosevelt administration's conflict with industry made US recovery from the Depression slower than that of other advanced nations.

6. During World War II Roosevelt adopted inclusive and flexible policies and chose effective leaders, resulting in the most rapid industrial and technical growth in US history. For the first time, scientists and engineers played an integral role in cooperation with industry and government (and the military).

7. After the war a new science paradigm created unbalanced emphasis on basic (discretionary) research, fueling growth of semidetached disciplinary studies in universities.

8. Turbulent events in the 1960s, including four assassinations of major political figures (starting with President Kennedy) and preoccupation of the Johnson White House, resulted in failure to deal effectively with environmental deterioration. Antibusiness trends fostered rise of tougher-minded, less open business leadership.

9. The Santa Barbara oil spill and ensuing crisis brought about revolutionary new environmental laws that were brilliantly designed for political effectiveness and passed with overwhelming bipartisan support. The laws promoted rapid progress at first, but their imbalances and rigidities created a rift in society between environmentalists and industry.

10. The intrusive laws opened the floodgates for serendipitous Congressional intervention into societal activities, allowing thousands of uncoordinated, poorly vetted bills to contend for political favor. Environmental NGOs and litigiousness grew.

11. The rift widened in the Carter and especially first Reagan administrations. It expanded into partisanship. Environmentalists were able to block industry initiatives but industry in turn blocked development of consensual or longer-term energy policies. The Reagan administration relegitimized business enterprise – but supply siders fostered overemphasis "on the bottom line," neglecting long-term stability and more thoughtful and inclusive governance.

12. After 1987 gridlock closed down most amendment or update of major environmental laws, except for the Clean Air Amendments of 1990 in the G.H.W. Bush administration. Lack of communication moved both environmental NGOs and industry toward radicalization.

13. The Clinton administration began with divisive EPA policies; however, the Reinventing Government program under Vice President Gore helped bring administrative reforms at EPA and more proactive, problem-solving approaches.

14. Despite superficial progress in terms of GDP, stock market indices, etc., underlying deterioration of the US infrastructure, industrial base, external trade balance continued.

15. Accentuated polarization developed in the George W. Bush administration. The United States remains unique among advanced nations in its rigid, compartmentalized environmental and regulatory structure; it and the NIMBY fostered by the current interpretations of NEPA (National Environmental Policy Act) contrary to founders' intent are blocking growth of commercial-scale renewable energy technologies. United States remains outside Kyoto.

16. In contrast to the United States the EU embraced sustainable development (i.e., integrated environmental and economic policies instead of primacy of environment). The consistent application of full consultation and intensive impact analysis before initiation of new environmental policies has fostered stability. Trust-based environmental advances, especially in Scandinavian nations have led to continuous innovation in both environmental policies and economic and industrial developments.

17. The EU as a whole has continued systematic development relying on intensive consultation and preparation to implement stronger environmental changes like those in the REACH (Registration, Evaluation and Authorisation of Chemicals) (chemical) laws that expand screening of potential toxic chemicals.

18. Conflict and polarization may be too great to permit near-term cooperative policies in the United States and reforms in environmental laws. But depoliticization of EPA could serve as a valuable first step, which would allow the agency to foster wider range planning and better cooperation.

19. US industry cannot be expected to compete with the environmental activist community in terms of open public communication and politicking. Industry will continue to lobby and network. To build trust and balanced policies, industry as well as all other concerned groups must be systematically included in the preparation of any new policy. It is easy to scoff at such ideas, but without balanced policies the United States will tend to continue partisanship, gridlock, and failed programs.

References

Academy of Finland. (2008). Academy of Finland: Generating scientific energy knowledge. *Energy and Enviro Finland*, 27 February, from http://www.energy-enviro.fi/index.php?PAGE=1503.

Ackerman, B. A. (2005). *The failure of the founding fathers and the rise of presidential democracy*. Cambridge, MA: Belknap Press, Harvard University Press.

Adam v. Norton et al. (2006) (U.S. Court of Appeals, Ninth Circuit, Northern California 1996), Docket 04-17365, from http://www.peer.org/docs/gs/06-11-12-reply-brief.pdf.

Adams, S. B. (1997). *Mr. Kaiser goes to Washington: The rise of a government entrepreneur*. Chapel Hill NC: University of North Carolina Press.

Albright, M. (2006). *The mighty and the almighty*. New York: Harper-Collins.

Alrøe, H. F., & Kristensen, E. S. (2002). Towards a system[at]ic research methodology in agriculture: Rethinking the role of values in science. *Agriculture and Human Values, 19*(1), 3–23.

American Association of Petroleum Geologists (AAPG). (2006). *Climate change policy*, from http://dpa.aapg.org/gac/papers/climate_change.cfm.

American Association of Petroleum Geologists (AAPG). (2007). *AAPG position statement, climate change* Retrieved Nov. 25, 2007, from http://dpa.aapg.org/gac/statements/climatechange.cfm.

American Civil Liberties Union (ACLU). (2005). *Science under siege*: ACLU.

American Geological Institute (AGI). (2007). *Oceans policy. 19 November 2007*. Retrieved 15 March 2008, from http://www.agiweb.org/gap/legis110/ocean.html.

American Lung Association (ALA). (2002). *Milestones in air pollution history* Chicago, IL: American Lung Association.

American Petroleum Institute (API). (1980). *Two energy futures: A national choice for the 80s*. Washington, DC: American Petroleum Institute.

American Philosophical Society (APS). (2006). Thomas Jefferson examines languages. Treasures of the APS (*American Philosophical Society Archives*) from http://www.amphilsoc.org/library/exhibits/treasures/vocab.htm.

American Society of Civil Engineers (ASCE). (2008). *Reportcard for America's infrastructure*, 2008, from http://www.asce.org/reportcard/2005/actionplan07.cfm.

American Wind Energy Association (AWEA). (2008). *U.S. wind energy projects*, 2008, from http://www.awea.org/.

American Wind Energy Association Siting Committee. (2008). *Wind energy siting handbook*. Washington, DC: American Wind Energy Association.

Anderson, T. L. (1997). *Breaking the environmental policy. gridlock*. Stanford, CA: Hoover Institution Press.

Andrews, R. N. L. (1999). *Managing the environment, managing ourselves*. New Haven, CT: Yale University Press.

Andrus, C. D., & Connelly, J. (1998,). *Cecil Andrus: Politics western style*. Seattle, WA: Sasquatch Books.

Annenberg. (2007, June 18–19, 2007). *Ceasefire! Bridging the political divide*. Retrieved 20 December 2007, from http://www.bridgingthepoliticaldivide.org/.

Appiah, K. A., & Gates, H. L. J. (2003). *Africana.* Jackson, TN: Perseus Publishing.

Applegate, J. (2002). *Engaged graduate education: Seeing with new eyes,* from http://www.aacu. org/pff/PFFpublications/engaged/research.cfm.

Armistead, T. F. (2008). *DOE drops clean-coal plant to focus on carbon capture,* from http://enr. construction.com/news/powerIndus/archives/080206a.asp.

Arnold, R., & Gottlieb, A. (1994). *Trashing the economy.* Bellevue, WA: Free Enterprise Press.

Auduc, A., Meissel, R., Pautremat, F. L., & Galaup, F. (1998). *Constitution et citoyenneté aux Etats-Unis, textes fondateurs et études de cas.* Paris: Adapt.

Axelrod, R. S., Downie, D. L., & Vig, N. J. (Eds.). (2004). *The global environment: Institutions, law and policy* (2nd ed.). Washington, DC: Congressional Quarterly Press.

B_B_C_News. (2007). *A timeline of the EU,* from http://news.bbc.co.uk/2/hi/africa/default.stm.

Bakan, J. (2007). *The corporation: The pathological pursuit of profit and power.* New York, NY: Free Press.

Baker, J. (1983, October 31, 1983). *The BLM was trashed.* Reprinted from *High Country News* (HCN).

Baltimore, D. (2007, February 19, 2007). *Building a community on trust,* from http://mitworld. mit.edu/video/64/.

Bamberger, R. L. (2003). *Energy policy: The continuing debate* (No. Report I B10116). Washington, DC: Congressional Research Service.

Bardach, E., & Kagan, R. A. (Eds.). (1982). *Social regulation: Strategies for reform.* San Francisco, CA: Institute for Contemporary Studies.

Baron. (2007). *Legal information mesothelioma case results.* Retrieved 11 January 2007.

Barry, J. M. (1997). *Rising tide.* New York: Simon and Schuster.

Barton, J., Jenkins, R., Bartzokas, A., Hesselberg, J., & Knutsen, H. (2007). Environmental regulation and industrial competitiveness in pollution-intensive industries. In S. Parto & B. Herbert-Copley (Eds.), *Industrial innovation and environmental regulation: Developing workable solutions.* Tokyo: United Nations University Press.

Beattie, D. A. (1997). *History and overview of solar heat technologies.* Cambridge, MA: MIT Press.

Bedard, R. (2005). *Final summary report, project definition study, offshore wave power feasibility demonstration project.* Palo Alto, CA: Electric Power Research Institute.

Beder, S. (2006a). *Free market missionaries: The corporate manipulation of community values.* London: Earthscan.

Beder, S. (2006b). *Suiting themselves.* London: Earthscan.

Bedini, S. A. (2002). *Jefferson and science.* Charlottesville, VA: Thomas Jefferson Foundation.

Belanger, D. O. (1998). *Enabling American innovation: Engineering and the National Science Foundation.* West Lafayette, IN: Purdue University Press.

Bell, D. (1973). *The coming of post industrial society: A venture in social forecasting.* New York, NY: Basic Books.

Bellona. (2005). *Carbon dioxide storage: Geological security and environmental issues.* Oslo Norway: Environmental Foundation Bellona.

Bennett, W. A. (2005.). Critical assessment of the delta smelt population in the San Francisco Estuary, California. *San Francisco Estuary and Watershed Science, 3*(2), 1–71.

Bertrani, R. (2007). World geothermal generation in 2007. *GHC Bulletin, 2007*(September), 8–19.

Bezdek, R. H., & Wendling, R. M. (2002). A half century of long-range energy forecasts: Errors made, lessons learned, and implications for forecasting, vol. 21,. *Journal of Fusion Energy,* (Nos. 3/4), 155–172.

Binder, S. (2003). *Stalemate: Causes and consequences of legislative gridlock.* Washington, DC: Brookings Institution Press.

Bingaman, J. (2008). Securing our energy future. *National Academies summit on America's energy future.* Washington DC: (National Academy of Sciences): National Academies Press.

Bird, K. J. & Houseknecht, D. W. (1998). Arctic National Wildlife Refuge, 1002 area, petroleum assessment. U. S. Geological Survey Open-File Report 98-34, Version 1.0 [CD-ROM]. Reston, VA.

Birur, D. K., Hertel, T. W., & Tyner, W. E. (2007). The biofuels boom: Implication for world food markets, *Food Economy Conference.* The Hague, Netherlands: Purdue University

Bogenschneider, K. (2002). *Family policy matters*. Mahwah, NJ: Lawrence Erlbaum Associates.

Bond, G. C., Showers, W., Elliot, M., Evans, M., Lotti, R., Hajdas, I., et al. (1999). The North Atlantic's 1-2 KYR climate rhythm: Relation to Heinrich events, Dansgaard/Oeschger cycles and the Little Ice Age. In P. U. Clark, R. S. Webb, & L. D. Keigwin (Eds.), *Mechanisms of global change at millennial time scales* (Geophysical Monograph 112, pp. 59–76). Washington, DC: American Geophysical Union.

Bond, J. R., & Fleisher, R. (2000). *Polarized politics congress and the president in a partisan era*. Washington, DC: Congressional Quarterly.

Boston City Council. 1856. Charter and ordinances of the City of Boston. In Boston City Council, p. 262. Boston, MA: Moore and Crosby City Printers.

Bower, T. (1992). *Maxwell–The book Maxwell tried to stop*. London: Viking.

Bradley, R. (2008). *Telephone discussion*. Fairfax, VA: World Resources Institute.

Braeman, J. (1994). *Roosevelt uses business leaders for World War II planning* (eNotes.com, 2006 ed.): Hackensack, NJ: Salem Press.

Bridis, T. (2006). *DP world: No plan to sell Miami Port ops*, from http://www.breitbart.com/article. php?id=D8GB0HJGC&show_article=1.

Briggs, M. (2004, August 2004). *Widescale biodiesel production from algae*, from http://www. unh.edu/p2/biodiesel/article_alge.html.

Bromley, D. A. (1994). *The President's scientists: Reminiscences of a White House science advisor*. New Haven, CT: Yale University Press.

Bromley, D. J. (1991). *Environment and economy: Property rights and public policy*. New York: Basil Blackwell.

Brooks, D. J., & Geer, J. G. (2007). Beyond negativity: The effects of incivility on the electorate. *American Journal of Political Science*. 51(1), pp. 1–16.

Brossard, E. B. (1983). *Petroleum–politics and power*. Tulsa, OK: PennWell Publishing Company.

Brown, L. (2006). Ethanol could leave the world hungry. *August 16 2006: 5:39 AM EDT (Fortune Magazine) CNNMoney.com*.

Brown, L. (2007). *Exploding U.S. grain demand for automotive fuel threatens world food security & political stability*, from SustainableBusiness.com, http://sustainablebusiness.com/ index.cfm/go/news.feature/id/1390/.

Brown, L., & Lewis, J. (2008, April 22, 2008). Ethanol's failed promise. *The Washington Post*.

Brundtland, G. H. (2004). Healthy people, healthy planet. *Nobel Peace Prize Forum*, http://www. stolaf.edu/news/speeches/bruntland/html.

Brundtland, G. H. (Ed.). (1987). *Our common future*. New York: Oxford University Press.

Bureau of Economic Analysis. (2008). *International economic accounts*, from http://www.bea. gov/international/bp_web/simple.cfm?anon=68941&table_id=1&area_id=3.

Burroughs, R. D. (Ed.). (1961). *The natural history of the Lewis and Clark Expedition*. East Lansing, MI: Michigan State University Press.

Bush, V. (1945). *Science the endless frontier*. Washington, DC: U.S. Government Printing Office.

Byrnes, P. (1998). *Environmental pioneers*. Minneapolis MN: Oliver Press.

Cain, L. P. (1978). *Sanitation strategy for a lakefront metropolis, the case of Chicago*. Chicago, IL: Northern Illinois University Press.

Cain, L. P. (2005). *Sanitation in Chicago: A strategy for a lakefront metropolis*, from http://www. encyclopedia.chicagohistory.org/pages/300017.html.

Caldwell, L. K. (1963). Environment: A new focus for public policy? *Public Administration Review, 13*(3), 132–139.

Caldwell, L. K. (1998). *The national environmental policy act : An agenda for the future*. Bloomington, IN: Indiana University Press.

Cape Wind Associates, L. (2005). *Cape wind energy project draft environmental impact statement*. San Francisco, CA: U.S. Army Corps of Engineers.

Carhart, A. H. (1959). *The national forests*. New York: Alfred Knopf.

Caro, R. (1975). *The power broker*. New York City: Vintage.

Carroll, S. J., Hensler, D. R., Gross, J., Sloss, E. M., Schonlau, M., Abrahamse, A., et al. (2005). *Asbestos litigation* (Monograph). Santa Monica, CA: RAND Corporation.

Carson, R. (1962). *Silent spring*. New York: Houghton Mifflin.

Carter, J. (1977). National address on energy. *Vital Speeches of the Day, XXXXIII*(14), 418–422.

Carter, J. (2000). Why we need an Arctic Refuge National Monument. *Defenders of Wildlife Magazine*, The Arctic Refuge and its coastal plain, Part II, the debate, from http://www.uconn.edu/ANWR/anwr.debateindex.htm.

Cassidy, P. (2008, April 23, 2008). Wind farm generates more than 40,000 comments. *Cape Cod Times*, http://www.capecadonline.com/aps/pbcs.dll/article?AID=20080423/NEUS/80423033.

CBS. (2003, Sept. 26, 2003). *Cheney's ties remain*. Retrieved 25 January 2008, from http://www.cbsnews.com/stories/2003/09/26/politics/main575356.shtml.

Cendrowicz, L. (2008, May 8 2008). *Europe grapples over biofuels*, 2008, from http://www.time.com/time/business/article/0,8599,1738434,00.html.

Center for Biological Diversity (CBD). (2006). *Emergency Petition to list the Delta Smelt (Hypomesus transpacificus) as an endangered species under the Endangered Species Act*: Center for Biological Diversity, The Bay Institute, The National Resources Defense Council.

Center for Marine Research and Environmental Technology (CMRET). (2005). *Center for Marine Research and Environmental Technology*, 2005, from http://www.olemiss.edu/depts/mmri/programs/cm_over.html.

Center for Sustainable Production (2008). *REACH – The new EU chemicals strategy: A new approach to chemicals management* 2008, from http://www.chemicalspolicy.org/reach.shtml.

Central Intelligence Agency (CIA). (1986). *The world factbook*, from https://www.cia.gov/library/publications/the-world-factbook/geos/us.html

Charleston Gazette. (2004, June 29, 2004). Wind farms: Mollohan and Rahall are wise to ask questions now [Editorial]. Charleston, WV: Charleston Gazette.

Cheldelin, S., Druckman, D., & Fast, L. (2003). *Conflict*. New York: Continuum.

Chertow, M., & Esty, D. C. (Eds.). (1997). *Thinking ecologically – the next generation of environmental policy* (paperback ed.). New Haven, CT: Yale University Press.

Clean Ocean and Safe Tourism Anti-Drilling Act, Senate (2005).

Clinton, W., & Gore, A. (1995). Reinventing environmental regulations. National Performance Review (2.2.6 ed.,). Washington, D.C.: U.S. Government Printing Office.

Cloud, D. (1873). *Monopolies and the People*. Davenport, IO: Day, Egbert and Fidlar.

Cohen, I. B. (1994). *Interactions: Some contacts between the natural and the social sciences*. Cambridge, MA: MIT Press.

Cohen, M. (2006). The roots of sustainability science: A tribute to Gilbert F. White, *Sustainability, Science, Practice, and Policy* (Vol. 2). Reston, VA: National Biological Information Infrastructure.

Cole, L. (1994). Civil rights, envronmental justice and the EPA: The brief history of administrative complaints under Title VI. *Journal of Environmental Law and Litigation, 9*, 309–398.

Commoner, B. (1972). *The closing circle: Nature, man, and technology*. New York: Alfred Knopf.

Congress, U. S. (1966). Marine Resources and Engineering Development Act of 1966. In U. S. Congress (Ed.) (Vol. 33 U.S.C. §§ 1101–1108): U.S. Congress.

Congressional Quarterly Service. (1973). 1969 Chronology. *Congress and the Nation: A Review of Government and Politics, 3*(1969–1972), 748–749.

Connor, S. (2007). If we fail to act, we will end up with a different planet. *The Independent*. 1 January 2007, from http://www.independent.co.uk.

Costle, D. (1985, 2006). *Ash council and creation of EPA*, from http://www.epa.gov/35thanniversary/publications/costle/14.htm.

Coulter, A. (2004). *How to talk to a liberal (If you must): The world according to Ann Coulter*. New York: Crown.

Cox, J. (2003, Thursday, Oct. 2, 2003). Deep trouble: A less-than Superfund. *Naples Daily News*.

Craig, M. G. (2008, May 5, 2008). *Defending the Reagan legacy: Rejecting revisionist history*. Retrieved 5 May 2008, from http://www.ashbrook.org.

Crain, M. (2005). *Impact of regulatory costs on small firms*. Washington, DC: U.S. Small Business Administration.

Cropper, M. L., & Oates, W. E. (1992). Environmental economics: A survey. *Journal of Economic Literature, 30*(June), 675–740.

Czech, B., & Krausman, P. R. (2001). *The endangered species act: History, conservation biology, and public policy.* Baltimore, MD: The Johns Hopkins University Press.

Dalton, K. (2007). *Theodore Roosevelt; a strenuous life.* New York: Random House.

Davis, A. C. (2008). California's budget deficit grows to $16 billion. Associated Press.

Davis, R. B. (1964). *Intellectual life in Jefferson's Virginia 1790–1830.* Knoxville, TN: University of Tennessee Press.

Day, J. N. (2001). Safe drinking water-Safe sites. In P. S. Fischbeck & R. S. Farrow (Eds.), *Improving regulation: Cases in environment, health, and safety* (pp. 17–42). Washington, DC: Resources for the Future.

de Tocqueville, A.-C.-H. C. (1835). *Democracy in America* (appearing in two volumes: 1835 and 1840).

Delay, T. (2006). *Rep. Tom Delay delivers his farewell address,* from http://www.washingtonpost.com/wp-dyn/content/article/2006/06/08/AR2006060801376.html.

Department of Energy (DOE). (2005). *National energy policy status report: Implementation of NEP recommendations.* Washington, DC: Department of Energy.

Department of Energy (DOE). (2008a). *20% wind energy by 2030-increasing wind energy's contribution to U.S. electricity supply.* Washington, DC: Department of Energy.

Department of Energy (DOE). (2008b). *Wind powering America wind resource maps,* from DOE 20% wind energy report, from http://www.20percentwind.org/20p.aspx?page=Report. Washington, DC: Department of Energy.

Department of Environment, Food, and Rural Affairs (UK) (DEFRA). (2007). *Surveys to help protect submerged habitats and heritage,* Retrieved 15 February 2008, from http://www.defra.gov.uk/marine/science/pshh.htm.

Det Norske Veritas. (2007). *Pipeline damage assessment from Hurricanes Katrina and Rita in the Gulf of Mexico* (Technical Report). Oslo, Norway: Det Norske Veritas.

Deutch, J., & Moniz, E. J. (2007). *The future of coal.* Cambridge, MA: Massachusetts Institute of Technology.

Devaney, E. E. (2006). *Testimony of the Honorable Earl E. Devaney, Inspector General of the Department of the Interior before Subcommittee on Energy and Resources, House of Representatives,* 13 September 2006.

Devine, R. S. (2004). *Bush versus the environment.* New York, NY: Anchor.

Dewey, S. H. (2000). *Don't breathe the air: Air pollution and U.S. environmental politics, 1945–1970.* Houston, TX: Texas A&M Press.

Diamond, J. (1997). *Guns, germs, and steel.* New York: W.W. Norton

Dickson, B., & Clooney, R. (Eds.). (2005). *Biodiversity and the precautionary principle: Risk and uncertainty in conservation and sustainable use.* London: Earthscan, James and James.

Dickson, D. (1988). *The new politics of science.* Chicago, IL: The University of Chicago Press.

Dirksen, E. M. (1964). Acceptance speech for receipt of good government award by American Good Government Society. Dirksen Center, from http://dirksencenter.org/print_end_quotes.htm.

Dobyns, H. (1983). *Their numbers became thinned.* Knoxville, TN: University of Tennessee Press.

Dolin, E. J. (1990). *Dirty water/clean water: A chronology of events surrounding the degradation and cleanup of Boston Harbor.* Cambridge, MA: Massachusetts Institute of Technology Sea Grant Program.

Donald, D. H. (1996). *Lincoln.* New York: Touchstone Simon & Schuster.

Dunlop, B. N. (2000). *Clearing the Air–How the people of Virginia improved the state's air and water despite the EPA.* Arlington, VA: Alexis de Tocqueville Institution.

Dupree, A. H. (1957). *Science in the Federal Government: A history of policies and activities to 1940.* Cambridge, MA: Harvard University Press.

Durant, R. F. (1987). Toward assessing the administrative presidency: Public lands, the BLM, and the Reagan revolution. *Public Administration Review, 47*(2), 180–189.

Durant, R. F. (1993). Hazardous waste, regulatory reform, and the Reagan revolution: The ironies of an activist approach to deactivating bureaucracy. *Public Administration Review, 53*(6), 550–560.

Easton, R. O. (1972). *Black tide: The Santa Barbara oil spill and its consequences*. New York: Delacorte Press.

Easton, T. A., & Goldfarb, T. D. (Eds.). (2003). *Clashing views on controversial environmental issues* (10th ed.). Guilford, CT: McGraw Hill/Dushkin.

European Commission (EC). (2003). Directive 2003/30/EC of the European Parliament and of the Council on the promotion of the use of biofuels on other renewable fuels for transport. Brussels, Belgium: European Commission.

Energy Efficiency and Renewable Energy (EERE). (2003). *Commonly used data*, from https://apps3.eere.energy.gov/ba/pba/analysis_database/docs/excel/common-items.php.

Energy Efficiency and Renewable Energy (EERE). (2008a). *Geothermal technologies program*, 2008, from http://www1.eere.energy.gov/geothermal/overview.html#direct_use.

Energy Efficiency and Renewable Energy (EERE). (2008b). *Labor and environmental groups launch a Green Jobs campaign*. 16 April 2008, from http://apps1.eere.energy.gov/news-detail. ofm/news_id=11709.

Eilperin, J. (2008, April 2, 2008). Environmental laws to be waived for fence. *The Washington Post*, p. A4.

Eldridge, M., Elliott, N., Prindle, W., Ackerly, K., Laitner, J. S., McKinney, V., et al. (2008). *Energy efficiency: The first fuel for a clean energy future–future resources for meeting Maryland's electricity needs*. Washington, DC: American Council for an Energy-Efficient Economy.

Elliott, R. N., & Spurr, M. (2008). *Combined heat and power: Capturing wasted energy*, from http://www.aceee.org/pubs/ie983.htm.

Energy Information Administration. (2007). *International energy outlook 2007*, 2008.

Engler, R. (1961). *The politics of oil: A study of private power and democratic directions*. Chicago, IL: Phoenix Books.

Environmental Protection Agency (EPA). (1990). *The environmental protection agency: A retrospective*, 29 Nov. 1990, from http://www.epa.gov/history/topics/epa/20a.htm.

Environmental Protection Agency (EPA). (1997, 10/21/97). *NTC New report shows clean air benefits significantly outweigh costs*, from http://yosemite.epa.gov/opa/admpress.nsf/8b75cea41 65024c685257359003f022e/ef494b0c753c5863852565370069dca4!OpenDocument.

Environmental Protection Agency (EPA). (2007a). *National Pollutant Discharge Elimination System (NPDES)*, from http://cfpub.epa.gov/npdes/per.cfm.

Environmental Protection Agency (EPA). (2007b). STORET, from http://www.epa.gov/storet/.

Estes, R. J. (1993). Toward sustainable development: From theory to praxis. *Social Development Issues, 15*(3), 1–29.

Esty, D. C., Levy, M. A., Kim, C. H., de Sherbinin, A., Srbotnjak, T., & Mara, V. (2008). *Environmental performance index*, New Haven, CT: Yale Center for Environmental Law and Policy, from http://research,yale.edu/envirocenter, July 2008.

European Union (EU). (2005). *The contribution of good environmental regulation to competitiveness*. White Paper by Network of heads of European Environmental Protection Agencies. Prague, Czech Republic: European Environmental Agency.

Euractiv.com. (2003). Waste electrical and electronic equipment legislation.

EVOSTC. (2005). *Oil spill facts*, from http://www.evostc.state.ak.us/facts/index.html.

FactCheck. (2004, September 30, 2004). *Kerry ad falsely accuses Cheney on Halliburton*. Retrieved 25 January 2008, from http://www.factcheck.org/kerry_ad_falsely_accuses_cheney_on_halliburton.html.

Fahrenthold, D., & Mufson, S. (2007). Drilling for oil off Virginia's shores. *The Washington Post*, 28 April, pA01b.

Federal Register. (2007). Proposed rules. *Federal Register (online), 72*(78), 20314–20317.

Feeney, C., & Neumann, D. (2004). *Removing the barriers to solar energy: The goal of the Santa Barbara County Million Solar Roofs Partnership*. Santa Barbara, CA: Community Environmental Council.

Fehrenbacher, K. (2008). *15 Algae startups bringing pond scum to fuel tanks*, from http://earth2tech.com/2008/03/27/15-algae-startups-bringing-pond-scum-to-fuel-tanks/.

Federal Energy Regulation Commission (FERC). (2007a). FERC hydrokinetic energy project policy. In FERC (Ed.). (Docket No. PL08-1-000). Washington, DC: Federal Energy Regulatory Commisssion.

Federal Energy Regulation Commission (FERC). (2007b). FERC issues first license for Hydrokinetic Energy Project. In FERC (Ed.). (Docket No: P-12751-000). Washington, DC: Federal Energy Regulatory Commission.

Feller, I. (1970). *The application of science and technology to public programs.* Cambridge, UK: Cambridge University Press.

Finnish Environment Institute. (2008). Five Finnish municipalities will show how to curb climate change. *Energy and Enviro Finland* 13 April 2008, from http://www.energy-enviro.fi/index. php?page=1673.

Fiorina, M. P., & Abrams, S. J. (2008). Political polarization in the American public. *Annual Review of Political Science, 11,* June 2008, from http://papers.ssrn.com/s013/papers.cfm?abstract_id=1141503.

Fiorina, M. P., Abrams, S. J., & Pope, J. C. (2005). *Culture war? The myth of a polarized America.* New York : Longman.

Fiorino, D. J. (2006). *The new environmental regulation.* Cambridge, MA: MIT Press.

Fischer, D. W. (1989). Returns to society from offshore hard mineral resource development: Special interests are seeking to monopolize public land under the sea by doing away with the 1953 act. *American Journal of Economics and Sociology, 48*(1), 31–46.

Fischer, G. (1964). The Nobel Prize in Physiology or Medicine 1948. In *Nobel lectures, physiology or medicine 1942–1962.* Amsterdam: Elsevier Publishing Company.

Fitzgerald, M. (2007). In wake of infrastructure tragedies, ASCE offers insight and leadership. *ASCE News,* September 2007.

Follansbee, R. (1994). *A history of the water resources branch, U.S. Geological Survey.* Denver, CO: U.S. Geological Survey, from http://water.usgs.gov/history.html

Foner, E. (2005). *Reconstruction* (Francis Parkman Prize Edition ed.). New York: History Book Club.

Foreman, D. & Haywood, B. (1993). *Ecodefense: A field guide to monkeywrenching* Chico, CA: Abzug Press.

FoundLocally.com (2008). *Sudbury history – The post-war years,* 2008, from http://www.foundlocally.com/Sudbury/Local/Info-CityHistoryPostWarYears.htm.

Friedemann, A. (2007, May 30, 2007). *Energy in a nutshell,* from http://www.energyskeptic.com/ Energy_In_A_Nutshell.htm.

Frieden, B. (1979). *The environmental protection hustle.* Cambridge, MA: MIT Press.

Friedmann, S. J. (2007). Geological carbon sequestration. *Elements,* 3(3), 179–184.

Fukuyama, F. (1995). *Trust.* Tampa, FL: The Free Press.

FundingUniverse. (2008). *Outokumpu Oyj,* from http://www.fundinguniverse.com/company-histories/Outokumpu-Oyj-Company-History.html.

Gardiner, D., & Portney, P. R. (1999). Does environmental policy conflict with economic growth? In W. E. Oates (Ed.), *The RFF reader in environmental and resource management* (pp. 21–26). Washington, DC: Resources for the Future.

Gates, S. M., & Leuschner, K. J. (Eds.). (2007). *In the name of entrepreneurship? The logic and effects of special regulatory treatment for small business.* Santa Monica, CA: Kauffman-Rand Institute for Entrepreneurship and Public Policy.

General Accounting Office (GAO). (1981). *Issues in leasing offshore lands for oil and gas development:* (Report to the Congress by the Comptroller General of the United States No. B 201745). Washington, DC: General Accounting Office.

General Accounting Office (GAO). (2004). *Report to congressional requesters: Individual fishing quotas:* (GAO Report 04–277). Washington, DC: General Accounting Office.

Geoscience Information Society. (2007a). Geoscience information society: News about the USGS libraries: Information for advocates. Retrieved 1 January 2008, from http://www.geoinfo.org/ GSIS_News_USGSLibraries.htm.

Gillilan, D. M., & Brown, T. C. (1997). *Instream flow protection: Seeking a balance in western water use.* Washington, DC: Island Press

Glater, J. D. (2005, October 9, 2005). *The tort wars, at a turning point.* New York: *The New York Times,* Business, p. 1 ff.

Goodell, J. (2007). *Big coal: The dirty secret behind America's energy future.* New York: Mariner, Houghton Mifflin.

Gore, A. (1992). *Earth in the balance: Ecology and the human spirit.* Boston, MA: Houghton Mifflin.

Gore, A. (2001, Jan. 12, 2001). *A brief history of Vice President Al Gore's national partnership for reinventing government during the administration of President Bill Clinton 1993–2001,* from http://govinfo.library.unt.edu/npr/whoweare/historyofnpr.html.

Gore, A. (2006). *An inconvenient truth.* Emmaus, PA: Rodale.

Gozan, J. (1992). *Solar eclipsed, multinational monitor,* from http://multinational.org/issues/1992/04/mm0492.07.html.

Graham, F. (1970). *Since silent spring.* Boston, MA: Houghton Mifflin Co.

Graham, L. R. (1993). *Science in Russia and the Soviet Union a short history.* Cambridge, MA: Cambridge University Press.

Graham, M. (1999). *The morning after Earth Day.* Washington, DC: Brookings Institution Press.

Gramling, R. (1996). *Oil on the edge: Offshore development, conflict, gridlock.* Albany, NY: State University of New York Press.

Green, R. J., & Lepkowski, W. (2006). A forgotten model for purposeful science. *Issues in Science and Technology* 2006 (Winter), 69–73.

Greenberg, D. S. (2001). *Science, money and politics: Political triumph and ethical erosion.* Chicago, IL: University of Chicago Press.

Greenpeace. (2007). Greenpeace investigations. Retrieved 29 January 2008, from http://research.greenpeaceusa.org/.

Greider, W. (1982). *The education of David Stockman and other Americans.* New York: E.P. Dutton.

Greve, M. S., Smith, F. L., & Wilson, J. Q. (1992). *Environmental politics: Public costs, private rewards.* New York. Praeger.

Grodsky, P. (1982). Some limitations of social scientists in affecting public policy decisions. In F. S. Sterrett & B. L. Rosenberg (Eds.), *Science and public policy II* (Vol. 387, pp. 47–56). New York: New York Academy of Sciences.

Grunwald, M. (2006). DeLay pulls no punches in final speech to House. *The Washington Post.* 9 June 2006, pA03.

Gunningham, N. A., Thornton, D., & Kagan, R. A. (2005). Motivating management: Corporate compliance in environmental protection. *Law and Policy, 27*(2), 289–316.

Gurian, P., & Tarr, J. A. (2001). The first federal drinking water quality standards and their evolution, 1914–1974. In P. S. Fischbeck & R. S. Farrow (Eds.), *Improving regulation: Cases in environment, health and safety* (pp. 43–69). Washington, DC: Resources for the Future.

Halliday, T. (1986). Six score years and ten: Demographic transitions in the American legal profession, 1850–1980. *Law and Society Review, 20,* 53–78.

Hannerz, M., & Palmer, C. H. (2003). Swedish forest research. *Scandinavian Journal of Forestry Research News and Views, 5,* 387–390.

Hansen, J. (1988). *The Greenhouse effect: Impacts on current global temperature and regional heat waves.* Washington, DC: Committee on Energy and Natural Resources, United States Senate.

Hansen, J., Fung, I., Lacis, A., Rind, D., Lebedeff, S., Ruedy, R., et al. (1988). Global climate change as forecast by Goddard Institute for Space Studies three-dimensional model. *Journal of Geophysical Research, 93,* 9341–9364.

Harbaugh, W. H. (1975). *Power and responsibility: The life and times of Theodore Roosevelt.* New York: Octagon Press.

Hardman, J. B. S. (2005). *Rendezvous with destiny: Addresses and opinions of Franklin Delano Roosevelt* (reprint of 1944 ed.). New York: Kessinger Publishing Co.

Harisalo, R., & Stenvall, J. (2004). Trust as capital, ch. III. In M.-L. Huotari, & M., Livone, (Eds.), *Trust in knowledge management and systems in organizations* Hershey, PA: Idea Group Publications, (pp. 51–80).

Harris, W. C. (2004). *Lincoln's last months.* Cambridge, MA: Belknap Press of Harvard University Press.

HarrisPoll. (2007, Nov. 1, 2007). *Oil, pharmaceutical, health insurance, managed care, utilities and tobacco top the list of industries that many people think need more regulation,* from http://www.harrisinteractive.com/harris_poll/index.asp?PID=825.

Hart, D. M. (1998). *Forged consensus: Science, technology and economic policy in the United States, 1921–1953.* Princeton, NJ: Princeton University Press.

Harvey, D. (2001). *U.S. aquaculture – Current status and future potential.* Washington, DC: U.S. Department of Agriculture, Appendix E-2, Western Region Aquaculture Industry.

Hayward, S. F. (2008). The United States and the environment: Laggard or leader? *AEI Online-Environmental Policy Outlook,* from http://www.aei.org/Publications/pub10.27548/pub-detail.asp.

Hazilla, M., & Kopp, R. J. (1990). *Social cost of environmental quality regulations: A general equilibrium analysis* (Vol. 98, pp. 853–873). Chicago, IL: The University of Chicago Press.

Heartland. (2007). *Heartland Institute.* Retrieved 24 November 2007, from http://www.heartland.org/PolicyBotSearch.cfm

Heatherly, C. L. (Ed.). (1980). *Mandate for leadership.* Washington, DC: Heritage Foundation.

Heinz, G. H., & Hoffman, D. J. (1998). Methylmercury chloride and selenomethionine interactions on health and reproduction in mallards. *Environmental Toxicology and Chemistry, 17,* 139–145.

Hendrick, B. K. (1937). *Bulwark of the Republic: Biography of the Constitution.* Boston, MA: Little Brown & Co.

Hess, H. D., et al. (1981). *Program feasibility document, OCS hard minerals leasing (Executive Summary).* Washington, DC: U.S. Department of the Interior.

Hickel, W. (2003). *Oral history, secretary of the interior series.* Center for the American West, from http://www.centerwest.org/prajects/secretaries/bio.php. Retrieved 1 October 2008.

Hickel, W. J. (1971). *Who owns America?* Englewood Cliffs, NJ: Prentice-Hall.

High Country News. (1983, October 31, 1983). Watt's ignorance of coal proved fatal. *High Country News,* from http://www.hcn.org/specialcollections/watt.jsp#watt.

Hinish, J. E. J. (1981). Regulatory reform: An overview. In C. Heatherly (Ed.), *Mandate for leadership* (pp. 695–707). Washington, DC: Heritage Foundation.

Hirst, E., & Moyers, J. C. (1973). Efficiency of energy use in the United States. *Science, 179*(4080), 1299–1304.

Hobson, K. (2006). Environmental responsibility and the possibilities of pragmatist-oriented research. *Social & Cultural Geography, 7*(2), 283–298.

Hoogenboom, A. (2006). *Rutherford Birchard Hayes: American President.* An online reference source, from http://millercenter.virginia.edu/academic/americanpresident/hayes.

Hopson, E. (1982). Watt's oil lease plan challenged: Natives, environmentalists join states in suit. *The Arctic Policy Review* (August, 1982).

Horn, R. E. (2001). Knowledge mapping for complex social messes. *Foundations in the knowledge economy.* Los Altos, CA: David and Lucile Packard Foundation.

Hornig, D. F., & Tukey, J. (1965). Restoring the quality of our environment. *Environmental pollution panel.* The White House: Report by the President's Science Advisory Committee.

Hughes, J. (2002). *The Manhattan project: Big science and the atom bomb.* Reading, UK: Icon Books, Ltd.

Hunt, L. (2002). *Against presentism. Perspectives, 2002 (May),* from http://www.historians.org/perspectives/issues/2002/0205/pre1.cfm. (21 Sept. 2007)

Inhofe, J. (2005). Climate change update Senate floor statement, *U.S. Senate Speech.* 22 Dec 07.

Inslee, J., & Hendricks, B. (2008). *Apollo's fire: Igniting America's clean-energy economy.* Washington, DC: Island Press.

Intergovernmental Panel on Climate Change (IPCC). (2007, April 6, 2007). *Climate change 2007.* Retrieved April 6, 2007, from http://www.ipcc.ch/.

Isenson, R. S. (Ed.). (1969). *Project Hindsight* (Final ed.). Washington DC : Office of the Director of Defense Research and Engineering.

Isidore, C. (2005). *Big oil CEOs under fire in Congress.* CNNMoney, 9 Nov. 2005, from http://money.cnn.com/2005/11/09/news/economy/oil_hearing/index.htm.

Jackson, K. T. (1995). *The encyclopedia of New York City* (pp. 1376). New Haven, CT: Yale University Press.

Jacobs, J. (1961). *Death and life of great American cities.* New York: Random House.

James, W. (1956). *The will to believe, human immortality.* Mineola, NY: Dover Publications.

Jasanoff, S. (1998). *The fifth branch: Science advisers as policymakers.* Cambridge, MA: Harvard University Press.

Jefferson, T. (1782). *Notes on the State of Virginia.* Electronic Text Center, University of Virginia Library, 4 Oct. 2007 from http://etext.virginia.edu/etcbin/toccer-new2?id=JefVing. sam&images/modeng&data=/texts/english/modeng/parsed&tag=public&part=front.

Johnson, J. (1939). *Peter Anthony Dey: Integrity in public service.* Iowa City, IA: State Historical Society of Iowa.

Kagan, R. A. (2003). *Adversarial legalism.* Cambridge, MA: Harvard University Press.

Kaiser, J. (1997). Science bureaucracy: USGS chief resigns after tough tenure. *Science, 277* (5033), 1755.

Kempthorne, D. (2008). *Secretary Kempthorne announces decision to protect polar bears under endangered species act; Rule will allow continuation of vital energy production in Alaska.* Washington, DC: U.S. Department of the Interior, from http://www.doi.gov/news/08_News_Release/0805142.html.

Kendrick, J. W., & Grossman, E. S. (1980). *Productivity in the United States: Trends and cycles.* Baltimore, MD: John Hopkins University Press.

Kennedy, J. F. (1960). Remarks of Senator J.F. Kennedy, Fond du Lac, Wisconsin. February 17, 1960. John F. Kennedy Presidential Library and Museum.

Kennedy, R. F., Jr. (2004). *Crimes against nature.* New York: Harper Collins.

Kleinman, D. L. (1995). *Politics on the endless frontier: Postwar research policy in the United States.* Durham, NC: Duke University Press .

Knopman, D., & Fleschner, E. (1999, May 1, 1999). *Second generation of environmental stewardship: Improve environmental results and broaden civic engagement,* from http://www.ppionline.org/ppi_ci.cfm?knlgAreaID=116&subsecID=150&contentID=767.

Knudson, T. (2004, September 3, 2004). Measure urges state to buy California wood. State enforces strict forest rules but imports most of its wood, sponsor says. *Sacramento Bee,* p. A1.

Koch, E., & Peden, W. (Eds.). (1944). *The life and selected writings of Thomas Jefferson.* New York: Random House Modern Library.

Kraft, M. E. (2004). *Environmental policy and politics* (3rd ed.). New York: Pearson Longman.

Kramer, T. (1982). Comment: Attorney fee awards to nonprevailing parties under the Clean Air Act. *University of Cincinnati Law Review, 51,* 635.

Krugman, P. (1994). *Peddling prosperity: Economic sense and nonsense in the age of diminished expectations.* New York: Norton.

Krupp, F., & Horn, M. (2008). *Earth: The Sequel. The race to reinvent energy and stop global warming.* New York: W.W. Norton & Co.

Kuhn, T. R. (1996). *The structure of scientific revolutions* (3rd ed.). Chicago, IL: University of Chicago Press.

Kumins. (1997). *OCS leasing moratoria, in oceans & coastal resources: A briefing book,* (No. Report 97-588 ENR). Washington, DC: Congressional Research Service.

Ladd, E. C., Jr., Palmer, D. O., & Shannon, W. W. (1978). Federal funding of academic marine science. Woods Hole, MA: *JOIDES Journal, 4*(3).

Lamb, H. (2004, March 27, 2004). *High-price gas blues? blame the greens,* 2004, from http://www.worldnetdaily.com/news/article.asp?ARTICLE_ID=37772.

Landy, M., & Dell, K. D. (1998). The failure of risk reform legislation in the 104th Congress. *Duke Environmental Law and Policy Forum, 9*(113), 1–20.

Larstad, T., & Gooderham, R. E. (Eds.). (2004). *Facts 2004.* Oslo: Ministry for Petroleum and Energy.

Lash, J. (2008). *Jonathan Lash,* 2008, from http://www.wri.org/profile/jonathan-lash.

Lash, J., Gillman, K., & Sheridan, D. (1984). *A season of spoils: The Reagan administration's attack on the environment.* New York: Pantheon.

Lautenberg, F. R. (1997). *Statement of Senator Frank R. Lautenberg, EPW committee, superfund hearing, MARCH 5, 1997,* from http://epw.senate.gov/105th/lau_3-5 htm.

Layton, E. (1971). Mirror-image twins: The communities of science and technology in 19th Century America. *Technology and Culture, 12*(4), 562–580.

Lazarus, R. J. (2004). *The making of environmental law.* Chicago, IL: University of Chicago Press.

Lear, L. (1997). *Rachel Carson: Witness to nature.* New York: Henry Holt & Co.

Lehmann, J., Gaunt, J., & Rondon, M. (2006). Bio-char sequestration in terrestrial ecosystems – a review. *Mitigation and Adaptation Strategies for Global Change, 11*, 403–427.

Leopold, A. (1949). *Sand County Almanac.* New York: Oxford University Press.

Lieberman, J. (2007). America's Climate Security Act (ACSA), press release (18 Oct. 2007), Washington, DC, from http://lieberman: senate.gov.

Lindall, W. (1974). Down to the sea in triplicate: Coastal zone management in Maine, a legal perspective. *FCCC Newsletter* (June 1974).

Lindstrom, M. J., & Smith, Z. A. (2001). *The National Environmental Policy Act.* University Station, TX: Texas A&M University Press.

Lipset, S. M. (1990). The work ethic – then and now. *Public Interest, 1990 (Winter),* 61–69.

Lipset, S. M. (1997). *American exceptionalism.* New York: Norton.

Logomasini, A. (2007, August 2007). *The green regulatory state,* from http://www.heartland.org/Article.cfm?artId=22838.

Lomborg, B. (2001). *The skeptical environmentalist.* Cambridge, UK: Cambridge University Press.

Lomborg, B. (2007). *Cool it: The skeptical environmentalist's guide to global warming.* New York: Knopf.

Lowenthal, D. (2003). *George Perkins Marsh: Prophet of conservation.* Seattle, WA: University of Washington Press.

Lucy, W. H., & Phillips, D. L. (2006). *Tomorrow's cities, tomorrow's suburbs*: American Planning Association, Chicago, IL: Planners Press.

Ludwig, U., & Schmid, B. (2007). Burning the world's waste. *Spiegel Online International,* 3 Dec. 2007, from http://www.spiegel.de/international/spiegel/0.1518.467239,00.html.

Luther, M. (1966). *Luther's house postils* (E. Klug, Trans. Vol. 1). Grand Rapids, MI: Baker Books.

MacDonald, J. B. (1990). Riparian doctrine. In K. R. Wright (Ed.), *Water rights of the fifty states and territories* (pp. 19–22). Denver, CO: American Water Works Association.

Mackenzie, G. C. (2001). *Nasty & brutish without being short: The state of the Presidential appointment process,* Washington, DC: Brookings Review.

MacLeod, C. (1963). *Use of pesticides.* Washington, DC: The White House.

Malmgren, B. (2005). Workers' struggle and environmental justice. The legacy of capitalism in Butte, Montana. April 2005 *Socialism and Liberation Magazine* 5(4), from http://socialismandliberation.org/mag/index.php?aid=339.

Manheim, F., Buchholtz ten Brink, M., Hastings, P., & Mecray, E. (1999). *Contaminated sediment database development and assessment in Boston Harbor.* Washington, DC: U.S. Geological Survey.

Manheim, F. T. (1986). Marine cobalt resources. *Science, 232,* 600–608.

Manheim, F. T. (1998). *Lake Pontchartrain basin: Bottom sediments and regional scientific and educational resources.* Washington, DC: U.S. Geological Survey.

Manheim, F. T. (2004). U.S. offshore oil industry: New perspectives on an old conflict. *Geotimes, 49,* 26–26.

Manheim, F. T. (2006). A new look at mining and the environment: Finding common ground. *Geotimes, 51*(4), 18–22.

Manheim, F. T. (2008). A review of Jay Inslee's Apollo's fire. *CHOICE Reviews Online* (May 2008).

Manheim, F. T., Buchholtz ten Brink, M. R., & Mecray, E. L. (1998). Recovery and validation of historical sediment quality data from coastal and estuarine areas: An integrated approach. *Journal of Geochemical Exploration, 64,* 377–393.

Manheim, F. T., & Hayes, L. (2002). *Lake Pontchartrain Basin: Bottom sediments and related environmental resources* (CD). Washington, DC: U.S. Geological Survey.

Manheim, F. T. & Manheim, E. (2005). Rock: The role and future of electronic "Beat" music. In F. Baron, D. N. Smith, & C. Reitz (Eds.), *Authority, culture, and communication: The sociology of Ernest Manheim,* Heidelberg, Germany: Synchron.

Mankiw, N. G., & Swagel, P. (2006). *The politics and economics of offshore outsourcing* (Working Paper No. 12398) Cambridge, MA: National Bureau of Economic Research.

Mann, C. S. (2006). *1491: New revelations of the Americas before Columbus* (2nd ed.). New York: Vintage Books.

Marco, G. J., Hollingsworth, R. M., & Durham, W. (Eds.). (1987). *Silent spring revisited*: American Chemical Society.

Markowitz, M. (2003, April 28, 2003). *NYC subway token, 1953–2003*. Retrieved Feb. 8, 2008

Marsh, G. P. (1864). *Man and nature*. Cambridge, MA: Harvard University Press.

McCartan, L., Menzie, W. D., Morse, D. E., Papp, J. F., Plunkert, P. A., & Tse, P.-K. (2006). Effects of Chinese mineral strategies on the U.S. minerals industry. *Mining Engineering, 58*(3), 37–42.

McCurdy, K. (2006). Earth Science influence in non-renewable resource policy: An evolutionary process, Geological Society of America, SE Section Meeting Abstracts, p. 13.

McGarity, T. O. (1983). Media-quality, technology, and cost-benefit balancing strategies for health and environmental regulation. *Law and Contemporary Problems, 46*(3), 159–233.

McGarity, T. O. (1991). *Reinventing rationality: The role of regulatory analysis in the federal bureaucracy*. New York: Cambridge University Press.

McIsaac, D. O., & Hansen, D. K. (2005). *Testimony before Committee on Council operations and reauthorization of the Magnuson-Stevens Fishery Conservation and Management Act* (MSA) (October 24, 2005 ed.). Portland, OR: Pacific Fishery Management Council.

McKibben, B. (1989). *The end of nature*. New York: Random House.

McKibben, B. (2007). *Deep economy: The wealth of communities and the durable future*. New York: Henry Holt & Company.

McLaughlin, B. (2007). *Minneapolis bridge collapse indicates extent of cracks and fissures in national infrastructure*, from http://www.associatedcontent.com/article/341899/minneapolis_bridge_collapse_indicates.html

McLaughlin, J. (1988). *Jefferson and Monticello*. New York: Henry Holt and Co.

Meadows, D. H., Meadows, D. L., Randers, J., & Behrens, W. W. I. (1972). *Limits to growth*. Washington, DC: Potomac Associates.

Meads, R., & Ballentine, B. (2002, Monday, 22 April, 2002). *Better regulation in the European Union: The role of regulatory impact analysis*. Retrieved 12 February 2007, from http://www.euractiv.com/en/Search?term=RIA&simple=true&combine=and&Search=Search

Measuringworth. (2008). *What was the U.S. GDP then?*, 2008, from http://www.measuringworth.com/datasets/usgdp/result.php.

Meine, C. (1991). *Aldo Leopold: His life and work* . Madison, WI: Univ. of Wisconsin Press.

Menard, H. W. (1971). *Science: Growth & change*. Cambridge, MA: Harvard University Press.

Metinko, C. (2007). Supervisors settle Altamont bird suit. 12 January 2007. Oakland, CA: *Oakland Tribune*.

Metropolitan District Commission. (1968). *Report of the Director of the sewerage division and chief sewerage engineer*. Boston, MA: Metropolitan District Commission, Sewerage Division.

Miller, T. (1999). *The 60's communes: Hippies and beyond*. Syracuse, NY: Syracuse University Press.

Mine_Safety_and_Health_Administration. (2008). *History of mine safety and health legislation*, from www.msha.gov.

Minerals Management Service (MMS). (1987). *Proposed marine mineral lease sale in the Hawaiian Archipelago and Johnston Island exclusive economic zones* (No. MMS 87-0004). Herndon, VA: Minerals Management Service and the Department of Planning and Economic Development.

Minerals Management Service (MMS). (2000). *Manteo exploration unit OCS leases offshore North Carolina*. Retrieved 14 March 2008, from http://www.gomr.mms.gov/homepg/offshore/atlocs/manteo.html.

Minerals Management Service (MMS). (2003). *U.S. offshore milestones,* Washington, DC: Minerals Management Service.

Minerals Management Service (MMS). (2004). *Deepwater Gulf of Mexico 2004 America's expanding frontier* (OCS Report No. 2004-021): Washington, DC: Minerals Management Service.

Minerals Management Service (MMS). (2005). *Hurricane Katrina and Rita research.* Retrieved 14 March 2008, from http://www.mms.gov/tarprojectcategories/hurricaneKatrinaRita.htm.

Mirovitskaya, N., & Ascher, W. (2001). *Guide to sustainable development and environmental policy.* Durham, NC: Duke University Press.

Mouawad, J. (2007). In Alaska's far north, two cultures collide. 4 Dec. 2007, New York: *The New York Times.*

Moyers, B. (2005). *Drilling on the front range.* Washington, DC: PBS WNET.

Mumford, L. (1971). *The Brown decades: A study of the arts in America. 1865–1895* (2nd Revised ed.). New York: Dover Press.

Murmann, J. P. (2003). *Knowledge and competitive advantage–The coevolution of firms, technology, and national institutions.* Cambridge, UK: Cambridge University Press.

Næss, A. (1989). *Ecology, community and lifestyle.* Cambridge, UK: Cambridge University Press.

Nakagawa, L. (2006). *Toxic trade: The real cost of electronics waste exports from the United States.* Washington, DC: World Resources Institute.

Nakamura, R. T., & Church, T. W. (2003). *Taming regulation.* Washington, DC: Brookings Institution Press.

Nash, A. E., Mann, D.E., & Olsen, P.G., 1972. *Oil pollution and the public interest: A study of the Santa Barbara oil spill.* Berkeley, CA: UC Institute of Governmental Studies.

National Academy of Sciences (NAS). (2003). *Oil in the sea III: Inputs, fates, and effects.* Washington, DC: National Academy Press.

National Academy of Sciences (NAS)/National Research Council (NRC). (1992). *Assessment of the U.S. Outer Continental shelf environmental studies program: III. Social and economic studies (1992).* Washington, DC: National Academy Press.

National Association of Clean Water Agencies. (2006, 2007). *Peak performance awards Platinum– Milwaukee Metropolitan Sewerage District, WI. Jones Island wastewater treatment plant,* from http://nacwa.org//index.php?searchword=Milwaukee&option=com.search&Itemid=.

National Center for Public Policy Research. (1993). *Environmental scientist: David Foreman.* Washington, DC: National Center for Public Policy Research.

National Institute for Occupational Safety and Health (NIOSH). (2007). *Distribution of cases of occupational illness in mining, 2005,* 2008.

National Oceanic and Atmospheric Administration (NOAA). (2007, August 6, 2007). *Annual commercial landing statistics,* from http://www.st.nmfs.noaa.gov/pls/webpls/MF_ANNUAL_ LANDINGS.RESULTS.

National Research Council (NRC). (2005). *Superfund and mining megasites: Lessons from the Coeur d'Alene river basin.* Washington, DC: National Academy Press.

National Science Foundation (NSF). (1968). *The Eighteenth Annual Report of the National Science Foundation.* Washington, DC: National Science Foundation.

Natural Resources Defense Council (NRDC). (2001). *Environmentalists fight general electric's latest evasion of PCB cleanup.* Retrieved 2008, from http://www.nrdc.org/media/ pressreleases/010406.asp.

Nelson, C. (2008). John Wesley Powell. In N. Koertge (Ed.), *New dictionary of scientific biography* (Vol. 6, pp. 149–152). Detroit, MI: Scribner's Sons.

Nelson, C. M., Rabbitt, M. C., & Fryxell, F. M. (1981). Ferdinand Vandeveer Hayden: The U. S. Geological Survey years, 1879–1886. *Proceedings of the American Philosophical Society, 125*(3), 238–243.

Nesson, F. L. (1983). *Great waters: A history of Boston's water supply.* Hanover, NH: University Press of New England.

New Jersey Public Interest Research Group. (2005). *The environmental case for wind power in New Jersey,* 2008, from http://www.njpirg.org/NJ.asp?id2=16596&id3=NJ&

Norton, B. G. (1991). *Toward unity among environmentalists.* New York: Oxford University Press.

Norwegian Petroleum Directorate. (2006). *Norwegian petroleum directorate facts: The Norwegian petroleum sector.* Oslo, Norway: Ministry of Petroleum and Energy.

National Parks Conservation Association (NPCA). (2007). *Alaska National Interest Lands Conservation Act.* Retrieved 19 March 2008, from http://www.npca.org/media_center/fact_sheets/anilca.html.

Natural Resources Defense Council (NRDC). (2008, 6/19/07). *Move over, gasoline: Here come biofuels,* from http://www.nrdc.org/air/transportation/biofuels.asp.

O'Connor, J., Major, J., & Grant, G. (2008). The dams come down: Unchaining U.S. rivers. *Geotimes, 53*(3), 22–28.

O'Toole, R. (1988). *Reforming the forest service.* Washington, DC: Island Press

Oakley, G. (1983, October 31, 1983). James Watt left his mark on Idaho. Paonia, CO: *High Country News,* 22 April 2005.

Office of Technology Assessment. (1984). *Oil and gas technologies for the arctic and deepwater.* Washington, DC: U.S. Congress, Office of Technology Assessment.

offshore-technology.com. (2007). Surviving the storm: offshore-technology.com.

Ognibene, P. (1975). *Scoop: The life and politics of Henry M. Jackson.* New York: Stein and Day.

Oliver, C. (2006). Managing our Nation's fisheries conference-focus on the future. Washington DC: North Pacific Fishery Management Council.

Organisation for Economic Cooperation and Development (OECD). (2007). *OECD key environmental indicators,* from http://www.sourceoecd.org/rpsv/statistic/516_about.htm?jnlissn=16081234.

Osborne, E. (2004). Measuring bad governance. *Cato Journal, 23*(3), 402–433.

Osiel, M. J. (1990). Review: Lawyers as monopolists, aristocrats, and entrepreneurs, *The Harvard Law Review Association, 103*(8), 2009–2066.

Outotec. (2007). *Outotec-More out of ore,* 2008, from http://www.outotec.com/pages/Page____21653.aspx.

Pacala, S., & Socolow, R. (2004). Stabilization wedges: Solving the climate problem for the next 50 years with current technologies. *Science, 305*(5686), 968–972.

Parker, M. (2008). North Carolina catfish producers eligible for grants. 15 May 2008, *Southeast Farm Press,* from http://Southeastfarmpress.com/news/catfish-production-0515/

Paturi, F. R. (1989). *Chronik der Technik* (3rd ed.). Düsseldorf, Germany: Bertelsmann Lexikon Verlag.

Pauly, D., & MacLean, J. (2003). *In a perfect ocean–The state of fisheries and ecosystems in the North Atlantic Ocean.* Washington, DC: Island Press.

Pearlstein, S. (2005, June 15, 2005). Think tank's leader's principled wisdom will be missed. *The Washington Post* p. D01.

Pelton, T. (2008). *Wind vs. nukes? you'd be blown away,* 2008, from http://weblogs.baltimoresun.com/news/local/bay_environment/blog/2008/04/wind_vs_nukes_youd_be_blown_aw.html.

Penny, T. J. (1993). Common Cents Deficit Reduction Act of 1993. Amendments to H. R. 3400 Defeated 20 Nov. 1993: The Government Reform and Savings Act of 1993. 103rd Congress.

Percival, R. V., Miller, A. S., Schroeder, C. H., & Leape, J. P. (2000). *Environmental regulation: Law, science, and policy* (3rd ed.). Gaithersburg, MD: Aspen Publishers Inc.

Petersen, E., Elbersen, B., Wiesenthal, T., Feehan, J., & Eppler, U. (2007, January 29, 2008). *Estimating the environmentally compatible bioenergy potential from agriculture,* from http://reports.eea.europa.eu/technical_report_2007_12/en.

Petroleum Today. (1976). Quotation from T. P. O'Neill. 1, p. 8.

Philibert, C. (2006). *Barriers to technology diffusion: The case of solar thermal technologies* (No. COM/ENV/EPOC/IEA/SLT(2006)9): Organisation for Economic Co-operation and Development International Energy Agency.

Phillips, D. G. (1964). *The treason of the senate.* Chicago, IL: Quadrangle Books.

Piper, J. R. (2005). *The major nation-states in the European Union.* New York: Pearson Longman.

Pombo, R. (2006). *Critics say money spent on California delta has produced little,* in section entitled News Roundup, The Westerner, http://thewesterner.blogspot.com/2006/02/news-roundup-questions.html.

Porter, M. E. (1990). *The competitive advantage of nations.* New York: The Free Press.

Powers, C. W., & Chertow, M. (1997). Industrial ecology. In M. R. Chertow & D. C. Esty (Eds.), *Thinking ecologically* (pp. 19–36). New Haven, CT: Yale University Press.

Presley, B. J., Wade, T. L., Santschi, P. H., & Baskaran, M. (1998). *Historical contamination of Mississippi River Delta, Tampa Bay, and Galveston Bay sediments*. NOAA Technical memorandum 127, Washington, DC: National Ocean Survey ORCA.

Price, D. J. d. S. (1963). *Little science, big science*. New York: Columbia University Press.

Price, D. J. d. S. (1965). Is technology historically independent of science? A study in statistical historiography. *Technology and Culture, 6*, 533–568.

Prochnau, W. W., & Larsen, R. W. (1972). *A certain Democrat: Senator Henry M. Jackson*. Englewood Cliffs, NJ: Prentiss-Hall.

Pullenger, M. (2008). *GenDis the impact of distributed generation of electricity networks from an SME perspective*, from http://www.cogen.org/projects/Gendis.htm.

Punke, M. (2006). *Fire and brimstone: The North Butte mining disaster of 1917*. New York: Hyperion Books.

Quick, G. R. (1989). Oil-seeds as energy crops. In A.Ashri, R. K. Downey, & G.Röbbelen (Eds.), *Oil crops of the world*. New York: MC Graw-Hill.

Rabbitt, M. C. (1980). Federal Science and mapping policies and the development of mineral resources in the United States during the first 25 Years of the U.S. Geological Survey, V. 2, 1879–1904. U.S. Department of the Interior, Geological Survey, Washington, DC: U.S. Government Printing Office.

Rabbitt, M. C. (1989). *The U.S. Geological Survey, 1879–1989*. U.S. Geological Survey Circular 1050, U.S. Department of the Interior, Geological Survey. Washington, DC: U.S. Government Printing Office.

Reagan, J. H. (1906). *Memoirs*. New York: The Neale Publishing Co.

Reddy, C. M. (2007). Oil in our coastal back yard. *Oceanus* posted 13 October 2004; accessed 22 Nov. 2007, from, http://www.whoi.edu/oceans/viewArticle.do?id=2471.

Redford, R. (2005). *An important message from Robert Redford* 12 October 2005, from http://hotspringsquy.com/2005/10/important-message-from-robert-nedford.html.

Reich, R. (2008). America's middle classes are no longer coping. *The Financial Times* 29 January 2008, from http://www.truthout.org/das_2006/01308B.shtml.

Reid, R. L. (2008). The infrastucture crisis. *Civil Engineering*. Special Report, from http://pubs.asce.org/magazines/CEmay/2008/Issue_01-08/article1.htm.

Reilly, W. K. (1991, January February, 1991). *The new clean air act: An environmental milestone*, 15 March 2007.

ResponsibleWindpower.org. (2006). *Wind farms: Mollohan and Rahall are wise to ask questions now*, from www.responsiblepower.org.

Rittel, H., & Webber, M. (1973). Dilemmas in a general theory of planning. *Policy Sciences, 4*, 155–169.

Roberts, M. (2003). *Walter J. Hickel*, from http://www.headwatersnews.org/p.hickel.html#confirmation.

Røkke, N. (2008, Feb. 5, 2008). *CCS activities in Norway-R&D and deployment plans*, from http://akseli.tekes.fi/opencms/opencms/OhjelmaPortaali/ohjelmat/ClimBus/fi/Dokumenttiarkisto/Viestinta_ja_aktivointi/Seminaarit/CCS_Seminar_050208/Nils_Rokke.pdf.

Roosevelt, F. D. (1941). *Day of infamy speech*, from http://historymatters.gmu.edu/d/5166/.

Roosevelt, T., & Davidson, D. J. (2003). *The wisdom of Theodore Roosevelt*. New York: Citadel Press.

Rosenbaum, W. A. (2005). *Environmental politics and policy* (6th ed.). Washington, DC: Congressional Quarterly Press.

Rosenbaum, W. A. (2007). *Environmental politics and policy* (7th ed.). Washington, DC: Congressional Quarterly Press.

Rosenthal, E. (2008). Europe, cutting biofuel subsidies, redirects aid to stress greenest options, 22 Jan. 2008, *The New York Times*.

Ross, B. (2008). *Spitzer quits but other public figures may be linked to escort ring*: New York: American Broadcasting Corporation.

Rossman, M. (1969). *On learning and social change*. New York: Random House.

Rowthorn, R., & Ramaswamy, R. (1997). Deindustrialization–Its causes and implications. *International Monetary Fund Economic Issues 10*, Washington, DC: International Monetary Fund.

Rycroft, R. W., & Kash, D. E. (1999). *The complexity challenge: Technological innovation for the 21st Century*. London: Pinter.

Ryther, J. H., Dunstan, W. M., Tenore, K. R., & Vaisnys, J. R. (1975). Controlled eutrophication–Increasing food production from the sea by recycling human waste. *Bioscience, 22*, 144–152.

Sample, I. (2008). Ocean currents may offset global warming over coming decade. *guardian.co.uk*.

Sauri-Pujol, D. (1992). Review of environment and property rights. *Economic Geography, 68*(4) 436–439.

Save-our-wild-salmon. (2008). *Congress urges federal government to base salmon recovery plan on sound science and economics*, 2008, from http://www.wildsalmon.org/pressroom/press-detail.cfm?docid=740.

Saxe, J. L. (1970). *Defending the environment: A strategy for citizen action*. New York: Borzoi Books.

Schleede, G. R. (2005). *"Big Money" discovers the huge tax breaks and subsidies for "Wind Energy" while taxpayers and electric customers pick up the tab*, from http://www.mnforsustain.org/windpower_big_money_discovers_the_money_schleede_0405.htm.

Schlesinger, A. M. (1946). *The age of Jackson*. Boston, MA: Little Brown.

Schlosberg, D. (1999). *Environmental justice and the new pluralism*. Oxford, UK: Oxford University Press.

Schreurs, M. A. (2002). *Environmental politics in Japan, Germany, and the United States*. New York: Cambridge University Press.

Schroeder, C. H., & Steinzor, R. (Eds.). (2004). *A new progressive agenda for public health and the environment: A project of the center for progressive regulation*. Durham, NC: Carolina Academic Press.

Schulte, B. (2007). Fred Krupp–Environmentalist. 12 Nov. 2007, *U.S. News and World Report*.

Scott, E. (2003). In search of certainty: The hardrock mining industry presses for national mining policy. *The Professional Geologist, 40*(November), 10–11.

Scott, M. (1983, October 31, 1983). Conservation is now a partisan issue. *High Country News*.

Scott, P. (1999). *Edward Jenner and the discovery of vaccination*, from http://www.sc.edu/library/spcoll/nathist/jenner.html

Seely, B. E. (1999). The other re-engineering of engineering education, 1900–1965. *Journal of Engineering Education, 88*(3), 285–294.

Sellers, C. G. J. (1958). Andrew Jackson versus the historians. *The Mississippi Valley Historical Review, 44*(4), 615–634.

Shallatt, T. (1994). *Structures in the stream: Water, science, and the rise of the U.S. Army Corps of Engineers*. Austin, TX: University of Texas Press.

Shapley, D. (1977). McKelvey ousted as Director of Geological Survey. *Science, 197*(4310), 1264.

Sherwin, C. W., & Isenson, R. S. (1966). *First interim report on project hindsight (summary)*. Washington, DC 20301: Office of the Director of Defense Research and Engineering.

Sherwin, C. W., & Isenson, R. S. (1967). Project Hindsight. *Science, 156*, 1571–1577.

Shulock, N. (1999). The paradox of policy analysis: If it is not used, why do we produce so much of it? *Journal of Policy Analysis and Management, 18*(2), 226–244.

Sierra Club. (1975). *Offshore petroleum exploration: Position adopted by the Sierra Club Board of Directors, 1975: The Sierra Club believes that no offshore petroleum exploration should occur unless and until the following conditions are met*, from http://www.sierraclub.org/ca/coasts/.

Singer, S. F., & Avery, D. T. (2007). *Unstoppable global warming: Every 1,500 years*. New York: Rowman and Littlefield.

Sissine, F. (2008). *Energy efficiency and renewable energy legislation in the 110th Congress* (No. RL 33831) Washington, DC: Congressional Research Service.

SITRA. (2007). The national action plan for environmental business in Finland. *Energy and Enviro Finland*, 27 February 2007, from www.sitra.fi.

Smaal, A. C. (2002). European mussel cultivation along the Atlantic coast: production status, problems and perspectives. *Hydrobiologia Springer Netherlands, 484*(1–3), 89–98.

Smith, B. L. R. (1990). *American science policy since World War II*. Washington DC: Brookings Institution Press.

Smith, E. (2002). *Energy, the environment, and public opinion*. Boulder, CO: Rowman & Littlefield.

Stahel, P. F., Clavien, P.-A., Smith, W. R., & Moore, E. E. (2008). Redundant publications in surgery: A threat to patient safety? *Patient Safety in Surgery, 2*(6), from http://www.pssjournal.com/articles/browse.asp.

Starobin, P. (2004). *The angry American*, from http://www.theatlantic.com/doc/200401/starobin.

Stern, S., Porter, M. E., & Furman, J. L. (2007). The determinants of national innovative capacity (2007, updated from 2000 (original) ed., Working paper w7876). Cambridge, MA: National Bureau of Economic Research.

Stockman, D. (1986). *The triumph of politics*. New York, NY: Harper and Row.

Stokes, D. E. (1997). *Pasteur's Quadrant: Basic science and technological innovations*. Washington, DC: Brookings Institution Press.

Stroup, R. L., & Meiners, R. E. (2000). *Cutting green tape: Toxic pollutants, environmental regulation and the law*. New Brunswick, NJ: Transaction Publishers.

Sullivan, P. (2004a, July 22, 2004). Anne Gorsuch Burford. *The Washington Post*, p. B6.

Sullivan, P. (2004b). Anne Gorsuch Burford, 62, dies; Reagan EPA Director. *The Washington Post*, p. 1.

Sun, S. X., & Hobbs, R. (2008). *power plant emissions to biofuels*. Presentation at National Renewable Energy Laboratory – Air Force Office of Scientific Research Workshop in Algal Oil for Jet Fuel Production, February 2008, from http://www.nrel.gov/biomass/algal_oil_workshop.html.

Swedish Forest Agency. (1994). *The Forestry Act*, from http://www.SVO.se/episerver4/templates/SNormalPage/aspx?id=11303.

Swedish Ministry of the Environment. (2008). *The Government's climate policy*, from http://www.sweden.gov.se/content/1/c6/10/33/84/63708a83.pdf.

Tarr, J. (Ed.). (2004). *Devastation and renewal: An environmental history of Pittsburgh and its region*. Pittsburgh, PA: University of Pittsburgh Press.

Tassava, C. J. (2008, Feb. 10, 2008). *The American economy during World War II*, from http://eh.net/encyclopedia/article/tassava.WWII

Theodore_Roosevelt_Association. (2005). *Conservationist: Life of Theodore Roosevelt*, 2008, from http://www.theodoreroosevelt.org/life/conservation.htm.

Thomas. (1995). Advanced Bill Summary & Status Search for the 104th Congress (1995–1996). The Library of Congress, from http://thomas.loc.gov/bss/2104quary.html.

Thompson, D. (2006, Feb. 28, 2006). Critics say money spent on California delta has produced little. Associated Press, cited in *San Francisco Chronicle* 28 February 2006.

Tichi, C. (1979). *New world, new earth: Environmental reform in American literature from the Puritans through Whitman*. New Haven, CT: Yale University Press.

Tietenberg, T. (1999). Regulatory reform in air pollution control. In W. E. Oates (Ed.), *The RFF reader in environmental and resource management* (pp. 69–80). Washington, DC: Resources for the Future.

Tillion, C. (2002). *Testimony before the North Pacific Living Marine Resources Panel*. Washington, DC: U.S. Commission on Ocean Policy.

Time. (1970). *Nader's Raiders strike again*. Retrieved 3-16-08, 2008, from http://www.time.com/time/printout/0,8816,942247,00.html.

TMpress. (2005). *Three decades of absolute power ... the story of CEO VP Dick Cheney ... and Iraq*. Retrieved 9 November 2007, from http://www.williambowles.info/guests/2005/cheney.html.

Town, M., (2006). *Victory in Virginia – Governor Kaine rejects offshore drilling* 2008, from http://www.sierraclub.org/pressroom/releases/pr2006-04-07.asp.

Transparency International. (2005). *Corruption perception index*. Retrieved November 2007, from http://www.transparency.org/policy_research/surveys_indices/cpi/2005.

Transparency International. (2006). *Corruption Perception Index.* Retrieved November 2007, *from* http:// www.transparency.org/policy_research/surveys_indices/cpi/2006.

Transportation Research Board. (1994, June 1994). *Environmental regulatory process: Does it work?: Dredging U.S. ports.* Paper presented at the Environmental Regulatory Process: Does It Work?: Dredging U.S. Ports Conference.

Twain, M., & Warner, C. D. (1873). *The Gilded Age, a tale of today.* Online eBook, Project Gutenerg, from www.gutenberg.net.

Ui, J. (1992). *Industrial pollution in Japan.* Tokyo: United Nations University Press.

UN Conference on Trade and Development (UNCTAD). (2008). *UNCTAD statistics.* Retrieved 4 July 2006, from http://www.unctad.org/templatesPage.asp?inItemID=1584.

UNESCAP. (2008). *What is good governance.* Retrieved 12 March 2008, from http://www.unescap.org/pdd/prs/ProjectActivities/Ongoing/gg/governance.asp.

Union of Concerned Scientists. (2008). *Digging up trouble, the health risks of construction pollution in California,* 2008, from http://www.ucsusa.org/clean_vehicles/california_driving/digging-up-trouble.html.

University of Cambridge. (2006). *Stabilisation wedges: Solving the climate problem for the next 50 years with current technologies,* from http://www-g.eng.cam.ac.uk/impee/topics/stabilisationwedges/files/Stabilisation%20Wedge%20v1%20PDF%20WITH%20NOTES.pdf, Cambridge, UK: University of Cambridge.

Uriu, R. M. (1996). *Troubled industries confronting economic change in Japan.* Ithaca, NY: Cornell University Press.

U. S. Army Corps of Engineers (USACE). (2007, November 26, 2007). *U.S. Army Corps of Engineers, a brief history.* Washington, DC: USACE.

U. S. Census Bureau. (2007). LandView® 6: Washington, DC: U.S. Census Bureau Geography Division.

U. S. Chamber Institute for Legal Reform. (2007). *Fast facts on asbestos,* from http://www.instituteforlegalreform.com/issues/htmlfeature.cfm?issue=ASB&Doctype=HTML&id=7.

U. S. Coast Guard. (2006). *Report to Congress: Oil spill liability trust fund hurricane impact*: Washington, DC: U.S. Department of Homeland Security.

U. S. Congress (1906). The Pure Food and Drug Act of 1906, 59th Congress, from http://coursesa.matrix.msu.edu/nhst203/documents/pure.html. Washington, DC.

U.S. Court of Appeals. (2007). Resisting environmental destruction on indigenous lands, A project of the indigenous environmental network, Center for Biological Diversity, and Sierra Club. In U.S. Court of Appeals (Ed.) (Vol. DOI 2007-152, pp. 2).

U. S. Senate. (2005). Clean Ocean and Safe Tourism Anti-Drilling Act. Washington, DC: U.S. Senate.

Valette-Silver, N. (1992). *Historical reconstruction of contamination using sediments cores: A review.* Rockville, MD: National Oceanic and Atmospheric Administration, National Ocean Service.

Valtin, T. (2006, May/June 2006). *Offshore drilling moratorium threatened.* Retrieved 16 November 2007, from http://www.sierraclub.org/planet/200603/offshore.asp.

Veletinsky, I. (2006). Russia makes waves. *Trendline Russia.* (advertising supplement to *The Washingon Post,* 27 December 2006; p. 3.)

Vest, C. M. (2005). *Pursuing the endless frontier.* Cambridge, MA: Massachusetts Institute of Technology.

Vest, C. M. (2007). *The American research university from World War II to World Wide Web: Governments, the private sector, and the emerging meta-university.* Berkeley, CA: University of California Press.

Vig, N. J., & Kraft, M. E. (Eds.). (2005). *Environmental policy: New directions for the twenty-first century* (6th ed.). Washington, DC: Congressional Quartely Press.

Vogel, E. (1985). *Comeback, case by case: Building the resurgence of American business.* New York: Simon & Schuster.

von Meyer, E. (1895). Geschichte der Chemie. Leipzig, Germany: Veit.

Wagner, F. W. (2006). Virginia energy plan SB 262 Virginia General Assembly, 7 December 2007, from http://leg1.state.va.us/cgi-bin/legp504.exe?old+sum+SB262.

Wagner, M. (2003). *The Porter hypothesis revisited: A literature review of theoretical models and empirical tests*, 2008, from http://129.3.20.41/eps/pe/papers/0407/0407014.pdf.

Walker, M. (2007). *Does a flawed triumph define EU at 50*, from http://www.terradaily.com/reports/Does_A_Flawed_Triumph_Define_EU_At_50_999.html.

Wallace, D. D. (1969). *South Carolina, A short history*. Columbia, SC: University of South Carolina Press.

Wallace, R. E. (1996). *Earthquakes, minerals, and me* (Open-File Report No. 96-260). Reston VA: U.S. Geological Survey.

Washington Post. (2005). Business corporate ethics. 26 April 2005. Washington, DC: *The Washington Post*.

Water Environment Federation (WEF). (2006). *California issues*, from http://www.water-ed.org/cabriefing.asp#THE%20BAY%20DELTA.

Wätzold, F. (2004). SO$_2$ emissions in Germany: Regulations to fight Waldsterben. In W. Harrington, R. D. Morganstern, & T. Sterner (Eds.), *Choosing environmental policy* (pp. 23–40). Washington, DC: Resources for the Future.

Weber, J. A. (1993). *The economic feasibility of community based biodiesel production*. Columbia, MO: University of Missouri.

Weber, M. (1904). Die protestantische Ethik und der ‚Geist' des Kapitalismus. *Archiv für Sozialwissenschaften und Sozialpolitik, XX and XXI* (Began Nov. 1904, continued in Spring, 1905), 1920 in überarbeiteter Fassung [1921].

Weisman, J., & Cillizza, C. (2006, April 4, 2006). DeLay to resign from Congress: Associates say reelection fears, not criminal probe, led to Republican's decision. *Washington Post*, p. A01.

Wessel, D. (2007). How Fred Krupp's singular style serves business, environment well. Retrieved 23 March 2007, *Wall Street Journal Online*, from http://webreprints.djreprints.com/1912570891197.html.

Western Wood Products Association. (2008). *Western Wood Products Association*, 2008, from http://www.wwpa.org/.

Westheimer, F. (2008). *Frank Henry Westheimer 1912-*. Michigan State University Department of Chemistry, Minibiography.

White, I. L., Kash, D. E., Chartock, M. A., Devine, D., & Lenard, R. L. (1973,). *North Sea oil and gas*. Norman, OK: University of Oklahoma Press.

Wibe, S., & Carlén, O. (2002). Editorial – Environmental regulations in the Swedish forest industry. What is the real problem? *Journal of Forest Economics, 8*(3), 167–168.

Wikipedia. (2007, October 14, 2007). *Justus von Liebig*. Retrieved October 16, 2007, from http://en.wikipedia.org/wiki/Justus_von_Liebig#Biography.

Wildavsky, A. (1995). *But is it true? A citizen's guide to environmental health and safety issues*. Cambridge, MA: Harvard University Press.

Wilderness.net. (2008). *The national wilderness preservation system* 2008, from http://www.wilderness.org/Library/Documents/Conservation_Quotes.cfm.

Wilkerson, C., & Limerick, P. (2003). Inside Interior: Walter J. Hickel, *Headwater News*. 15 October 2003, accessed 27 April 2007.

Wills, G. (2003). *"Negro President": Jefferson and the slave power*. New York Houghton-Mifflin.

Wood, D. M., & Yesilada, B. A. (2004). *The emerging European Union* (3rd ed.). New York: Pearson Longman.

World Maritime News. (2001). Japanese shipbuilding giants cut their losses. *MarineLink.com*.

World Resources Institute. (2007). *GHG protocol initiative*, from http://www.wri.org/project/ghg-protocol

Worster, D. (2001). *A river running West*. New York: Oxford University Press.

Wright, G., & Czelusta, J. (2002). *Exorcizing the resource curse: Minerals as a knowledge industry, past and present*. Retrieved March 21, 2008. Stanford University Working papers, from http://www-econ.stanford.edu/faculty/workp/swp02008.pdf.

Yaffee, S. L. (1982). *Prohibitive policy*. Cambridge, MA: MIT Press

Yandle, B. (1997). Environmental protection: Lessons from the past and future protection. In T. Anderson (Ed.), *Breaking the environmental policy gridlock* (pp. 140–167). Stanford, CA: Hoover Institution Press, Stanford University.

Yergin, D. (1991). *The prize: The epic quest for oil, money & power*. New York: Simon & Schuster.

Yochelson, E. (1997). *Charles Doolittle Walcott, paleontologist*. Kent, OH: The Kent State University Press.

Yochelson, E. (2006). Best and brightest of his generation. *Science, 312*, 1601.

Zachary, G. P. (1999). *Endless frontier*. Cambridge: MIT Press.

Ziman, J. (1968). *Public knowledge*. Cambridge, UK: Cambridge University Press.

Zovod, S., & Skaggs, S. (2005). House passes new endangered species legislation. *Endangered Species Committee Newsletter, 8*(1) 1–4.

Notes

Page range (5–7)

1. Former Senator David Boren (D-OK), testimony during nomination hearings for Robert M. Gates as Secretary of Defense, December 5, 2006, cited from Federal News Service by The New York Times.
2. Secretary of State in the Clinton Administration, Madeleine Albright.
3. Some political scientists find positive values in incivility. For example, Brooks and Geer (2007) challenge the idea that incivility undermines our electoral processes or that we have a "civility crisis" in public life. They claim that polling experiments in which substance remains the same while the tone and negativity are varied find no relation between incivility and trust or political efficacy. "Negativity and labeling may create a colorful and exciting display that reminds voters that politics isn't dull."
4. Other evidence: the presence of Republican governors in strongly Democratic states and the reverse (e.g., Mitt Romney in Massachusetts, Kathleen Sebelius in Kansas) and the fact that half the states have divided legislatures (Fiorina & Abrams, 2008).
5. Now an Emeritus Professor at the School of Public Policy, George Mason University. Following a series of authoritative books on oil and gas and other energy policy development in the 1970s, Kash was nominated by President Carter and approved as the Chief of the US Geological Survey's Conservation Division, which had responsibility for offshore oil and gas leasing, in 1978. The book series included the following: Energy under the Oceans (Norman, 1973); North Sea Oil and Gas (1973; sponsored by the White House Council on Environmental Quality); Energy Alternatives (1975); and Our Energy Future, based on a project funded by the National Science Foundation's Directorate of Research Applied to National Needs (1976).
6. Discussions during my initial project at George Mason University, 2004.
7. Vice President Richard Cheney, a former five-term congressman from Wyoming and Secretary of Defense under President G.H.W. Bush, resigned his post as President of Halliburton Co., a leading oilfield service and construction company, in order to become the Republican candidate for Vice President in 2000.
8. National Energy Plan, *Reliable, affordable, environmentally sound energy for America's future* (White House, May 2001, http://www.whitehouse.gov/energy/).
9. Analysis by the nonpartisan Fact Check organization of the Annenberg Public Policy Center, University of Pennsylvania discredited the technical validity of widely distributed reports about Cheney's continuing financial ties to Halliburton (e.g., TMpress, 2005; CBS, 2003; FactCheck, 2004).
10. In his book, Kennedy did not offer disclaimers separating his positions from that of NRDC, nor have there been official reservations by NRDC known to me regarding the book's contents.

Page range (9–15)

11. From 1959 to 1977, API published an informative journal, *Petroleum Today*, that reported on technical and policy developments (see also note 18). In 1980, API published a major policy study, *Two Energy Futures* that covered not only oil and gas, but also coal and nuclear energy and peripherally, other mineral resources (API, Washington D.C., 166 pages, with 9 pages of references and database sources). From 1979 to 1989, a cooperative report series on oil spills was published, after which significant external publications seem to have largely ceased.

12. For many years, until its merger with the Exxon Corporation in 1999, Mobil Oil Co. published brief policy notes in leading newspapers.

13. "The oil and natural gas industry shares a keen interest in the policy issues arena. As demand for energy to keep our homes, vehicles, and businesses running continues to increase, so does our advancement in technology, allowing us to provide safe, reliable, and affordable energy. While serious challenges face our nation on a variety of fronts, oil and natural gas industry representatives remain actively engaged with government leaders to ensure informed decision making so the energy needs of tomorrow are met." Website for the American Petroleum Institute, Policy Issues, http://api-ec.api.org/policy/, September 7, 2007; accessed December 20, 2007.

14. "For all its unique beauty, the Arctic National Wildlife Refuge remains under assault. Year after year, Big Oil's high-priced lobbyists push legislation to industrialize this national treasure. Isn't it time that this special place got the lasting protection it deserves?" Website of Defenders of Wildlife, headlined "Help Save the Arctic National Wildlife Refuge," April 2006, accessed January 24, 2008, http://www.savearcticrefuge.org/.

15. State PIRGS were founded in 1972 by Ralph Nader.

16. During Senate confirmation hearings for Wilson as President Eisenhower's Secretary of Defense, Wilson was asked whether he could make a decision adverse to the interests of General Motors. Wilson answered yes but added that he could not conceive of such a situation "because for years I thought what was good for the country was good for General Motors and vice versa." That statement was subsequently widely quoted as "What was good for General Motors was good for the country."

17. President of Union Oil Co. Fred L. Hartley: "I don't like to call it a disaster, because there has been no loss of human life. I am amazed at the publicity for the loss of a few birds."

18. The quote, in *Petroleum Today* (1976) cites O'Neill: "From an environmental viewpoint, our interest is not in restricting petroleum development, but insuring its compatibility with two of our major industries – tourism and fishing."

19. "White spent most of his life chairing countless taskforces – several of them under White House or Congressional auspices – to mitigate the flood damage caused by these ill-advised and seemingly inexorable pursuits (Cohen, 2006)."

20. The total number should not be taken as a number of blogs, since the same blog may yield multiple hits.

21. NASA Goddard Institute for Space Studies, Columbia University in 1989; now Chief Scientist for NASA.

22. First published in Danish as Verdens Sande Tillstånd in 1998.

23. Lomborg resigned from this Institute in 2004 to return to research at the University of Aarhus (Wikipedia, 2007). Lomborg has also published several new books (Lomborg, 2007).

24. Another indication of the split in the earth science community is offered by the American Association of Petroleum Geologists. In 2006, the Association issued a climate policy statement that showed skepticism: "Human-induced global temperature influence is a supposition that can be neither proved nor disproved (AAPG, 2006)." However, responding to new revelations and growing pressures in the wake of the IPPC report of 2007, the AAPG replaced the former statement on their website with a new one: "Certain climate simulation models predict that the warming trend will continue, as reported through NAS, AGU, AAAS, and AMS. AAPG respects these scientific opinions... ...." before going on to suggest qualifications (AAPG, 2007).

Page range (15–24)

25. The petition, circulated in April 1998, implied backing of the National Academy of Sciences (NAS); NAS quickly denied any relationship with OISM and further indicated that most credentialed scientists viewed global climate change as real, and human contribution to global warming through greenhouse gas emissions as significant.
26. Beder has charged that "The new associations present a united front for their corporate members and assert the power of large corporations in political forums…. They have also created an array of front groups that achieve their political goals whilst appearing to be independent of the founding corporations or associations (Beder, 2006a)."
27. An indication of the information barriers was provided in a conversation with a federal agency employee on loan to the Republican staff of the House Resource Committee. The staff member noted a meeting of Republican committee staff in 2006, where the conversation turned to the award-winning film by former Vice President Al Gore. All in the room had not only not seen *An Inconvenient Truth* but affirmed that they did not intend to see it. It might be noted that the Heartland Institute's home web site featured a hot link to a videotape debunking the Gore film's "snow job," built around scripts from the Gore film interspersed with clips of critical comments by skeptical experts.
28. A graphic example of the contrast between European and American environmental organizations' attitudes was experienced in Hamburg during an interview with German Greenpeace representatives. My contact, in charge of an energy policy project, had visited US Greenpeace headquarters in New York. After a discussion of energy policy, the German Greenpeace leader mentioned his desire to meet an American oil company executive. The American Greenpeace representative responded with some emphasis,"We don't talk to those people!"
29. The Sierra Club has been represented by a subsidiary, the Legal Defense Fund, which became Earthjustice in 1997, as well as by its own lawyers. It also joins other environmental organizations in suits.
30. Railway stations serving regional traffic were vital civic centers. Long-distance train stations (e.g., Union Station in Kansas City) lent the same kind of semiexotic atmosphere as international airline traffic lends to airports today. A popular radio show, Grand Central Station, presented human dramas that took place with the backdrop of New York's main railway hub. In the background one could hear crowd and steam locomotive noises, arriving trains, "all aboards," and the other sounds of the long-distance travel hub for the city.
31. Robert Moses was a political scientist by training – rather than an engineer.
32. I have not tried to research this complex story. A Wikipedia article on "The Great American Streetcar Scandal" is editorially flagged with questions about objectivity and inadequate referencing. This suggests that the subject needs more definitive treatment, which may offer an opportunity to a resourceful Masters or Ph.D. candidate.
33. Jacobs discussed the history and change in urban society due to market-led growth supported and tempered by civic values and good governance.
34. No kin to the G.H.W. Bush family.
35. Born in 1890, Bush's career prior to 1944 is shown by his biographer, George Zachary (Zachary, 1999) to be that of an engineer's engineer. After his Ph.D. in electrical engineering from MIT/Harvard (he completed a 169 page dissertation in one year), he developed a concept for magnetic field detection of submarines that he pursued for the Navy under later Nobel Prizewinner, Robert Millikan, in 1917. Joining MIT's electrical engineering department in 1919, he staved off boredom by cowriting a respected textbook on electrical engineering in 1922, and cofounding a highly successful company renamed Raytheon in 1924. In 1928, he developed the differential analyzer, an analog forerunner of today's computers, and in 1932 was appointed Vice President of MIT by the incoming President, Karl Compton. Adding to science leadership roles under President Roosevelt, Bush became the organizer of the Manhattan project that created the atom bomb.

Page range (27)

36. Among major conclusions from the work of 13 diverse teams studying developments from 1946 (and earlier) to 1963 were the following:
The time frame between basic research developments and the military breakthroughs utilizing these concepts was often close to 20 years, and up to 60 years. This time frame was so long that valuable discoveries would become most valuable to countries prepared to exploit them, not necessarily those that created them.
The "basic science" involved in the advanced military technologies from which a series of developmental steps produced the final breakthrough was primarily available in standard advanced courses in physics, engineering, chemistry, handbooks, and other syntheses, not specialized research. The results confirmed the conclusions of science historian D.J. de Solla Price (Price, 1965) that "Technology seems to arise directly from the trunk and main branches of the tree of science, rather than from its growing twigs."
Of the 710 studied "events" 91% were technical and 9% were scientific. Eighty-nine percent of the science and technology events utilized in military breakthroughs were achieved through systematically applied effort by diverse skills joined specifically to achieve(Hart, 1998) the desired technology. The contribution of undirected university research was only 0.3%.
37. In one regional conference opened by leading politicians like Kenneth Curtis, Governor of Maine, and Senator Ted Kennedy of Massachusetts, Paul Ylvisaker and Gustave Heningburg raised challenges in their opening address: "Are science and technology (or scientists and technicians) useful to public organizations?" "Are scientists and technologists willing to work in the value-laden environment of state and local government? (Feller, 1970)."
38. For example Myron Tribus, Director of the MIT Center for Advanced Engineering, was quoted: "The National Science Foundation, by its 'practices, procedures, protocols, power structure, philosophies, peer review, political constituencies, and personnel, simply was the wrong organization to be concerned with enhancing our ability to deploy technology' (Belanger, 1998)."
39. In 1979 "Frank Press as Science Advisor and Richard Atkinson as NSF director killed it without regret," cited by Green and Lepkowski (2006).
40. In a publication jointly sponsored by NSF and the American Colleges and University, James Applegate pointed out that advanced student preparation must get away from the well frog approach; "the well frog lives its life at the bottom of its disciplinary well. It believes the sky is limited to what it can see from the bottom of its well." Applegate also cites a book by D. Stokes to the effect that "the continuum between basic and applied research that has dominated much of higher education's thinking since World War II is not only an inaccurate historical description of research, but also totally inadequate as a policy framework to guide twenty-first century research (Applegate, 2002)."
41. "Basic applied research" formed a considerable part of my own earth science research. My colleagues and I served conscientiously and made significant scientific discoveries. But I knew something was wrong. USGS had in the past been an outstanding and widely appreciated service organization. But after 1965 the Geological Division also became influenced by the basic research trend. Many conscientious colleagues felt that their prime responsibility was to "do good science." Publications in peer-reviewed outside journals (or books) were the prime criteria for promotion. Whether or how these products would become translated into societal value was rarely considered. On the other hand, USGS products that served the public most directly, such as topographic maps, were often the least highly regarded in terms of professional advancement.
42. Vest was one of those impossibly bright scientist leaders who not only manage their own organizations but simultaneously sit on top advisory councils and corporate boards, testify in Congress, keep up with contemporary developments on a broad spectrum of subjects, and in his case, was also a decent family man. Vest was also known for his candor.

Page range (29–40)

43. Term given by Derek de Solla Price to an informal collegial association of peer professional scientists interested in given scientific disciplines or subfields, and accustomed to writing for common types of competitive scientific journals, seeking prestigious grants, and often having much closer relationships to peers in foreign countries than other members of their own academic department.

44. During a guest lecture at MIT, Dr. Klaus Weyrich, Director of Research for the giant Siemens Corporation of Germany, shared background on the processes by which innovative products are achieved by his firm – whose advanced medical equipment has gained wide application in hospitals throughout the United States, and various industrial uses. He emphasized that the advanced product development process requires a multifaceted search for new potentials that maintains oversight over market needs, the activity of competitors, and teams of scientific and technical people working in close conjunction with sales organizations.

45. W.E.B. DuBois, a visionary African American intellectual leader had referred to "the talented tenth" for the role that talented individuals played disproportionally in societal affairs (Appiah & Gates, 2003).

46. The main spillage of oil during the Katrina event was from ruptured storage tanks and refineries on land along the Mississippi River or tributary canals (offshore-technology.com, 2007). In November 2005, a double-hulled tank barge hit a platform that sank during the storm. The tank barge was ruptured and discharged at least 1,400,000 gallons of oil (U.S. Coast Guard, 2006).

47. The first smog inversions were noted in Los Angeles in 1943, leading to an Air Pollution Act signed in 1947 by Governor Earl Warren. In 1967, Governor Ronald Reagan signed the Mulford-Carrel Act, creating the California Air Resources Board, to manage all state air pollution control activities. The board set maximum levels for total suspended particulates, photochemical oxidants, sulfur dioxide, nitrogen dioxide, and carbon monoxide in 1969.

48. See Table 8.1 in Chapter 8; Frank T. Manheim and Greg Fuhs, unpublished database of US environmental laws, George Mason University, Fairfax, VA.

49. Documents retrieved with the assistance of archivists of the Lyndon Baines Johnson Library in Austin, TX found several task force studies and recommendations to the President, but few specific actions other than press releases and speeches (materials at the LBJ Library pertaining to natural resources and the Department of the Interior, NRFG145, June 1986).

50. Kennedy references in this section are based largely on papers made available through the Lyndon Johnson Library and Museum, March 3, 2008.

51. Personal communication with Russell Train, the Environmental Protection Agency Administrator under Richard Nixon, May 27, 2008.

52. Personal communication with William A. Radlinski, who was Associate Director of the US Geological Survey, 1969–1979, 2004.

53. A Department of Science, Space, Energy, and Technology was formally proposed by Representative Tim Penny (D WI) in 1993 (Penny, 1993). Representative Penny's amendment was a significant departure from the previously envisaged departments, in that it would have dismantled or privatized many of the federal science agencies as a part of its estimated savings of $90.4 billion. The basic concept behind a Department of Natural Resources was widely respected (and became adopted by many states), but the complex problems involved in bringing together existing agencies with widely divergent origins and constituencies got in the way of implementation at the federal level. The most systematically developed and promising initiative for at least partial federal agency restructuring was undertaken by the Stratton Commission, described below.

54. The Stratton Commission's report, Our Nation and the Sea, which was delivered to Congress in 1969, laid the foundation for the Coastal Zone Management Act and the Fishery Conservation and Management Act (American Geological Institute, 2007).

55. National Environmental Policy Act (NEPA).

56. Russell E. Train, in conversations during May 2008, told about the origin of the proposals. Train, a member of the Nixon White House staff under the direction of John Ehrlichman in 1969–1970, indicated that Ehrlichman (later implicated in the Watergate scandals that brought down Nixon) and he were "tree huggers" by inclination. President Nixon's interests were more aligned with the business community, but he was keenly aware of the importance of the environmental movement. In typical fashion, he wanted to take leadership in emerging developments. He therefore charged Ehrlichman to assembling a White House team that included input from Public Health Service's air pollution staff (HEW) to develop a plan for air pollution control. John C. Whitaker (later Undersecretary of Interior after William T. Pecora) was named to head the developments, and Christopher DeMuth, now President of the American Enterprise Institute, among leaders in the work.

57. Richard E. Ayres was a founding member of the NRDC in 1969; Leon G. Billings was the Administrative Assistant to Senator Muskie and a key person involved in writing the CAA and subsequent laws. The five senators referred to were Edmund Muskie (D-ME), Chairman of the Senate Air and Water Pollution Committee; Thomas Eagleton (D-MO); J. Caleb Boggs (R-DE), who had helped write the Water Quality Acts of 1965 and 1970; John Sherman Cooper (R-KY); and Howard Baker (R-TN). Baker had an engineering degree before going to law school.

58. According to Leon Billings, the provision for citizen litigation was inspired by proposals for public interest litigation by Philip Hart (D-MI) and a book on public litigation by Joseph Saxe (Saxe, 1970). Senator Thomas Eagleton became a special advocate for this provision, arguing that without it the law might be passed but nothing might happen.

59. I recall an overfilled auditorium at the Woods Hole Oceanographic Institution for a guest lecture by Barry Commoner in the early 1970s. Student listeners, many with hair or clothing signaling their sympathy with challenges to the status quo, sat in the aisles and stood in the back of the hall. In the course of his talk, which reviewed the crisis of the cities (see his book, *The Closing Circle*), Commoner, perhaps half seriously, suggested that automobile manufacturing had gone out of control, and society should consider going back to the Model T Ford.

60. This and other statutes are discussed in more detail in Chapter 4.

61. Nixon vetoed the Clean Water Act Amendments, which had been passed unanimously in the Senate, in large part because of its budget; Congress overrode the veto by a large margin (Russell Train, personal communication, June, 2008); Train was the second EPA Administrator, 1973–1977. Nixon turned away from the environmental movement because of its opposition to the Amchitka Island nuclear tests (John C. Whitaker, personal communication; June 2008; Whitaker was Undersecretary of the Interior, 1973–1975).

62. The results of negotiations between Interior Secretary Morton and Governor Askew over offshore leasing in the Mississippi–Alabama–Florida (MAFLA) area included initiation of comprehensive baseline studies of the offshore area.

63. A key issue in the campaign between Andrus, a former Idaho state representative, and his incumbent Republican opponent, Don Samuelson, was the ASARCO Company's plans to open a large open-pit molybdenum mine in the White Cloud Mountains. Andrus opposed it while Samuelson supported it. Idaho had a history as a mining state with a conservative political bent. Andrus's astute campaign strategies won the governorship by 11,000 votes (Andrus & Connelly, 1998). Andrus carried only 13 of Idaho's 44 counties but these included the most populous areas. Having served one term as Idaho governor, Andrus was nominated for Secretary of the Interior by Jimmy Carter on December 18, 1976, and was confirmed on January 20, 1977.

64. Unpublished correspondence from the Interior Department, Office of the Secretary through the Freedom of Information Act.

65. When Reagan offered her a minor post on a federal advisory commission, Gorsuch, responded with a now famous quote calling the nation's capital "too small to be a state but too large to be an asylum for the mentally deranged" (Sullivan, 2004b).

Page range (50–55)

66. William Bettenberg, personal communication, 2005.
67. Bureau of Land Management, Office of Surface Mining, US Geological Survey, Bureau of Reclamation, Fish & Wildlife Service.
68. William Bettenberg, personal communication, 2005.
69. http://www.doiu.nbc.gov/orientation/mms2.cfm.
70. Individual bills ultimately placed so much of the US EEZ under moratorium for offshore leasing that a consolidated moratorium was reluctantly declared by former President G.H.W. Bush in 1990. It was extended by President Clinton to 2012 (W. Clinton, 1998, http://www.ems.org/energy_policy/response.html); see Kumins, 1997, OCS Leasing Moratoria, in Oceans-Coastal Resources: A Briefing Book, Congressional Research Service Report 97–588 ENR.
71. As a Presidential candidate, the senior Bush expressed the desire to be "the environmental President" and worked hard to promote compromise and progress in contentious areas. However, the distance between contending forces eventually wore him down. His Administration's Clean Air Act Amendments of 1990 were the last major revisions to be achieved for the core 1970s laws up to the present.
72. Litigation by oil companies continued regarding the North Carolina offshore leases. On June 26, 2000, the Supreme Court, by an 8:1 margin, ruled that the Government (Congress through a 1990 bill and the Interior Department) had broken the 1981 lease contracts and should repay front-end bid money to the companies (Minerals Management Service, 2000). As a result of a similar suit, the Interior Department in 2002 agreed to buy back leases in the Destin Dome area (Eastern Gulf of Mexico), as well as producing properties in the Florida Everglades.
    In existing lease areas new technologies continued to advance. These included "bright-spot" seismic surveying that for the first time permitted remote identification of potential petroleum accumulations, horizontal drilling, and ultradeep drilling, enhanced oil production in the Gulf of Mexico. A new record deep water drilling depth of 10,011 feet was set (MMS, 2004).
73. Even "revisionist" political scholars who make unsentimental critiques of the Founding Fathers (Ackerman, 2005) mainly propose reinterpretation of the Constitution to fit new circumstances. They do not propose rewriting of the document, a step that Thomas Jefferson initially thought might be necessary every 30 or 40 years. In fact, thoughtful people of all political persuasions have tended to recognize that the Founders' realistic weighing and design of political systems to recognize fundamental human needs and desires, as well as recurring tendencies and sources of problems in human relations, came during a fortunate window in US history.
74. At the time of the Declaration of Independence, there had been no major nations governed by representative systems since the Roman Republic (509–44 BC) and the Athenian Democracy (500–322 BC). The founders of the new nation went back more than 2,000 years to research political models potentially relevant to the new republic. Particular interest focused on the classical philosophers and politicians' debates about the strengths and weaknesses and historical outcomes of democratic models. Open writings about such topics under royalty were difficult even where parliaments were in place – as in eighteenth century Britain. The framers of the Constitution were especially concerned about the cycle: democracy – mobocracy – chaos – reestablishment of authoritarian rule – that had repeatedly occurred in antiquity.
    James Madison was the preeminent scholar of Classic literature, combing it for material relevant to the task of constructing the Constitution. He fortunately combined encyclopedic knowledge with an attitude of service and promoting good judgment, rather than egotism or display of erudition. Through his historical perspectives he was often able to move debates from unproductive directions, head off appealing but flawed statements and avoid articulations that would become dated with time.
    After the signing of the Constitution in 1787, broad acceptance of the new document by the states was not yet assured. In September 1787, Alexander Hamilton published several

letters supporting the Constitution in a New York newspaper under the pen name, "Publius." That series expanded to a series of essays now known as the Federalist Papers. It served as something like a modern interactive web site for political leaders and citizens of the time. It fostered discussion, allowing the airing of objections or concerns about interpretation of practical applications of the still controversial document. James Madison contributed more than a quarter of the material in the Federalist Papers, describing and explaining the various function of government with notes on historical precedents (Davis, 1964).

75. Did leaders of American Revolutionary era live in a simpler time when people could more calmly set about solving political challenges? That complexity and human foibles were alive and well in the Revolutionary war period is indicated by numerous examples: General Benedict Arnold initially served the revolutionary cause gallantly, but then betrayed it for money and because of an oversized ego that was skillfully exploited by British agents. Thomas Conway, an Irish officer appointed Brigadier General in the Continental Army, and General Horatio Gates, victor at the Battle of Saratoga, attempted to foment a plot to undermine General George Washington during the lowest ebb of the Revolution. Washington barely restrained unpaid, angry soldiers who were allowed to languish by Congress after the war. Later, in President Washington's administration, conflicting political philosophies of influential members of his cabinet brought the work of the new government close to the breaking point.

76. Although individual senators or other politicians at times might seek discussions or accommodations, none of the partisanship has limited dialog and creative resolution on a larger scale.

77. In the 1770s slavery was generally regarded as an evil institution that should be eliminated. The invention of the cotton gin in 1793 was among factors that by 1820 led South Carolinians to no longer apologize for slavery, but to defend it (Wallace, 1969).

78. Even in his last speech, which advocated Negro voting rights and incited John Wilkes Booth to his assassination plan (April 14, 1865), Lincoln proceeded with discretion (Foner, 2005). Speculations about what Lincoln might have done had he lived are offered by (Harris, 2004). Lincoln's "with malice toward none" policy, his political skill, and personal support in Congress for otherwise potentially controversial gestures of generosity to the south, as well as high standing with African Americans (giving him influence in controversies) would surely have moderated the movement by white southerners to suppress the black population. The reascension to political power by the southern whites in turn influenced national policy.

79. Roosevelt made the subtle but powerful point that accusing a guilty man of more than he had done could be more damaging than falsely accusing an honest man. How could this be true? An honest man might be vindicated when false accusations were challenged. This might be harder in the case of a man who was already under suspicion. If distortions or skewed information remained uncorrected, they could adversely influence understanding of problems or issues and lead to mistaken and damaging policy.

80. Roosevelt had come from affluent family circumstances. His home, now a historic site in Oyster Bay, Long Island, served as the locale where Roosevelt mediated an end to the Russo-Japanese War of 1905. In spite of this background, he had publicly chastised men who sought to corner markets, build cartels, and otherwise gain advantages by dubious means. A series of scandalous accusations published in Cosmopolitan magazine in 1906 sparked widespread feelings against some 23 wealthy members of the US Senate. Believing that exaggerated accusations were arousing undesirable class hostilities, Roosevelt made a speech in which he referred to the journalistic articles as "muckraking," the muck rake being a long-handled rake used on manure piles or solid waste in cesspools. Roosevelt condemned attacks of this kind while at the same time supporting many of the reforms advocated by the investigative journalists ("Pure Food and Drug Act of 1906").

81. Accepting the award of the American Good Government Society in 1964, Senator Everett McKinley Dirksen defined good government as follows: "It cherishes and practices a strict respect for the Constitution which brought this very government into being; it through

Page range (59–61)

its officers and agents maintains a wholesome and respectful regard for the people from whom government derives its powers; it exhibits courtesy and good manners in all of its dealings with the people at home and the nations abroad; it is not moved to hasty and ill-advised action by the emotions of any given moment; it in the language of the ancient law shows restraint and does not follow a multitude to do evil; it charts a course calculated to be beneficial now and in the future to all of its citizens; it exalts the dignity of human-personality; it observes the golden rule in all of its dealings with other governments in the family of nations; it stands firm for a right and for equal justice under the law; it is ever mindful that the blessings of liberty spring from the everlasting covenant between the instant generation and those generations who have gone before and those who will come after our day and time (Dirksen, 1964).

82. A curious alignment regarding slavery is mentioned by historian Burton Hendrick: "Conspicuous among the apologists for slavery in Philadelphia were the Yankee delegates, while the state that took first place in denouncing slavery and insisting on its exclusion from the Constitution was Virginia." Connecticut formed an axis with Georgia and South Carolina to push the "Connecticut Compromise" in accommodating slavery (Hendrick, 1937).

83. Personal communication from James Bryson, Palo Alto, CA, 2007, from genealogical research in the Middle Atlantic.

84. Were there alternatives to the slavery compromise? If the antislave states had rejected slavery in the Constitution, and the three southern states had carried out their threat to separate (North Carolina's politics were closer to those of Virginia and it might have remained in the Union), this would have avoided the ensuing reactionary laws and Supreme Court decisions. It would have forestalled the later proslavery drift in Virginia and Maryland. It could have prevented expansion of slavery into Tennessee, Kentucky, Missouri, and Texas, and would have isolated the slave states in an increasingly anachronistic system. Stressed by escape of slaves to northern sanctuaries, internal opposition may have accelerated bloodless voluntary abolition of slavery by the breakaway states, followed by their rejoining the union.

85. Jackson attracted strong feelings, usually either for or against him but rarely in the middle. His first major biographer, James Parton, is cited as describing him as "one of the greatest of generals, and wholly ignorant of the art of war. A writer brilliant, elegant, eloquent, without being able to compose a correct sentence.... The first of statesmen, he never framed a measure. He was the most candid of men, and was capable of the profoundest dissimulation. A most law-defying, law-obeying citizen. A stickler for discipline, he never hesitated to disobey his superior. An urbane savage ..... (Sellers, 1958)."

86. Under the pressure of changing political policies, growing commerce and population, major revisions in town ordinances took place in 1833 and 1854, transferring authority from the former Board of Health to the discretion of political bodies (Board of Aldermen, appointed by the City Council) and the Chief of Police (!). As noted from an excerpt of the 1854 ordinances, the Aldermen could authorize connection of privy vaults to the sewer system for fees.
    "Section G. The superintendent of sewers, under the direction of the board of Aldermen, is authorized to permit, under suitable restrictions, and on the payment of such sum, not to exceed 30 dollars, as they shall deem expedient, the construction of sufficient passageways or conduits under ground, for the purpose of conveying the contents of any of the vaults aforesaid into any common sewer or drain." From ("Charter and Ordinances of the City of Boston," 1856).

87. Comment on Jefferson's measurement of temperatures, pressures, and relation between forest cover and climate (Bedini, 2002).

88. This kind of water use is referred to as riparian water law.

89. (Hannerz & Palmer, 2003).

90. Genealogy Quest, 2007, http://www.genealogy-quest.com/glossaries/epidemics.html.

91. Except for awards of land to soldiers who had seen service in wars. The Homesteading Act of 1862 was passed after seccession of the southern states.

Page range (61–68)

92. J.W. Powell, second Director of the US Geological Survey, was an early advocate of federal ownership or management of water resources (see Section 8.3).
93. Mark Twain and Charles Dudley Warner *The Gilded Age: A Tale of Today* (1873); a novel describing fictional but realistic characters tied to bribery and a railroad subsidy scheme. The term "gilded age" had its origin in this book.
94. The Union Pacific Railroad, in a famous scandal involving its financial subsidiary, the Credit Mobilier, inflated the costs of proposed railway expansion to in a cost-plus operation from an estimated $30,000 per mile to $60,000 per mile, thereby pocketing huge sums.
95. An example is the North Butte mining disaster of 1917 with its subtext of callous and manipulative corporate managers against some of the nation's most powerful unions (Punke, 2006).
96. In the 1770s, inoculation was a procedure whereby infectious material from persons with smallpox was allowed to weaken with time and was then introduced into uninfected persons in a cut through the skin. Not without risk, inoculation caused a milder case of smallpox, which then conferred immunity to future infection. A true vaccine was first produced from cowpox by Edward Jenner in England in 1796 (Scott, 1999).
97. "He most adored the high Alps. He climbed the Becca di Nona (Pic di None) south of Aosta, among pastures so hard of access that goats were pulled up to them by ropes. He scaled the Schilthorn and the Faulhorn in the Bernese Alps. He crawled through the Col de la Traversette in the Alpine crest north of Monte Viso. Mesmerized by glaciers – Zermatt, Aletsch, Grindelwald – he risked life and limb studying ice flow structure" (Lowenthal, 2003, p. 256). Whether or not such strenuous exploits were wise for an overweight sexagenarian, he returned from summer climbs ready to tackle "brainwork with renewed vigor." "As I am getting so strong at 63," he boasted, "I suppose I shall climb the Himalayas at 100 (Lowenthal, 2003, p. 256)."
98. In recent years sharp global temperature drops, called Dansgaard-Oeschger events, have been interpreted from ice core data; they have been linked with interruptions of the Gulf Stream by massive fresh water flows from North America. Concern about such an event has been raised in conjunction with the possibility of the melting of Greenland's ice cap (Bond et al., 1999).
99. Roosevelt made trips to western sites with both John Muir and John Burroughs (Dalton, 2007).
100. Editors' review of Mann's book in the Amazon.com web site.
101. Humphreys' preparations for the report included 18 months of study of European and Middle Eastern deltaic rivers and consultation with the greatest European hydraulic engineers. The report included critical analyses of all relevant scientific literature, and "hundreds of pages of drawings, graphs, and raw data on sandbars, riverbanks, levees, on every imaginable river phenomenon." On the strength of the report and his service as a General in the Civil War, Humphrey was appointed Chief Engineer of the US Army Corps of Engineers in 1866. According to Barry, thereafter the unyielding military man took over from the scientist/engineer. Among other things, Humphrey tried to have Eads' bridge removed on grounds that it obstructed river traffic.
102. The breathtaking nature of Eads' achievement has been described in John Barry's book about the great flood of 1927 (Barry, 1997). The open-hearth process for steelmaking had only been developed in 1867. Not only was a steel bridge of such magnitude considered unfeasible, but constructing it meant designing the equipment to manufacture components, like multistrand cable wire. Equipment was designed to test materials to heretofore unheard-of tolerances. Testing did not just involve random samples, but was applied to every piece that went into the bridge.
103. Peter Anthony Dey began as an engineer for the New York and Erie Railroad in 1846. He became known for his expertise in conducting surveys for future railroad lines. The recommendations he made for an extension of routes from Iowa to Great Salt Lake became the

basis for Union Pacific's plans, and in 1864 he became Union Pacific's Chief Engineer. However, Dey resigned "the best position in my profession this country has ever offered to any man" when he recognized that the construction charges specified in a contract with *Crédit Mobilier*, an American branch of a renowned French investment bank, were inflated. They nearly doubled his own estimates. Congress was subsidizing the creation of the westward railway extension, and had appointed oversight committees for the railroad projects. As described in a contemporary book (Cloud, 1873), Dey's resignation prompted the discovery that the government overseers and a number of Congressmen were implicated in a huge fraud. Dey became an important witness during Congressional hearings on the case in 1873.

104. Legal battles continued until 1906, when the Supreme Court finally ruled in favor of Illinois. St. Louis would have to filter its water to avoid the risk of infectious organisms (Cain, 1978, 2005).

105. Stearns had participated in the design and construction for the Boston sewage system. For the Board's 1893 annual report, he had completed a 220-page systematic analysis of drinking waters for all towns in Massachusetts, that included computations of water yield from rainfall records and other data, and painstaking measurements of turbidity, color, taste, and pollution (Nesson, 1983).

106. Design achievements produced by engineers of the New York City Water Development Board in 1905 resolved the water supply crisis that was limiting growth of the Greater New York City metropolitan area. The engineers developed an ingenious system of interlinked water reservoirs in the Catskill Mountains on the border of New York and New Jersey, with pure water from these reservoirs transported to the city by gravity through long-distance aqueducts.

107. Bruce E. Seely, Professor of History and Chair of the Department of Social Sciences at Michigan Technological University (http://www.social.mtu.edu/people/bseely.htm), and Henry Petroski, Professor of Civil and Environmental Engineering (and also Professor of History) at Duke University (http://www.cee.duke.edu/fds/pratt/ccc/faculty/petroski), are leading experts on the history of American engineering. Their curriculum vitae on the web pages provide extensive lists of publications relevant to this topic.

108. Bruce Seely, oral communication, April 5, 2008.

109. According to Henry Petroski (oral communication, March 2008), Alfred Nobel was an engineer, and in his will he specified that the prize money should be used for applied science. His two executors were also engineers. However, lawyers associated with translation of Nobel's will and funds into action apparently associated with basic scientific circles, and ultimately Nobel prizes came to be awarded for pure scientific achievements, not necessarily with any applied result.

110. In the concluding chapter, I argue that an organization's (or a societal sector's) effectiveness is to a significant extent dependent on its ability to attract talented, open minded, and responsible leaders. The negative image often applied to the Corps in the last decade may or may not have validity. It certainly cannot be expected to attract top talent to future leadership.

111. http://en.wikipedia.org/wiki/Atomic_Energy_Commission.

112. The US Commission of Fish and Fisheries was created in 1871, the first federal natural resources agency. In 1903, it was attached to the Department of Commerce and Labor (which became the Department of Commerce and the Department of Labor in 1913; commercial fish stayed with Commerce and, since 1970, has been part of the National Oceanic and Atmospheric Administration. The Department of Agriculture (established in 1862) expanded and renamed its Division of Economic Ornithology and Mammalogy in 1885, the Bureau of Biological Survey.

113. Briefly called the National Biological Service.

114. Ventilation and banning of miners under the age of 12 were the subjects of the first mine safety law (1891). Fifty years passed (1941) before inspectors were allowed in coal mines, and another

few years (1947) before the first modest regulations were codified. In 1952, the Federal Coal Mine Safety Act finally mandated annual inspections and penalties for noncompliance of some orders. In 1966, noncoal mines were included in the regulations. The "Coal Act" of 1969 increased the reach and enforceability of the previous laws and regulations. Increasing awareness of health and the environment in the early 1970s prompted the Secretary of the Interior to create a new agency in 1973, the Mining Enforcement and Safety Administration (MESA); he moved the health and safety functions from the Bureau of Mines to MESA, ostensibly to avoid potential conflict of interest between the functions of mineral resource development and mine safety and health. Congress took this trend further in 1977 with the Mine Act, which amended the Coal Act. MESA was moved to the Department of Labor and renamed the Mine Safety and Health Administration (MSHA). Mining fatalities dropped from 272 in 1977 to 86 in 2000, indicating both good oversight by MSHA and cooperation by mining companies and employees. However, several preventable fatal and near fatal mining accidents have occurred since 2000. During this period, MSHA has had significant cuts in funding and employee ceilings, apparently reducing the agency's ability to adequately inspect mining operations. This statement is based on information from MSHA's web site (Mine_Safety_and_Health_Administration, 2008) and Goodell's book, *Big Coal* (Goodell, 2007).

115. Hickel had actively supported the planning development. Hickel was fired after his letter to President Nixon cautioning a restrained approach to youthful antiwar protesters was prematurely released to the press. The cooperative task force efforts relating to the organization of a comprehensive Department of Natural Resources were described to me by an active leader and participant in the effort, William A. Radlinski, who was at the time an Associate Director of the US Geological Survey (William A. Radlinski, oral communication, April 2008).

116. PL 91–190 42 USC 4321–4347. It was signed by President Nixon on January 1, 1970.

117. A monograph on NEPA (Lindstrom & Smith, 2001) indicates that Senator Muskie, who chaired the Air and Water Pollution Subcommittee of the Senate Public Works Committee, was concerned that Jackson's Senate 1075 bill "would debilitate the existing environmental goals and programs of his Air and Water Pollution Subcommittee." Muskie and other members on his committee are cited as thinking that federal agencies "could not be trusted to follow the new course."

118. A key modification was that, besides preparing an impact statement and observing other requirements, agencies had to comply with specific environmental standards. Senator Muskie was familiar and sympathetic to the problems faced by local industries in Maine, and had earlier opposed national standards and more stringent requirements that would overly encroach on states' independence in managing environmental issues. However, in 1970, he became a leading Democratic Presidential candidate. He was attacked by a book sponsored by Ralph Nader's new Center for the Study of Responsive Law, *The Morning After Earth Day*, charging that he had "failed the nation in the field of air pollution control legislation" (M. Graham, 1999). Facing President Nixon, Muskie promoted drastic new pollution laws.

     William van Ness, Jr., Chief Counsel for Senator Jackson's Senate Interior and Insular Affairs Committee and currently President of the Jackson Foundation, confirmed the switch in policy on the part of Muskie. With respect to later discussion of Jackson's disappointment about excessive litigation based on NEPA, van Ness nevertheless felt that the NEPA law had "the virtue of getting all the facts out into daylight. All relevant issues are brought forward about whether a project is just and valid. Everything is made accessible to the public, and then a decision is made (William van Ness, Jr., oral communication, April 23, 2008)."

119. Jackson had also expressed his irritation with opponents of economic development, then prevalent in society (Ognibene, 1975).

120. US federal laws are designated by PL numbers, which delineate the Congress (e.g., 91st Congress 1970) and number in order of passage, and code. The federal code number is assigned after the new law is compiled into the Code of Federal Regulations (CFR), which comprises 50 titles. For example, Title 43 is Public Lands and Title 50 is Wildlife and Fisheries. The

last number refers to the section of the code. All laws are reported in the Federal Register. The code for the OSHA, the Occupational Safety and Health Act (December 29, 1970), is 29 USC 651 et seq.

121. The CAA of 1970, at only 50 pages, was transitional to much longer acts by 1972. The 1990 Amendments ballooned to 800 pages (see also Reilly, 1991).

122. Billings was Staff Director, Senate Committee on Public Works' Subcommittee on Air & Water Pollution, 1966–1978.

123. The additional expense involved in cleaning sites that involved no groundwater aquifers or other potable water supplies to standards associated with drinking water greatly increased cleanup costs, unnecessarily burdening organizations assigned responsibility, and reducing the number of sites that could be cleaned (Day, 2001).

124. EPA developed a huge database system, called STORET, designed to accommodate the vast production of data (U.S. Environmental Protection Agency, 2007b).

125. (Percival et al., 2000); This aspect of the laws had strong advocacy by Ralph Nader and his Center for the Study of Responsive Law (supported by hosts of student volunteers) and other leaders of the consumer rights and public litigation movement. It had also been advocated in an influential book by Joseph Saxe, *Defending the Environment: A Strategy for Citizen Action* (Saxe, 1970).

126. According to Prochnau and Larsen (1972), the CEQ concept was added to NEPA at the urging of Senator Muskie.

127. http://www.epa.gov/swerosps/bf/

128. That is, the loss of alternative opportunities not chosen.

129. PL 104–142 Mercury-Containing and Rechargeable Battery Management Act, May 13, 1996, USC Sec. 103 b 3. (3) "On each rechargeable consumer product containing a regulated battery that is not easily removable, the phrase 'CONTAINS NICKEL-CADMIUM BATTERY. BATTERY MUST BE RECYCLED OR DISPOSED OF PROPERLY' or 'CONTAINS SEALED LEAD BATTERY. BATTERY MUST BE RECYCLED,' as applicable."

130. Unpublished database, in progress, F.T. Manheim.

131. In the late 1990s, EPA threatened to take the City of New Orleans to court over violation of performance standards for wastewater treatment. Treated New Orleans sewage effluent was normally discharged to the Mississippi River. Under conditions of high rainfall holding capacity for combined sewers and the treatment system was exceeded, and raw sewage and industrial effluents would be discharged to Lake Pontchartrain (a large estuary that forms New Orleans' backyard) through canals. Fears and claims about toxic bottom sediments in Lake Pontchartrain from Save Our Lake, an influential citizen organization, posed major potential complications for New Orleans. At the time I led a comprehensive database compilation project at the US Geological Survey for all available analytical data on bottom sediments in Lake Pontchartrain. A preliminary version of the database, including EPA's own data, was quickly published (Manheim, 1998) and made available to EPA. This showed that metal and other contaminants in the bottom sediments were far below toxic levels, and allowed the EPA-New Orleans negotiations to focus on realistic, rather than feared conditions. A more comprehensive environmental assessment of bottom sediments with much ancillary data was published in 2002 (Manheim & Hayes, 2002).

132. A rule of thumb: cities are typically governed more efficiently than states, and states more efficiently than the federal government; that is, government closest to those governed, and more in touch with the problems and issues works most efficiently.

133. Basis for adversarial legalism was laid in the laws of the 1970s.

134. Scenic Hudson Preservation Conference v. FPC, 354 F.2d 608 (CA2). Page 407 U.S. 926; bhttp://supreme.justia.com/us/407/926/case.html.

135. The aggrieved African American plaintiffs, led by Thurgood Marshall, successfully claimed that they had been subjected to abuse of power by government, for which they petitioned for redress to the Supreme Court.

Page range (95–102)

136. A striking example of the clout of the citizen litigation provisions of the CWA is offered in a vignette about litigation by a recent graduate of the Northwestern University Environmental Law program against the Milwaukee's Sewage Treatment Authority (Chapter 8). The lawyer, with a few volunteer supporters and modest donated funds, won a round in federal district court after a series of failed court suits alleging technical violations of the CWA on the part of both the Authority and EPA. This litigation, in my judgment, resulted from the lawyer's literal interpretation of the CWA combined with inadequate scientific background about municipal sanitation systems and lack of concern about holistic and creative solutions to costly problems. It nevertheless generated a long series of articles in the Milwaukee newspapers that portrayed the young lawyer as a heroic David against the institutional Goliath – which had received national awards for superior performance of its sewage treatment department for 5 years in a row. Such results can discourage qualified, highly motivated staff from taking important public service positions – to be replaced by people willing to put up with harassment for the sake of a secure public employment job.

137. The concept that economic expansion is fundamentally at odds with biodiversity continues to be held by many environmentalists. For example, authors of a recent monograph on the Endangered Species Act, wildlife biologists Brian Czech and Paul Krausman, conclude that that there are only two options left: "abandon ESA or abandon economic growth" in the United States (Czech & Krausman, 2001).

138. One might note that the expenditures for war and national emergencies like Katrina are included in the figures that make up Gross Domestic Product, even though the "value" may be transferred abroad, or the outlays temporarily replace losses rather than being part of normal productive enterprise.

139. "This paper maintains that deindustrialization is primarily a feature of successful economic development. Advances in the service sector, rather than in the manufacturing sector, are likely to encourage the growth of living standards in the advanced economies … .. (Rowthorn & Ramaswamy, 1997)."

140. China's consumption of steel for building its infrastructure and durable goods has increased demand for US steel; some small iron mines have been reopened in order to supply raw material to US steel mills (McCartan, Menzie et al., 2006).

141. O'Toole (1988) pointed out that in the 1950s and earlier, Forest Service policy was to selectively cut mature trees in ways that hardly altered the appearance of forests. Massive clear cutting not only responded to industry pressure but also boosted the Forest Service's discretionary budget. On the other hand, Carhart (1959) indicated that in the special case of Douglas fir, clear cutting was necessary and acceptable forest management because seedlings required sunshine to develop.

142. Unpublished database, F.T. Manheim and G. Fuhs, 2007.

143. Who could have predicted that a sweet-voiced church singer, Janis Joplin, would have embraced the sudden evolution of a drug-laced hardrock lifestyle and die of an overdose in 1970? Rapid transformations within the academic community are illustrated by the "before and after" images of Michael Rossman on the cover of his book *On Learning and Social Change* (Rossman, 1969), Fig 2.4. US influence also extended abroad, leading Mick Jaggar, a three-piece suit-attired student at the London School of Economics in 1961, to embrace the role of a countercultural rock star in 1962, once the success of the Rolling Stones was assured (Manheim and Manheim, 2006, p. 267).

144. In addition to the variable definitions and criteria that go into calculating benefit/cost (BC) assessments, which led to EPA's report of BC estimates in which the Clean Air Act was supposed to have created benefits 40 times larger than its costs ($22 trillion dollars of value for Clean Air Act Amendments from 1971 to 1980). Discusing BC assessments has become something of a cottage industry among economists (Rosenbaum, 2008, p. 145). Rosenbaum (2008, p. 153) also points out the consistent tendency of the USACE to develop inflated estimates of proposed economic projects.

Page range (102–112)

145. See Section 1.2; reforms are discussed in Chapter 7.
146. By focusing only at the current militancy of the environmental movement and the ethical lapses of big business leaders, one easily overlooks the evolution of confrontation and alienation in these developments.
147. Industries have grown around benefit cost analysis, which can be accessed by queries on Google Scholar.
148. Except in some of the policy arguments of environmental organizations criticizing inadequate environmental enforcement or opposing developments that are perceived to involve risk to the environment.
149. For example, the price fluctuations of commodities produced in the United States are much lower than those that must be imported, which are subject to a variety of political and economic variables in the world market. This applies to both metal and mineral commodities and petroleum. Unpredictable variations in commodity prices have a destabilizing effect on industry operation. Most recently these have resulted in several regional air carriers filing for bankruptcy. The powerful indirect influence of US policies on corn production and biofuels has helped raise world food prices and other adverse developments (Brown, 2006). Failure to anticipate environmental regulations for safer, cleaner products, and operations led most manufacturers of water-based basecoats in the United States "to rely on technologies developed by European suppliers" (Gardiner & Portney, 1999).
150. According to Hazilla and Kopp (1990), curtailed job growth, lower saving rates, reduced capital formation; and the reallocation of resources reduce GDP by 5.8%.
151. An in-depth evaluation by Wagner (2003) indicated that the Porter hypothesis outcome is most likely to occur under "efficient" regulation that encourages innovation. Much of US regulation was designed exclusively for environmental protection and deliberately excluded flexibility or discretion that could allow economic influence.
152. The mass of anecdotal evidence gathered by the conservative activist, Ron Arnold (*Trashing the Economy*; Arnold & Gottlieb, 1994), about the adverse effects of environmentalists' campaigns and their perceived obliviousness or hostility to economic activity and the livelihoods of ordinary people is discussed in Section 8.8, Guerrilla Warfare.
153. Barton and others concluded that factors affecting competitiveness are difficult to compare from industry to industry (Barton, Jenkins, et al., 2007). While environmental regulation promotes innovation, it also has immediate costs that add to the price of commodities. In the short run, regulation may depress profits, but in the long run, it may lead to changes that improve competitiveness (see Wagner's critique of the Porter hypothesis, 2003). Gunningham, Thornton, and Kagan (2005) conclude that most businesses comply with environmental regulations, although for different reasons and in different ways. Few were found to be intent on skirting the regulations for economic reasons.
154. "Overall, ASCE gives the nation's infrastructure a grade of D and estimates that $1.6 trillion will be needed over a 5-year period to improve it sufficiently to meet today's demands" (Fitzgerald, 2007).
155. Regulatory details that follow are condensed from Robert Kagan's influential book about US adversarial legalism (Kagan, 2003, where not otherwise cited).
156. James Hansen, Chief Scientist for NASA and a pioneering climate investigator and modeler, has recently reported evidence for more rapid increases in global temperature than had been included in earlier models. He argues that if the United States does not take vigorous action domestically and use its influence globally to curb greenhouse gases, accelerated warming will set in motion irreversible changes toward "a different planet" (Connor, 2007).
157. Global climate change and US energy policy: "The US must act." The Intergovernmental Panel on Climate Change (*4th Assessment Report, 2007*), the *Stern Report* of 2006, the National Academy of Sciences, and other prestigious scientific and technical advisory books and bodies have concluded that global warming due to greenhouse gas emissions has already raised land and ocean temperatures by more than 0.7 degree Celsius during the past

century. Unabated emissions of carbon dioxide to the atmosphere will accelerate temperature increases, causing destabilizing effects on the Earth. The most severe effects are expected on the world's poorest people, owing to their locations in tropical, drought-prone areas, or low-lying coastal regions.

158. Gro Harlem Brundtland is a former Norwegian Prime Minister who chaired the UN Commission report of 1987 that launched the concept of sustainable development in the European Union. In her lecture at the Nobel Peace Prize forum of 2004, Brundtland recounted her experience as Norway's Environmental Minister in 1977 (Brundtland, 2004). The blowout at the giant Ekofisk oil platform and subsequent events triggered new aware-ness. Brundtland noted that until economic dimensions became part of the discussion, "this debate was mainly limited to those with special interests." "I realized that you cannot make real changes in society unless the economic dimension of an issue is fully understood." In other words, separating environmental from economic policy cannot achieve success.

159. In a recent panel discussion at the annual EPA Environmental Partnership Summit, Baltimore, May 20, 2008, Malachy Hargadon, environmental representative for the Euro-pean Commission Washington Delegation commented on swings of movement within the EU. At times, beginning with the European Commission chairmanship of Manuel Baroso, economic development was emphasized, with special focus on Regulatory Impact Analysis (RIA). At other times the movement was toward environmental concerns, especially global climate change. At a meeting of the EU Competitiveness Council, Dromoland Castle, Ireland in 2004, concerns about excessive bureaucracy and regulation were claimed to be inhibiting intra EU trade and export development.

160. EU tariffs on industrial products are among the lowest in the world, however energy subsidies through tax exemptions, rebates, and selective trade restrictions have been used to aid in the growth of new industry, smooth energy price fluctuations, and otherwise help domestic industries. This is especially true for the coal industry, which in some nations pays no tax, and for which support is allowed until 2010. Whereas subsidies over the past decade were concentrated on biofuels, which was critical in assisting the growth of the German biofuel industry, similar to subsidies currently aiding the US ethanol industry, European nations are now beginning to divert the biofuels subsidies to other renewable energy resources (Rosenthal, 2008).Total EU biofuel subsidies in 2006 were 3.7 billion euro.

161. Representative Jay Inslee suggested that the United States is a fountainhead of innovation that can be put to work in new energy initiatives. This idea seems confirmed by my 2006 review of ocean and tide energy. Although the United States had only been involved for a few years, a list in an article summarizing international initiatives (see Section 5.3.4 and 6.5.1) found 19 in the United States, 10 in the United Kingdom (which has been at it longer), and lesser numbers in other countries. The study found connecting experimental operations to the local grid a more difficult problem.

162. Congressional legislative staffers indicate that a majority of the bills are submitted mainly to signify concern, interest, or activity, with no further expectations. Some bills may consoli-date ten or more other bills, or represent considerable effort in preparation. Only a minority will have bipartisan sponsorship.

163. One should make a qualification here. The original framers of the CAA in 1970, including members of the Clean Air Act task force in the Nixon White House (Russell E. Train, per-sonal communication, May 27, 2008), and Senator Muskie, knew what they were doing and what they wanted to achieve. But that was a very specific time and place, whose conditions have little in common with ordinary legislative environments.

164. Evidence of mass filings of dubious claims against the US asbestos industry was revealed by discovery processes described in the decision of Judge Janis Jack (see details in endnote 169).

165. I emphasize that although conservatives sometimes lump academics and environmentalists, equat-ing the two seems to me (from my personal experience) wrong and misleading. It is true that some academics are environmental activists or even radicals on their own time, and a few are radicals

on the other side. They may let advocacy show in lectures or presentations. But within the social science as well as the natural science community, peer-reviewed journals or other media that carry high standards want authors to play by scientific rules and not give their media a bad name by pushing advocacy at the expense of facts. In fact, they may accept work by controversial authors whom they would rather reject if excluding well-prepared research would expose them to criticism for letting their own preferences govern content. I also suspect that personal campaigns against dissident critics of global warming models, for example, as "deniers" does not come from the mainstream scientific community. My criticism of academic research is not about quality - which is often high. It is about the goals of scientific work and the use of the nation's scarce scientific talent.

166. For example, the Union of Concerned Scientists reported detailed research on health hazards of highway and construction equipment in California (Union of Concerned Scientists, 2008). Its research offered extensive documentation and monetization of health costs of outdated highway equipment, for example, $8 million+ per premature death, and lesser amounts for hospitalizations, school absences, etc. It showed minimal interest in the economic and infrastructural aspects of the fixing the problem - or exploring cooperative and innovative approaches to dealing with issues in holistic fashion. The long-term consequences of such one-sided emphases are well established. The results create delays or obstruction in infrastructural development, and limit new initiatives to essentials (i.e., discouraging innovative and future-oriented initiatives). Environmental critics gain favorable images as public-spirited agents, while those who deal with the economic side of activities get unfavorable public images. Bright, idealistic, technically oriented young people tend to be discouraged from careers in these "grubby" operational sectors of society. This problem may be less severe in California than in other states. California has traditionally demanded a high level of performance from state agencies, has offered better salaries, and has attracted higher caliber personnel.

167. In one of his editorials and columns for the Wall Street Journal, Jude Wanniski, an early economic advisor to the Reagan Administration and influential journalist (who coined the term "supply-side economics"), advocated deficit creation by the Reagan administration because this would inhibit future Democratic administrations from expanding government. Deficit creation, according to David Stockman (1986) was clearly not Reagan's objective, but was due to overoptimistic political assumptions and supply-side economic doctrines not backed by economic analysis.

168. Rigid US building codes typically do not permit state of the art green solutions in building construction, although such innovations are common in Europe and exist in pockets in the United States; for example, the use of fly ash in concrete (http://www.greenbuilder.com/sourcebook/Flyash.html), earth-concrete mixtures in low-cost, earthquake-resistant dome homes (http://www.calearth.org/), and water-free toilets (http://www.envirolet.com/). Similarly, clothes dryers and hot water storage tanks – which waste a large amount of energy – are not used in European homes; they dry clothes outdoors or on indoor racks, and employ highly efficient flash (on-demand) water heaters instead.

169. The ruling of Judge Janis Graham Jack sent shockwaves through the media and legal professions. An extensive discovery effort revealed in a 249-page decision showed that legal firms had trolled for participants in class action suits through large-scale advertisements. Many persons included in the suits had no long-term exposure to asbestos and were "diagnosed" by special doctors who mass produced questionable diagnoses by the hundreds of thousands. An account of the history of asbestos litigation indicates that not only lawyers but affected companies also bear some responsibility for the lucrative appeal of class action suits (Glater, 2005). Many companies did not fight unreasonable suits, but instead paid large sums or otherwise made deals that would limit ultimate liability (but did not always do so). Only when, with some hesitation, companies made a stand and had an objective judge who had had experience in human health issues (Judge Jack) was the massive fraud revealed.

170. In the mining industry, including both extraction and processing. These are the occupations in which workers are most likely to encounter respirable (breathable) particles of asbestos and silica.

171. In contrast to these numbers, the firm of Baron and Budd, specializing in asbestos-related suits, recently reported 68 cases with recoveries to individual clients of from 1 to $10 million after legal fees. It is not certain when the cases were settled (Baron, 2007). It may be a sign of the times that a history of asbestos litigation is available online from the firm, "Hugesettlements.com."

172. In a New York Times Op-Ed article of December 16, 2005, Robert F. Kennedy Jr. begins his attack on the project "As an environmentalist, I support wind power, including wind power on the high seas. I am also involved in siting wind farms in appropriate landscapes, of which there are many. But I do believe that some places should be off limits to any sort of industrial development." Kennedy's article can be regarded as NIMBY and the art of the environmental lawyer raised to impassioned heights. He evokes the uniqueness of the area, for example, "fishing villages immersed in history and beauty," and "dark clouds of terns and shorebirds descending over the thick menhaden schools," and then turns to the menacing destructiveness of the project, "Hundreds of flashing lights … will steal the stars and nighttime views" "..noise of turbines" .. "40,000 gallons of potentially hazardous oil" … "loss of up to 2,355 jobs because of the loss of tourism" … "[fishermen's] gear fouled in a spiderweb of cables." Kennedy advocates moving the project to deeper water (where the economics would kill the project) where "thousands of turbines could be placed." Independently, Senator Ted Kennedy critiques the project and makes it clear that making money on the placement of turbines is distasteful to him.

173. A Google search on "Cape Wind Project" yields interested readers 260,000 or more web sites (January 17, 2008).

174. Hydropower refers to land-based power stations that use falling water to turn turbine blades; hydrokinetic refers to wave-, tide-, and current-based water power technology, designed for use in oceans and rivers.

175. This section is based largely on the following references: (B_B_C_News, 2007; Piper, 2005; Wood & Yesilada, 2004).

176. Fiscal responsibility continues to be a major requirement at the time of initial application for membership to the EU. New applicant states undergo rigorous EU scrutiny and cannot join until they "get their houses in order." Specifically, the currency exchange rate, interest rate, and rate of inflation are examined, along with debt as a percentage of GDP.

177. Required sellers of electrical appliances, TV, etc. to accept items for recycling.

178. REACH act is 849 pages and covers about 30,000 chemicals.

179. Reminiscent of Theodore Roosevelt's admonitions. "I recognize the right and duty of this generation to develop and use our natural resources, but I do not recognize the right to waste them, or to rob by wasteful use, the generations that come after us" (Theodore Roosevelt, speech, Washington, DC, 1900).

180. This photograph replaced "Earthrise," taken in 1968 by NASA's Apollo 8 astronauts, which showed only part of the Earth. Both were taken 240,000 miles from Earth.

181. Differences in geography (size of country and latitude), population density, level of technological advancement, productivity, type of energy production, and lifestyle influence ranking systems; a potential distorting effect is population density. For example, though The Netherlands is considered among advanced nations for many aspects of environmental management, its high population density causes it to rank only 55th in the EPI index. The relatively sparse population for Sweden, Norway, Finland, and Colombia, on the other hand, is a factor in their favorable rankings, though the three named Scandinavian nations rate high in almost every ranking system. A significant point of view in the United States remains that America is a leader in efficient use of energy and has the right to use as much as it needs to maintain its standard of living (Hayward, 2008). Rankings for the United States and 148 nations in key environmental indicators can be found in Measuringworth (2008) and the OECD (2007).

Page range (148–165)

182. Schreurs (2002) noted a case in which a lower administrative court decided against expansion of a large German coal-fired power plant on environmental grounds. The decision was reversed by an upper administrative court on labor and economic grounds. In the United States, court decisions rarely give weight to economic considerations because the key environmental laws place exclusive or dominant emphasis on environmental protection. A noted environmental suit in Japan in the 1980s was brought against the US government, which planned to build a military housing unit in a forest but neglected to consult with the local government official. This reflects the cooperative aspect of Japanese processes.

183. Norway's plentiful North Sea oil is exported and the currency is banked for future needs (see Section 6.3.3).

184. As EU countries, Sweden and Finland agree to follow EU laws and regulations regarding the environment. They can go beyond the requirements, however, and frequently do. But even Sweden and Finland have been taken to the European Court of Justice for failure to adequately follow the environmental protection directives of the EU.

185. This count was made from an extensive database of US environmental laws (F.T. Manheim, unpublished, prepared with the assistance of Greg Fuhs, George Mason University, Fairfax, VA). Many but not all earlier laws have been superseded. The US code of laws contains a complex accumulation of parts of earlier laws, replaced titles or additions, and titles from newer laws. The laws include both designations of geographic areas, like National Forests, National Parks, and Wilderness areas, management, and laws of which part refers to environmental issues.

186. For an earlier research study of the Norwegian offshore oil and gas industry (Manheim, 2004), I expected to find regulations comparable to those issued by the US Mineral Management Service. My search had no success. Finally, at a special symposium on energy and the environment in Washington D.C., sponsored by the Norwegian Embassy, I encountered an official of the Norwegian Petroleum Directorate and an engineer with a Norwegian oilfield contracting company. Through these individuals I came to understand that there were no published regulations! The leasing supervisory authorities provided informal guidelines to participating oil companies as well as the Norwegian national oil company, Statoil, but the formal operation plans were worked out by cooperation and negotiation. The Directorate set high goals for environmental performance, targeting 2006 for achieving zero harmful discharge, and sub-seafloor oil well "completions" (i.e., connections from the borehole to the pipeline to shore facilities). This meant there was no hardware on the seafloor on which fish trawlers could snag nets.

187. Except off the Lofoten Islands, a spawning ground for cod.

188. I viewed this operation on a farm near Bassum, North Germany, which had been retained in my mother's family for more than 500 years.

189. The US Environmental Protection Agency issued rules eliminating the oxygen requirement as of May 8, 2006.

190. The tragic Atlantic haddock and other groundfish landings story shown in Fig. 7.1a is the result of 35 years of reliance on laws and "method" rather than professional judgement. The Pacific-Alaska halibut landings picture (Fig. 7.1b) reflects management guided by an international agreement under the leadership of knowledgeable and committed scientists and professional fisheries managers. In the Atlantic example, environmentalists, commercial fishermen, and consumers have lost, whereas Alaska's outcome is positive for all stakeholders.

191. I found that it was important to browse in peripheral areas, in order to find important perspectives that a logical and organized (or biased) mind would never think of. For example, Daniel Yergin's Pulitzer Prize-winning book about the international history and politics of oil, *The Prize* (Yergin, 1991), led me to the important story of Everett Lee DeGolyer (1886–1956), entrepreneur, geologist, and scholar. DeGolyer discovered the world's largest oil field (Potrero del Llano No. 4), which opened up the huge Gold Lane oil district in Mexico, while taking off a year from undergraduate studies in mining engineering at the University of Oklahoma! He later introduced the use of geophysics to the global oil industry,

Page range (167–192)

was the most famous oil expert in America, and became an advisor to President Roosevelt on oil supply during World War II. He also saved the *Saturday Review of Books* from bankruptcy, and guided it to a 20-fold increase in subscribership. His story is not just a colorful historical note. It sheds light on changes in the leadership of the American oil industry from the prewar period through the 1970s. Changes in types and attitudes of people among both environmentalists and industry have had underexamined influence on developments in the post-World War II period.

192. One RFF Board member, a CEO cited by Washington Post columnist Steven Pearlstein, was quoted in a column noting the departure of Paul Portney: "Their [RFF's] hearts are generally green, but they don't lie about the numbers, and they have an enormous amount of credibility on both sides" (Pearlstein, 2005). RFF's early staff, including long-time President Paul Portney, are sometimes given credit for initiating one of the major innovations in US environmental regulatory development, pollution permit (or emissions) trading. The technique is credited with providing up to 80% cost savings over alternative methods when utilized for reduction of sulfur dioxide emissions to the air in the 1991 Clean Air Act Amendments, and is now used as part of the Kyoto Protocol for reducing global CO_2 emissions. See Krupp and Horn, (2008).

193. Because of litigation owing to environmental opposition and NIMBY.

194. Reinventing Environmental Regulation provides ten principles. One of these principles is that environmental regulations must be "performance based" and must allow flexibility while requiring accountability in attaining goals. Another principle is that "market incentives should be used to achieve environmental goals, whenever appropriate." The document also includes "25 High Priority Actions." A section in which open-market air emissions, effluent trading in watersheds, and other topics are discussed (Clinton & Gore, 1995).

195. S.2191: Lieberman-Warner Climate Security Act of 2007.

196. Protestantism and the Protestant Ethic: Two key religious leaders who influenced the economic development of western culture were Martin Luther and John Calvin. Luther, an Augustinian friar who famously criticized Catholic Church hierarchy and set off the Protestant Reformation, believed that people served God through their work. An example of Luther's many writings on this theme is: "We have disdained such tasks and obedience, and have rushed into the monasteries, [thinking that we] can do it more excellently than the dear child Jesus… who in his youth gathered wood, picked up shavings, started the fire, fetched water, and did other household chores (Luther, 1966)." Weber wrote that although Luther believed that although work and the professions created differing social classes, each vocation had equal spiritual dignity (Weber, 1904). Luther's teachings have been less influential in the United States or on development of profit-oriented economic systems than the teachings of John Calvin because Luther disapproved of commerce as an occupation, and regarded accumulating or hoarding wealth as sinful (Lipset, 1990).

197. The issue remains an uncomfortable one for science policy theorists. It has perhaps been most diplomatically and artfully articulated by Donald Stokes (Stokes, 1997).

198. Judith Tegger Kildow, Associate Professor of Ocean Policy, Massachusetts Institute of Technology; now Distinguished Professor of Science and Environmental Policy, California State University at Monterey Bay, Seaside, CA.

199. In late 2007, researching for this book, Dr Richard Green, who had been a former manager of the RANN program during the time in question, kindly agreed to an interview. He confirmed the contract-like requirements of the program at that time.

200. Clarence King, in US Geological Survey 1st Annual Report, 1880, p. 4.

201. In the election of 1877, Democrat Samuel J. Tilden of New York had a plurality of 250,000 votes over Hayes, but the Election Commission awarded the four states, electonal votes to Hayes, which gave him 185 to Tilden's 184. A special Congressional Commission awarded the southern Electoral College votes to Republican Hayes. His furious opponents "dubbed Hayes 'Rutherfraud' and 'His Fraudulency;'" Ari Hoogenboom, "Rutherford Burchard Hayes (1822–1893)," American President An Online Reference Resource, http://millercenter.virginia.edu/academic/americanpresident/hayes.

202. Hayes' efforts, including the commissioning of a comprehensive survey of civil service systems in other nations, culminated in the Pendleton Act (1883), passed during the administration of Chester Arthur, who succeeded the assassinated James Garfield.
203. Schurz's was an "American dream" story. Carl Schurz was born in Liblar, near Cologne, Germany in 1829. He fought for democratic values as a lieutenant in the 1848 revolution that swept through Europe. On its defeat in Germany by the Prussians, he was sentenced to death for treason but fled, immigrating to America with his wife in 1852. By the age of 40, the immigrant Schurz gained a law degree and rose to become elected Senator from his adopted home state of Missouri. Schurz was a passionate progressive, a supporter of Lincoln's Republican Party and the antislavery movement, served as a successful general in the Civil War, and was an advocate of civil service reform. He was appointed Secretary of the Interior in the administration of President Rutherford B. Hayes. Schurz is widely regarded as having been one of the United States' most outstanding Secretaries of the Interior. Among his goals was to reform Interior's policies for Native Americans. He later opposed the Spanish American War (1898) and US expansion abroad. Schurz's idealistic models for public service are reflected in quotes from the web site of his namesake Schurz High School, Chicago, IL, http://www.schurzhs.org/carl_schurz.jsp?rn=4833003; the most famous are "My country, right or wrong; if right, to be kept right; and if wrong, to be set right" and "Ideals are like stars; you will not succeed in touching them with your hands, but like the seafaring man on the ocean desert of waters, you choose them as your guides, and following them, you reach your destiny."
204. Clifford Nelson, written communication, March 31, 2008; based on US Geological Survey Appointments Ledger 9–873 (1879–1890), Record Group 57, National Archives, College Park MD, see also (Nelson, Rabbitt, & Fryxell, 1981).
205. Garfield was assassinated in 1881; he was succeeded by Vice President Chester A. Arthur, who served out Garfield's term (1881–1885).
206. First water resources units within the USGS: 1894, Hydrographic Division (streamgaging); 1902, Hydrographic Branch (comprising the Division of Hydrography for streamgaging, the Division of Hydrology for ground water, the Division of Hydro-Economics for water quality, and the Reclamation Service, which later became the Bureau of Reclamation).
207. According to Mary C. Rabbitt, "Twice during that period, the USGS experienced a brain drain, as industry and the academic world raided the Survey's staff" (Rabbitt, 1989).
208. Mary C. Rabbitt (1989), Excerpts http://www.usgs.gov/125/.
209. The time from completion of a manuscript to USGS "Director's approval" is months to years; "Director's approval" is required prior to publication. Open file reports, often not cataloged in major library archiving systems, became popular in the early 1970s in part because they could be locally prepared and rapidly released. But their distribution was limited and they were not readily accessible except on request; they bore a disclaimer saying they had not been reviewed and revised in accordance with USGS standards. At the same time outside publications burgeoned (not shown in figure).
210. Personal interview with Joan Davenport in her office in the Interior Building, Washington, DC, December 1980.
211. While Director, Menard maintained a part-time office and research assistant at the Woods Hole Oceanographic Institution to continue his research interest in global tectonics and ocean island formation. He published two books on these subjects in 1986.
212. In the early 1970s, two colleagues and I wrote a letter to the Chief of the Conservation Division, noting informal conversations with oil company executives. The representatives acknowledged that earlier release of exploration data than the existing 10-year holding period would actually benefit oil companies, if done uniformly – though their company's public position opposed any liberalization of release as a standard policy. We noted the advantages gained by Canada through liberalized data release policies, which would avoid large expenses in unnecessary duplication of field exploration data, and provide the government and other researchers important data. We learned only decades later that the letter

never reached the Conservation Division. It had been circular-filed by Coastal and Marine Geology's office chief "to avoid political problems."

213. Personal interviews with W.S. Radlinski, Springfield, VA, 2004.

214. In the RIF of 1995, the Water Resources Division absorbed a number of individuals from other Divisions. The Geological Division, which before the RIF numbered about 2,200 people, absorbed most of the loss of 500 persons (before restorations as a consequence of the class action suits).

215. In a long departure message on September 10, 1997, Eaton reiterated the mood of Congress on his appointment and the need to "alter the bureau's external image,... .internal and external cooperation and communication and coordination among the divisions, program relevance.. and a perceived lack of responsiveness."

216. An informed source reported that at a breakfast with Interior Secretary Babbitt, Eaton had been given his walking papers, after which he announced his resignation with 20 days notice.

217. Based on biological reports on polar bears in Alaska, the US Fish and Wildlife Service recommended and Interior Secretary Dirk Kempthorne accepted the need for placing polar bears on a potentially endangered status (Kempthorne, 2008).

218. I had initiated comprehensive databases of contaminated sediment chemistry in coastal areas of the United States (Manheim, Bucholtz ten Brink, et al., 1999). Leaders of the EPA's contaminated sediment program indicated to me that the agency would be interested in the possibility of funding USGS to develop a nationwide database of contaminated sediments in the coastal environment. When I conveyed this opportunity to USGS managers, it got a cool reception, as did most suggestions that might involve commitments to serve other agencies that did not come through higher-level sources. Of course, any major commitment would need careful evaluation for suitability and to avoid overstretching capabilities. The issue is that internal preoccupations systematically inhibited the whole organization's openness to societal needs it had the ability and mandate to fulfill.

219. I myself engaged enthusiastically in discretionary research in my earlier USGS career. This included NSF projects together with leading academic research institutions. But I got my undergraduate science education before the Vannevar Bush Endless Frontier research era boomed, when there was none of the stigma on "applied" science that later developed. So although I participated in and enjoyed the heady chase after new knowledge for its own sake, I did not develop the consuming inner conviction that I saw in many younger colleagues. They seemed unable to shake off the belief that "real science" lay primarily in the passionate pursuit of discretionary investigations at the frontiers of knowledge, communicated to like-minded researchers through scholarly publication.

220. In 1970, California passed the California Environmental Policy Act, a parallel to the NEPA Act of 1969. By 1976 there were 4,000 Environmental Reports filed annually. By 1975 228 lawsuits invoking the Act had been initiated.

221. An appeal by MMSD to the US Supreme Court of the Wisconsin Appeals Court reversal was returned to the District Court for action. In 2008, the Wisconsin Judiciary closed action by assessing fines of $70,000 to the District, attorneys' fees of $170,000 to the Friends' plaintiffs, and costs of $275,000 to be applied to further specific improvements (oral communication, Jim Fratrick, Wisconsin Department of Natural Resources, June, 2008).

222. Frank Lautenberg (1997): "The bill also repeals an existing preference for cleaning up the pollution to protect future generations and the environment. Instead, S.8 would allow the materials to remain at sites, so long as there is a fence around them, even if the materials continue to pose health risks."

223. Instead of making a direct attack on a law regarded by environmental organizations as a bulwark for the protection of the nation's biological heritage, the new legislation stressed constructive modification and "strengthening" of the law. Unlike previous tactics, the reauthorization bill reflected considerable research on the ESA law, its purpose and operational history.

Page range (217–234)

224. TESRA weakened or removed a number of enforcement actions incorporated in ESA, but nevertheless claimed that the reauthorized statute could enhance rather than reduce species protection. It claimed that ESA had been inefficient and extremely expensive, and that its uncompromising provisions promoted hostility instead of constructive solutions. For example, because of the severe restrictions or penalties invoked by finding of endangered species or habitat on private land, landowners or leaseholders had a disincentive to seek or report presence of endangered species on their property, and might even do harm. These findings were described in an 81-page report by an extensive task force investigation launched by Representative Pombo prior to writing TESRA.

225. At a hearing in 1989 on the Magnuson Act appropriations before the Subcommittee on Fisheries and Wildlife Conservation and the Environment, Committee on Merchant Marine and fisheries, House of Representatives, Clem Tillion (retired President of the Alaska senate, former Chairman of the North Pacific Fisheries Management Council) stated, "The..... problem was an entrenched bureaucracy there in Washington, DC, where time after time they would override Council decision, often voted on and made by senior NMFS officials on the West Coast (Tillion, 2002)."

226. On March 12, 2008 Spitzer himself resigned as Governor of New York owing to personal ethical problems (Ross, 2008).

227. Examples of legitimate issues that may be buried in controversial tactics: (1) Use of heat generated in steel making may be inhibited by the "New Source Regulations" that the Bush Administration is alleged to have undermined; (2) On the other side, partisan campaigns by environmental organizations against offshore drilling include valid concerns (such as the absence of long-term planning to counter "boom and bust" economics developed in coastal communities which support offshore oil developers).

228. Dale Carnegie, author of *How to Make Friends and Influence People*, still a bestseller after 70 years, began his 1936 book with graphic examples of a key conclusion: condemnation and criticism of people produces negative response. Carnegie showed that even ruthless killers like Two Gun Crowley and Al Capone regarded themselves as persons with basically good intentions.

229. http://www.anncoulter.com/cgi-local/content.cgi?name=bookstore (Coulter, 2004).

230. Table 2 in the poll: 4%, 2003, 2004; 3%, 2005–2007, HarrisInteractive, (2007. The Harris Poll(r) #107, November 1, 2007 2007).

231. http://www.sourcewatch.org/

232. http://www.greenpeace.org/usa/campaigns/global-warming-and-energy/exxon-secrets.

233. A curious historical development revealed in these sites is that the majority of the largest private foundations, which were created by the owners of major industrial corporations like Sun Oil Co. (Pew Trust), Ford Motor Company (Ford Foundation), and Rockefeller Brothers Foundation (Standard Oil), have devoted significant proportions of their funding to activist environmental organizations that lead in attacking and filing lawsuits against industry.

234. These topics are discussed in more detail in Chapter 9.

235. Failure of a major geothermal energy project. An opportunity to develop a major new strategic minerals industry centered in Hawaii reached advanced stages in the late 1980s after joint German-U.S. research discoveries from the Pacific Ocean sea floor. The sides of seamounts, deep-ocean volcanoes that do not reach the surface of the ocean, were found to be encrusted with thick ferromanganese crusts with economically recoverable concentrations of strategic metals like cobalt, manganese, and platinum. The environments in question were largely biological deserts. They posed minimal biological, but significant engineering challenges for crust recovery. An unusual cooperative effort cosponsored by the US Minerals Management Service and the Hawaii Department of Economic Development produced an EIS for the planned lease sale (MMS, 1987). It not only included detailed analyses of environmental issues including fish, plankton, and infrastructural impacts for Hawaii. It included evaluations of potential engineering recovery methods and on-land processing systems for the crusts. Individual sections of the EIS were not pro forma. They included contracted chapters providing innovative technical

Page range (241–242)

adaptations by leading experts. Scientists and economists from the University of Hawaii also developed potential economic uses for residues from ore processing. Most important was the proposed use of geothermal energy from Hawaii's volcanic substrates in the beneficiation and processing of the ores. The Hawaiian government strongly favored the development, which would have provided an advanced industry, superior jobs to replace those lost from declining sugar and pineapple growing operations, and other benefits to the state.

All went well until the first hearing in Honolulu. Residents raised questions about the operations. These were written down by the MMS hearing coordinators, who responded with "Thank you," but did not respond to other than noncontroversial technical questions. As a USGS marine scientist leading a major analytical/research program on the seafloor crusts (Manheim, 1986), I knew the coordinators, who were well versed with scientific and policy detail, but probably did not regard themselves authorized to provide "official" response to questions that might have policy implications. The audience's irritation palpably increased as the hearing continued. Commentators openly accused MMS of running a charade – going through a pro forma hearing process and recording comments although the federal and State governments had already struck a deal. Tempers rose until voices reached angry shouting levels. The hearing and subsequent written commentaries together produced a negative result. Ultimately, leasing plans were suspended and never completed.

MMS had done conscientious job. There was no hidden agenda. But the structure of the leasing–EIS system did not allow the bureaucratic agency to achieve a trust relationship. This led to costly failures. None of the competitive MMS mineral leases from 1983 to the present have ever been successful, to my knowledge, though federal water areas were allocated to federal agencies like the US Army Corps of Engineers and US Navy for sand mining, and indirect benefits have been gained by the "hard minerals" leasing program (Fischer, 1989). For comparison, offshore sand and gravel mining has been successfully conducted by the UK since 1948, coordinated with fisheries authorities (DEFRA, 2007).

236. William James (1842–1910) was trained as a medical doctor at Harvard University and appointed as an instructor in medical subjects at the University. James gravitated toward studies of the mind and psychology in the formative period for that field. He became a Professor of Psychology and later Philosophy, known as the leading exponent for the philosophical school of pragmatism, which is sometimes expressed as "truth is what works" (James, 1956).

237. As told in a detailed biography (Bower, 1992), Maxwell started life in a poor Jewish family in a Ruthenian village on the border of Czechoslovakia and Romania. Through an extraordinary series of events, some still murky, he made his way through Vichy France and joined the Czech legion of the British Army. Succeeding in the unusual feat of transferring as private to the regular British Army, he distinguished himself in combat during his unit's battle with the Germans, and ultimately rose to the rank of Captain, changing his name several times in the process. (Later, as a publisher, Maxwell insisted on being addressed as Captain.)

After the war Maxwell made himself useful by devising skilful ways to assist German publishers to distribute stocks of scientific publications under military restrictions and other difficulties in the early postwar period. With no previous training in science, he somehow gained a sophisticated understanding of scientific publication, its economics and distribution. By the early 1950s he took over leadership of Pergamon Press, a small science publisher in the United Kingdom. Maxwell recognized that the rapid increase in growth of university and other scientific research had created a new market for journals to serve emerging scientific disciplines. Libraries had become a potentially significant market. By adroit networking, Maxwell arranged for leading scientists in different countries to become editors of new journals, for example, *Geochimica et Cosmochimica Acta* (in which I published my first major scientific paper). The journal would offer steeply discounted subscription rates for individual scientists whose institutions bought library copies.

Page range (246–258)

238. "Never before in all our history have all these forces been so united against one candidate as they stand today. They are unanimous in their hate for me – and I welcome their hatred (Hardman, 2005)."

239. When he became overwhelmed, a further moderating agency, the War Mobilization Board was added under James Byrnes. Research oriented activities were coordinated by flexible and dynamic leaders like Bush and Compton.

240. Blumer was born in Basel, Switzerland in 1923; he earned a Ph.D. at the University of Basel in 1949. He worked for Royal Dutch Shell Holland, Shell Houston, and CIBA, Basel. He served on the Coast Guard – National Academy of Sciences panel on pollution monitoring, testified before the US Senate Subcommittee on Air and Water Pollution in 1970, and died unexpectedly of cancer in 1976.

241. In an address before the Associated Industries of Georgia, Indiana Standard Oil's Robert E. Wilson "criticized businessmen for 'lack of courage in leading the fight against both demagoguery and dishonesty in city, state, and nation.' He called upon them to restore to American public life the high standards of ethical conduct that characterized their economic behavior (Engler 1961)."

242. Under President Robert O. Anderson, ARCO was one of the more foresighted and innovative US companies. It discovered the massive Prudhoe Bay oil field in Alaska, and was one of the first companies to produce lead-free gasoline in the early 1970s. On retirement Anderson became a philanthropist and Chairman of the Aspen Institute, which sponsored non-partisan policy conferences and research.

243. One of the spectacular results of this kind described was a massive campaign conducted by the Department of Agriculture and the State of Florida in 1957–8, using a new biological control technique against the screwworm fly. This fly had devastated livestock in the south. The joint undertaking, involved 20 aircraft, and release of 3.5 billion irradiation-sterilized screwworm flies over Florida. According to Carson (1962) this campaign completely eradicated the screwworm fly in Florida in 1959. Other methods described by Carson included chemisterilants, pheromone attractants, insect-attacking bacteria and insect predators. She accepted the use of minimum quantities of pesticides under carefully controlled conditions. Even though the Department of Agriculture had played a major role in launching chemical assaults on insect pests, and had tried to obstruct attempts to require consultation with state agencies before insect eradication campaigns, she refrained from attacking it. She recognized the Department's responsibilities for supporting agriculture, and gave full credit for the Department's efforts to explore alternative pest control methods (e.g., the screwworm initiative) once it became aware of the risks of the pesticide approach and options for alternatives.

244. Bush adminisration positions are critiqued in Chapter 4. It is the bald mixing of science and politics that indelibly marks the Gore book as "political" (however much science there is in it) rather than material to be weighed in scientific terms.

245. Everette Phillips, described in http://www.sierraclub.org/ca/coasts/cuchessi.asp.

246. The authors draw from the philosophic ideas of the nineteenth century American pragmatic philosopher, Charles Peirce, and Niels Bohr in his debates with Einstein and other authors, the idea of unitary science that communicates to society while it learns. They then apply the unitary science idea to organic farming.

247. However, one knowledgeable former colleague, a marine scientist in Finland, whom I had known for more than 30 years, remained highly skeptical about the "orthodox" scientific positions on human influence on global climate change, and I learned that the President of Czechoslovakia, Vaclav Klaus, was in this camp, as well.

Index